China's Medium and Long-Term Science and Technology Program

Zhenghong Chen

China's Medium and Long-Term Science and Technology Program

History and Philosophy

Springer

Zhenghong Chen
WMO Regional Training Centre Beijing
China Meteorological Administration
Beijing, China

ISBN 978-3-030-75145-6 ISBN 978-3-030-75146-3 (eBook)
https://doi.org/10.1007/978-3-030-75146-3

Translation from the language edition: 当代中国中长期科技规划历史与理念研究 (The History and Philosophy of Contemporary China's Medium and Long-Term Science and Technology Program) by Zhenghong Chen, © China Meteorological Press 2015. Published by China Meteorological Press. All Rights Reserved.

This Springer imprint is published by the registered company Springer Nature Switzerland AG
The registered company address is: Gewerbestrasse 11, 6330 Cham, Switzerland

Dedication

This book conducts a panoramic study on the history of China's Science and Technology which focus on Medium and Long-Term Science and Technology Program (MLSTP), in general speaking these Programs have a duration 5–30 years, over the past 60 years through an epochal assessment of the project's conceptual context. At this point, some basic points become clear: the historical evolution and philosophy development of China's MLSTP are the results of an amalgamation of political, economic, and social factors within distinct contemporary contexts. As a national action plan, MLSTP has incorporated many of the factors that go beyond the intentional factors of science and technology. MLSTP is not only a macro vision and blueprint for scientific and technological development; it is also a political act of realizing the national will. While ensuring the MLSTP builds on its great achievements, we should also reflect upon its deficiencies and disadvantages in order to better promote the advancement of science and technology in China.

This book comprehensively lays out the historical and theoretical dimensions. Our process of writing history has been guided by a clear vision of historical constructivism; we have compiled the MLSTP philosophical of different eras into a conceptual framework for this era, and used this framework to research and analyze the historical and philosophy evolution of MLSTP. Research on MLSTP is important for enriching contemporary studies in the history of science and the science and technology policy. Since the establishment of the People's Republic of China over 60 years ago until 2020, this country had enacted an MLSTP 10 times, this book separates the development of the MLSTP into three different historical eras: the era of economic planning, the era of economic transformation and the new century. Each historical epoch corresponds to a different MLSTP philosophy concept; therefore, we are able to study the conceptual evolution of MLSTP using historical research as our foundation. Through a summary of the entire book, we are able to draw the seven concrete conclusions.

Foreword

After several years of accumulation and revision, I am very happy to know that Dr. Chen's book will be published. I was shying to write the foreword of his book. Why? First, I rarely write a foreword for scholars, even for my students. I've only written three forewords of three books so far.

Second, Mr. Zhenghong Chen was a doctoral student of Prof. Liu Xiaoli when he was studying at Beijing Normal University since 2005. Professor Xiaoli entrusted me to help guide Zhenghong's doctoral studies When she was introduced as a high talent expert by the Renmin University of China in 2006. So that, I and Mr. Chen had a special reason in a sense of pairing phenomenology. As a matter of fact, this preface should be written by Prof. Xiaoli, but Zhenghong feels that the partial content of his manuscript seems to be closer to the STS (Science, technology, and Society) topic that I have paid close attention to in recent years. There is neither the problem of strengthening the appearance nor the problem of ostentation. Therefore, it is inconvenient for me to push his invitation again and again.

Mr. Zhenghong had done some education work in Zhejiang Province before he was recruited by the doctoral program of Beijing Normal University. While he was in the University, he not only studied hard and cared about current affairs but also obeyed the etiquette of teachers and students. In this way, we usually have a deep impression of him. In addition, I have done scientific research management work. His doctoral research direction was still in science and technology development planning. It was natural to discuss when we meet each other. Therefore, it was reasonable that Prof. Xiaoli asked me to assist him in his studies.

In fact, Mr. Zhenghong's original research on the subject of *the history and philosophy of China's medium and long-term scientific and technological program* was very pioneering and ambitious in 2006, and he has also achieved his goal well. In this regard, the experts of his doctoral dissertation defense committee gave a very high evaluation. In today's view, this pioneering not only gives a detailed description of China's science and technology development program based on rich historical materials, provides many little-known facts but also makes a more forward-looking international perspective for examination and judgment, and its enlightenment value is self-evident. Especially when the philosophy of science and technology of phenomenology was just popular in China, Zhenghong's *theory of ideas* research also contained the meaning of the Times. His penetrating understanding of the history of contemporary China's medium and long-term science and technology program not only outlines a miniature of China's science and technology development but also summarizes China's experience in contemporary science and technology competition from the special side.

Specifically speaking, as a research achievement, Dr. Chen not only used the large-scale perspective method of history but also used the small-scale anatomy technology; and not only paid attention to the focus of people and events but also paid attention to the clue extension and link transformation in different historical periods. He not only took note of the interaction between science and technology planning and social-political economy but also took note of science and technology planning on the evolution of autonomy. Furthermore, Chen not only studied the particularism of countries and regions but also the universalism of perspective and methodology. As far as I know, in order to write this book well, Zheng Hong used his one-year study at the University of California Berkeley to collect and consult countless valuable materials, participated in relevant international high-level conferences, and asked for advice from some high-level international scholars, including Nobel Prize winners, so as to change Chinese scholars long-term inability to see China from the perspective of the world tradition, all above which ensured the academic vision and standard of this book.

In addition, different from the general books, Zhenghong's this book had strict academic nature and special unique Qi and EMOTION, which not only showed Zhenghong's own ideological style and academic personality but also integrated with the topic selection properly. The so-called "Qi" means that there is an ideal and pursuit in the thesis, a realm, a responsibility, and an aura that takes the rise and fall of the country under the world as its own responsibility. The so-called "emotion" is not only the thesis devoted to sentiment but also expresses the author's yearning and hope for the beautiful things and his love for the great motherland. This was reflected in the layout of this book and the consideration of the text. If not, strictly speaking, Zhenghong is also a poet. He has the temperament of a poet. When he during the doctoral program, he occasionally showed his poetic talents at some parties. These reserves are now concentrated in his book. Therefore, this book is fluent, full of rhythm, good to read, and very durable.

Zhenghong's research in 2005 was timely and ahead of Time, and the publication of his book was lucky. The lucky is not meaning that he can finally publish the book as a souvenir, or as an important element for showing off and appraising. It's because he catches up with the rapid rise of Think Tanks in mainland China so that they can be put to practical use and have a wide audience. I don't know what people think of Think Tanks as "new things". But as far as I know, many people and institutions never seem to come up with an idea before, but today they have transformed themselves into "Think Tanks". However, in any case, if some Think Tanks do science and technology works, Zhenghong's this book should be as their reference book. On the one hand, reading the history of science can be learned from the past, and on the other hand, Zhenghong's attitude towards problems and foresight of insight is also worth learning. I believe that people who read this book will be inspired by it. This also shows that the publishing house has a very academic and strategic vision in publishing Zhenghong's book at this time.

Of course, that doesn't mean Zhenghong's book is absolutely perfect. Looking back today, his book may have three deficiencies. First, there is a "time gap" between the book and the description object due to the delayed publication of the book, for example, the situation in the last decade has changed rapidly but not yet been intensively examined, which will weaken the timeliness of the work to a certain extent. Second, the subject focuses on the official planning policies because of their own nature. However, it ignores the small folk things that may play an important role, which is inadequate for those readers who have genealogical and anthropological preferences. Third, there are few reflective especially critical opinions in the book. On this issue, when I talking with Zhenghong at the beginning, he expressed a lot verbally, but there was little left in the implementation of the text. In my opinion, the intention was inadequate, because the development of science and technology in China is moving from imitation to indigenous innovation. Without a profound critical reflection, China cannot complete this transformation as soon as possible. In recent years, China's high input and low output in science and technology, as well as its small effective contribution to economic and social development, are very illustrative. But Zhenghong is a cautious man and unwilling to speak out publicly, which may be a wise approach.

Looking through Zhenghong's manuscript, I have been immersed in the aftertaste of his past learning situation, especially in 2007 when he spent New Year's Eve in my home with Shi Wei and several other students in order to wait for his invitation letter from America. Some interesting things seem to have happened yesterday. Nowadays, after a flash of years, many things have become as history, society is becoming more and more Darwinian, and some new strange phenomena are hissing. Therefore, from Zhenghong and his book, I think of our disciplines, our teachers and students, and the busy but happy day and night. So, while congratulating Zhenghong on the publication of his book, I also brought some private appreciation by the way. I would like to commemorate the days when I worked with Profs. Wang Desheng, Liu Xiaoli, and other friends to take students with me. I think that's what Zhenghong missed.

July 2015

Xiaoting Liu
Professor of Beijing Normal University
vice president of the Chinese Society
for Dialectics of Nature/Philosophy
of Nature, Science, and Technology

Preface

What is the goal of scientific development? This involves the judge's own evaluation criteria. Many scholars have different opinions on whether it is to meet the interest of researchers or to achieve some utilitarian purpose, or both. And what is the goal of technological development? Although there are many debates, lots of these arguments trend are consistent to meet certain material needs and other purposes of human beings. After the modern scientific revolution, the boundary between science and technology is gradually blurred, and the tendency of integration of science and technology is constantly strengthened. If we want to ask what the goal of science and technology development is, there will be many different answers. After all, resources are limited. In the 20th century, China has made tremendous achievements. In this process, the state and government have their own understanding and answer to the goal of science and technology development. This book attempts to sort out the development goals of science and technology in this historical process, especially after 1949.

At first, I don't know how to start to complete such systematic work. In the past, many scholars have written a lot of profound and comprehensive works in this field. It is not easy to write a book with new ideas and depth. In 2007–2008, I had the opportunity to study in the Office for History of Science and Technology at the University of California, Berkeley, which has more than 20 Nobel Prize winners. Anyone can pass these great Masters of Science and even eat fast food at one table without being surrounded by young students like in China. To expand the gymnasium, Berkeley needs to cut down several trees, which were opposed by environmentalists, who lives on the trees for several years. There is no possibility as China's demolition speed. I studying abroad am short, and one of the biggest enlightenment is that if we want to understand the goals of China's science and technology development in the 20th century, we must consider historical accumulation and current reality in China.

From the perspective of historical accumulation, the Chinese nation has made great contributions to world civilization. Chinese are still in the process of National Rejuvenation, which shows that China today has not reached the realm of prosperity like in ancient China. In science and technology, it did not reach the point of ancient when the light for the world. Although there is controversy about the existence of Science in ancient China, the Needham Problem shows that contemporary Chinese science and technology need another model of great development. From the perspective of the current practice, China's resources are limited, but its development goals are magnificent and numerous. China doesn't need to argue about whether to develop a certain science and technology as foreign countries do, but how to develop and how to develop rapidly.

The brilliance of history in ancient China and the current embarrassment make the science and technology of China in the 20th world need to undertake more historical missions. After 1949, China's science, technology, Academy, and culture are in a huge system or network. The core of this system/network is the long-term planning and development of science and technology. Projects and scientists become part of this huge system and contribute to the common achievement of the established goals. Therefore, from the perspective of medium and long-term science and technology planning, it can be a useful attempt to elaborate the development goals of science and technology in China in the 20th century.

In the field of science and technology planning in China, science and politics are inseparable. In China, the meaning of politics seems to be more than that of other countries, and the understanding of ordinary people seems to be more complicated, which there is more tacit knowledge or no-coding knowledge. In order to deepen the research, I clearly defined that the word of politics in this book as a neutral word, just a word for analysis. In the preface, it is necessary to emphasize that we can understand the neutral context of "politics" by treating politics as science does.

In the development process of science and technology planning, its core philosophy is constantly changing and advancing with the times. The composition and color of technology planning are relatively strong, which ensures the measurability of science and technology planning as a national action and government decision-making. As a national action, it is bound to be accompanied by many "political" considerations. Therefore, planning text analysis, government decision-making documents, and various opinions, which seem to be insufficient references in the eyes of some rigorous scholars, become necessary references source.

The book analyzes the utilitarian purpose of science and technology planning from the perspective of commendation. Without this utility, there is no need for science and technology planning, and China's rapid development will be difficult. However, for science, especially basic science, the purpose of utility brings some results worthy of reflection. For example, the status of natural science is not commensurate with the economic status in China, and the Nobel Prize for natural science, which is recognized by the world, is often missing. However, the development of affairs is always wireless and complex. There are usually decades between making original achievements and winning the Nobel Prize. Today's science and technology planning may give birth to the Nobel Prize decades later. This is why we should look at and analyze the history and philosophy of medium and long-term science and technology planning from a positive and rational perspective.

The 12th Five-Year Plan was completed in 2015 and the 13th Five-Year plan will be completed in 2020, and the 14th Five-Year plan of science and technology will soon be launched. The research area of this book does not involve the 13th Five-Year Plan for science and technology for the time being. If the historical review and philosophical reflection of the past science and technology plan can help to see and specify the future science and technology development plan, and even other science and technology plans, such as meteorological science and technology plan, then the value of this book has been reflected, and the goal of science and technology development has been combed a useful attempt.

Beijing, China
December 2019

Zhenghong Chen

Acknowledgments

This book was written beyond the work from the doctoral program. In the twentieth century, China's scientific and technological development has made great achievements, and scientific and technological planning has made great contributions to that. If I want to study this field thoroughly and write it well, it's not easy to be practical. I can't help myself to do this work, and I hope to get the guidance and help from academia and colleagues. It is not easy to study this field thoroughly and write well, and I do this job with arduous effort, hoping to get help from the academic community and colleagues.

After working in the China Meteorological Administration from 2009, I wanted to publish an academic work on science and technology planning and write a book on science and technology development planning related to meteorology. Meteorology involves thousands of industries, is related to national interests and climate change, and is also a concern of all mankind. Looking back carefully, in fact, in the twentieth century, climate change was becoming more and more concerned and influential, which not only affect economic development but also affect the development of science and technology, such as energy science, environmental science, atmospheric science, wind energy, solar energy technology, etc., which affect the formulation and implementation of science and technology development plan. The medium and long-term scientific and technological planning described in this book has many special discussions on the contents of atmospheric science and related science and technology. Many experts in the meteorological field have also participated in the medium and long-term scientific and technological planning, contributing an important force to the development of national science and technology. It can be predicted that in the future, the layout of science and technology planning in terms of climate change will be more comprehensive. I even want to rename this book as like science and technology planning in the context of climate change. However, thinking over and over again, this is not only the issue of renaming but also the need for major changes in content, ideas, and main points of view. There is no big difference between rewriting a book and rewriting it. I have to stop for the time being. Maybe I can continue to work in this way in the future.

There are too many senior scholars and colleagues who need to thank for while the completion of this book. Thanks to Profs. Xiaoli Liu and Xiaoting Liu, their meticulous attitude, rigorous academic style, upright personality, and profound knowledge make me unforgettable. Thanks to Prof. Song Tian, for his substantial and very important help in my application for studying in the United States. During my stay in California, Prof. Tian took good care of me. His rich academic views and fresh and profound thinking inspired my different understanding of science and technology. Thanks to Prof. Jianhui Li, for his generous help in sponsoring my study abroad. No matter whether he is in an administrative position or not, Prof. Li was almost responsive to my request for help. Thanks to Prof. Chunyu Dong, for letting me share his application and experience in studying abroad. In Prof. Dong's class, I figured out the mystery and beauty of philosophy. Thanks to Prof. Cathryn Carson of the University of California, Berkeley for inviting me to study. The beautiful Berkeley campus and unselfish Chinese friends gave me a new understanding of the ultimate value of life.

Thanks to Mr. Bushi Cheng who was the former deputy chief engineer of Yunshi (the name of big-aircraft), for accepting my interview. Mr. Cheng's narration was helpful to my research. Thanks to Academician Yihui Ding for his guidance and help in this book. Academician Ding has experienced many science and technology planning and participated in the formulation and implementation of many major national science and technology projects. His guidance for me had made my further understanding of science and technology planning. Thanks to my wife Han for her support. As a natural science professor, she has a lot of good suggestions for my work. Thanks to the China Meteorological Press for its support and help. Thanks to Baochun Jiang, Kerin Bruestle, and others for their generous help in the process of translation of this book. Thanks for Springer International Publishing AG for inviting me to publish this book. In this process, Dr. Michael Lechner has given sincere and great help. Carmen Spelbos, Doris Bleier, Ambrose Berkumans, and other editors have given a lot of help. Thank you very much.

In the process of writing this book, I have received support from the project of China Association for Science and Technology, the project of meteorological science and technology history of China Meteorological Administration, and the National Natural Science Foundation of China. Thanks to my research team. The literature and references and so on in the book have been verified in many ways. If readers find any mistakes or problems, please criticize and update them. Your help and tolerance will help my future research work.

Beijing, China
December 2019 Zhenghong Chen

Contents

List of Figures

List of Tables

Introduction

1

Abstract

This chapter points out the theme of the whole book on the history and philosophy of the ten Medium and Long-Term Science and Technology Programs from 1949 to 2020 in China. There is no "Polanyi-Bernal debate" on whether or not to make science and technology planning in China, but how to make better science and technology planning and play a practical role. This chapter analyzes the research progress in this field at home and abroad, and shows the writing approach of this book based on historical facts and philosophical reflection.

Keywords

Medium and long-term science and technology program · Polanyi-Bernal debate · China's basic research · History of science in contemporary China · Science and technology resources · Reflection on planning

The development of science and technology (S&T) in the modern world is affected by many social factors. Since the end of World War II, a new scientific research model known as "big science" has been rising. In the meantime, as the Third Technological Revolution gradually unfolds, the world has seen nonlinear, uncontrollable technological developments accompanied by growing global awareness of the negative effects of scientific and technological developments. The increasing importance of scientific and technological advances in the game of nations is a major driving force behind the emergency of the concept of "S&T planning".

An S&T plan is a forward-looking programmatic plan of a country or government that establishes a guiding framework for the development of S&T in the country or jurisdiction of the government. In China, S&T plans have distinct Chinese characteristics and their corresponding era. The study of Medium- and Long-term S&T Programs issued by the Chinese government since the founding of the People's Republic China (PRC) is an important subfield of the history of contemporary S&T in China. The principles, philosophy, of an S&T program, refers to overall guidelines, fundamental views, and development goals that underpin the S&T program. The principles of a Chinese S&T program are usually reflected in the concisely written but meaningful guidelines and are at the core of the planning system. They are the foundation of the entire scientific and technological plan and reflect policymakers' understanding of the history of S&T and the art and science of planning. To gain insight into the strengths and weaknesses of China's S&T planning system, it is necessary to study the evolution of Chinese S&T plans and planning principles over the past few decades. It can help us gain a more holistic understanding of China's S&T system and history.

Z. Chen, *China's Medium and Long-Term Science and Technology Program*,
https://doi.org/10.1007/978-3-030-75146-3_1

Scientific and technological advancements in China take an inverted U-shape. China far outstripped Europe in S&T in ancient times. However, it gradually fell into a scientific and technological stagnation trap and "missed" the technological revolution entirely in the modern times, which is why China now has to rely heavily on imports of technologies. Since the founding of the PRC, the country has been pursuing scientific and technological advancements with renewed vigor and S&T planning has become an essential governing tool for the Chinese government to guide scientific and technological developments. Since the founding of the PRC, the Chinese government has developed ten national S&T programs (Fig. 1.1), but only nine of them came into force.[1] They are[2]:

1. *Science and Technology Development Program (1956–1967)*;
2. *Science and Technology Development Program (1963–1972)*;
3. *Science and Technology Development Program (1978–1985)*;
4. *National Medium- and Long-Term Science and Technology Development Program (1986–2000)*;
5. *National Medium- and Long-Term Science and Technology Development Program (1990–2000–2020)*;
6. *Ten-Year Program & the Eighth Five-Year Plan for the Development of Science and Technology in the People's Republic of China (1991–1995–2000)*;
7. *Ninth Five-Year Plan for National Science and Technology Development (until 2010)*;
8. *Tenth Five-Year Plan for National Economic and Social Development—Science, Technology and Education*;
9. *National Medium- and Long-Term Program for Science and Technology Development (2006–2020)*; and
10. *Eleventh Five-Year Plan for National Science and Technology Development.*

Every S&T program since the founding of the PRC was directly influenced by the planning philosophy. The evolution of the abovementioned S&T programs and planning philosophy reflects changes in China's political and economic environment and Chinese policymakers' understanding of factors driving scientific and technological advancements and the role of S&T in society. Studying the history and philosophy of Chinese S&T programs can help put the relationship between S&T planning and the development of China into perspective. In-depth analysis of the ten national S&T programs can shed some light on how S&T shape the society.

This book reveals the complex influences of and interactions between various factors that affect the development and implementation of Chinese S&T programs, and illustrating the evolution of these S&T programs with the times and the importance of underpinning philosophy. National S&T programs are at the core of China's S&T system.[3] In other words, the formulation and implementation of a national S&T program is closely linked to the development of China's entire S&T system. Therefore, studies of Chinese S&T programs can deepen our understanding of China's S&T system. The interests of different sectors and groups as well as national interests are reflected in national S&T programs. The study of the historical evolution of S&T programs has profound implications for our comprehension of some major theoretical issues.

In the international fora, there used to be a debate, known as the Polanyi-Bernal debate, over

[1]The fifth national science and technology plan (i.e., the *National Science and Technology Development Program 1990–2000–2020*), aborted after development for some reason, is included in the 10 programs due to its significance.

[2]In the face of these program studies later in this book, the program name may not be completely consistent with this list. It is a scientific and technological planning in the same period of time, with different titles.

[3]Science and technology are merged into one word in the Chinese language, "keji." Strictly speaking, science and technology are two concepts. However, given how frequently the term "keji" is used in Chinese government documents, we also use "keji" in the Chinese edition of this book to refer to science and technology. But it does not mean that we agree to the two concepts should be merged.

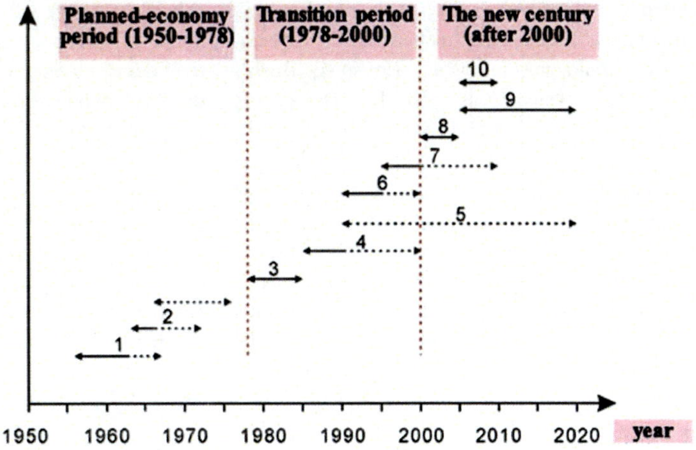

Fig. 1.1 Medium and long-term S&T programs since the founding of the PRC

the necessity of planning the development of S&T. The debate took place in the 1930s in the United Kingdom. At one side of the debate was John Desmond Bernal who believes that science is utilitarian. As a staunch advocate of central planning of scientific developments, Bernal emphasizes the social functions of science and calls for government intervention in and planning for the development of science. Bernal extolled the virtues of the centralized planning system of scientific research in the Soviet Union. At the other end of the debate was Polanyi who is firmly against planning the development of science. Polanyi divides science into pure science and applied science. He believes that centralized planning of pure science would result in a complete failure because scientific development itself is unpredictable. Polanyi argues that "a scientific community formed by spontaneous co-ordination is the most effective way to enhance knowledge. Any intervention in pure science is not only unhelpful but harmful. He believes that only applied science has social functions and pure science does not have any secular functions."[4] Polanyi's arguments have two starting points. First, according to Polanyi, the majority of a scientist's knowledge is personal knowledge, either tacit knowledge or skills that cannot be clearly and fully articulated. In nowadays terms, tacit knowledge is unencodable personal knowledge. The other starting point is that there is a division of knowledge among individual scientists. In other words, each individual scientist is only an expert within a narrow field, which is insignificant relative to the entire natural world. He contends that scientific knowledge is partial and scientific planning is impossible because the central planning institution cannot access personal knowledge or partial knowledge. The Polanyi-Bernal debate had significant implications for the development of post-war scientific plans and science policies.[5]

After World War II, Vannevar Bush (1890–1974) echoed Polanyi's scientific liberalism views in *Science the Endless Frontier*, calling for the separation of the scientific system from economic and social controls. He clearly stated that universities and research institutions need autonomy and independence and should be insulated from the influence of commercial interests. In 1950, the United States passed the *National Science Foundation Act* and established the famous National Science Foundation (NSF). In the meantime, it should be pointed out that Bush does not want S&T to develop completely without purpose. He believes that universities and research institutions should maintain autonomy but also need federal funding. As for how

[4]Fan (2002).

[5]Fu (2008).

science will advance after the provision of funding, it is scientists' matter and the government should not intervene. However, Bush's arguments were challenged by Democratic Senator Harley Martin Kilgore (1893–1956). Kilgore contends that scientific research is an important way to achieve social goals, economic growth, full employment, and health benefits and the state and society have an obligation to support scientific development. He believes special attention should be paid to the geographical distribution of scientific resources and scientific research has obvious social effects. The state should intervene in the direction of S&T development. In a nutshell, Kilgore agrees with Bernal.[6] The post-1945 science systems of the United States and the OECD countries were the results of compromises between the philosophies of Polanyi and Bernal. Considerable amounts of money were spent on science planning with specific purposes, especially in the military and health fields. In the meantime, independent scientific institutions also received large amounts of financial assistance from governments.

By the 1980s, Bernal's theory became the mainstream theory. The empirical evidence shows that knowledge drives economic growth and science plays an increasingly important role in propelling economic growth. The traditional role of scientists has become obscure, and the independence of decision-making relevant to scientific developments is constrained by a growing number of factors. In response to the rising demand for financial support for S&T, the state needs to make some trade-offs and prioritize scientific projects in different fields. Influenced by resource constraints, the "guns versus butter choice" is a central question that policymakers need to consider when planning the development of S&T. State intervention in promoting scientific and technological development seems an irresistible trend.

As for the influence of the Polanyi-Bernal debate in China, during the Republic of China era, Wu Xuezhou (1902–1983), a member of the Academia Sinica who participated in the formulation of China's national medium- and long-term development programs after the founding of the PRC, delineates the debate in his *Analysis of China's Post-war Scientific Research Programs* (1942), proposing that "S&T planning is necessary, but research institutions should also maintain a certain degree of autonomy."[7] The Polanyi-Bernal debate didn't have a direct impact on S&T planning after the founding of the PRC, but the ideas on both sides of the debate are worth pondering upon. In the modern world, the boundaries between pure and applied science as drawn by Polanyi, between science and technology, and between scientific, technological, economic and social domains, are increasingly blurred. Therefore, S&T planning is necessary in the modern world. S&T planning by the Chinese government is the embodiment of Bernal's philosophy. However, there is still great room for improvement with respect to the independence of scientific developments and the scientific community in China. Chinese S&T programs have obvious Chinese characteristics as reflected in the final decision-making authority of the government in S&T planning when confronted with tricky problems and overemphasis on the utility of S&T.

With respect to the division between pure science and applied science, Stokes Donald E. offers a different view. In his book *Pasteur's Quadrant: Basic Science and Technological Innovation*, he puts forth the concept of use-inspired basic research and maintains that the relationship between government and the scientific community should be rebuilt. His argument that the government must play a crucial role in S&T planning has a major impact on the development of planning principles in the S&T domain.[8] As for the role of government, Henry Etzkowitz proposes a triple-helix innovation regime, arguing that both statist and laissez-faire regimes are not conducive to the development of S&T, and the triple helix of university-industry-government is the inevitable form of social

[6] Based on the views of Dr. Li Bin, Ph.D., Institute for the History of Natural Science Chinese Academy of Sciences.

[7] Fu (2008).

[8] Stokes (1997).

organization in the future.[9] The theories of Stokes and Henry Etzkowiz offer more rational perspectives on S&T planning.

In addition, John A. Alic et al. mention in *Beyond Spinoff: Military and Commercial Technologies in a Changing World* that the U.S. system is highly fragmented and is not conducive to the implementation of specific and consistent technology policies, but the open, polyarchic U. S. political structure also brings opportunities for innovation and facilitates flexibility and creativity. This shows that, unlike China, the United States does not have a centralized science planning system, but government intervention does exist in some areas to guide the development of S&T. In the study of industrial development and innovation in China and the United States, Kathrine. A. Walsh points out that the government still plays a central role in the 11th Five-Year Plan, and China's current industrial R&D paradigm is strikingly similar to that in the United States. Richard P. Suttmeier, a researcher specializing in Chinese S&T policies, in his study of the evolution of the relationship between science and society in China, asserts that the *Science and Technology Development Program (1956–1967)* has both technological and political[10] functions. The majority of literature currently published on the subject has concentrated on discussing the overall development patterns of S&T. There is limited literature on China's medium- and long-term S&T programs and planning philosophy.

Many Chinese researchers have debated over S&T planning. For example, there was a fierce debate about the guidelines of the first national S&T program released in 1956. Researchers have also voiced different opinions during the formulation of subsequent scientific and technological programs. But these debates are not about whether China needs S&T planning, but how to effectively plan the development of S&T provided that planning is necessary. Before the release of the *National Medium- and Long-Term Program for Science and Technology Development (2006–2020)*, there was also a debate about whether technology planning is necessary and whether big science projects are worth it.

In 2004, eleven Chinese biological scientists sent a letter to the top officials of the government, setting off an online debate over big science projects in Chinese academic circles. In this debate, researchers raised two major concerns with big science projects.

The first concern is that science cannot be planned. Research, especially biological research, is basically exploring the unknown with unclear results. Devoting large sums of money to R&D in a specific field is quite risky. The second concern is whether the fairness and transparency in the allocation and use of public R&D funds can be ensured. Many researchers are worried about whether the 2.5 billion Yuan allocated to the new large-scale biological research project can be used reasonably. It is very likely that the project may be a waste of time and money, causing China to fall further behind in biotechnology."[11]

Professor Rao Yi, a well-known Chinese scholar, believes that different approaches should be taken to planning S&T. For technology, the state may establish a general direction and make plans for projects. For science, the state should make plans for entire fields rather than specific projects. The large-scale project known as "Two Bombs and One Satellite" (atomic and hydrogen bombs and man-made satellite) is a technology development plan, not a science development plan. Because science is all about pushing the boundaries of human knowledge and exploring the unknown, it will be improper and risky to plan the development of science. Chinese policymakers are putting the cart before the horse by making plans for specific scientific projects. In fact, China's science system still has many fundamental problems. Rao Yi is very concerned about China's S&T planning. After the release of the *National Medium- and Long-Term Program*

[9]Etzkowitz (2005) (trans: Zhou, C.).

[10]It should be noted that "political" used in this book is a neutral academic term, the same as "science and technology" and "planning", and doesn't have such connotations as when used in policy documents or literary writings.

[11]Yuan (2004).

for Science and Technology Development (2006–2020), he published an article entitled "National Science Planning should be Focused on Improving the Science System Rather than Planning Specific Projects" to voice his concerns.[12] The lack of standardization in the review of S&T programs is another issue that causes researchers to question China's big science projects.

However, there are also a large number of Chinese scientists supporting big science projects. Supporters argue that it is not easy for domestic scientists to carry out scientific research on a tight budget. The 2.5 billion Yuan is a long-term R&D investment, and only a small part of the fund is allocated to specific biological research projects.

Zhu Qingshi, a member of the Chinese Academy of Sciences and president of the University of Science and Technology of China, said that big science projects are necessary. Some Chinese biologists said that there were no big science projects in the life sciences sphere in the past, and now the country has begun to attach importance to the development of life sciences.

Whether it is in China or the United States, the establishment of large-scale projects is a complex process, and thorough pre-project review and analysis are needed. The bigger the project, the more arguments it will set off. What really needs improvement are the processes of argumentation and decision making, which should be open and democratic. "A big science project should not be linked to the interests of an individual or a group. In the case of limited funding, government decisions will ultimately play a decisive role."[13] Unlike the Polanyi-Bernal debate, these debates were not theoretical but they reflected the evolution of Chinese policymakers' understanding of S&T planning. They were ended in a non-academic manner with no agreement reached. The particularity of China's S&T planning system has once again enhanced the significance of this research field.

There are many master's theses and doctoral dissertations on China's S&T policies, but there is limited published literature on Chinese S&T programs after the founding of the RPC. In the *Study of the Science and Technology Policies of the Communist Party of China since the Founding of the People's Republic of China*, Cui Luchun elaborates on the close relationship between China's S&T policy and the political line of the Communist Party of China and the details of the *Science and Technology Development Program (1956–1967)* and the *Science and Technology Development Program (1978–1985)*, but there is limited discussion of the planning principles and guidelines. Wang Huizhen maintains that there are two critical periods in the history of national S&T planning in China: the period from the 1950s to the late 1970s, during which the focus of planning was on defense, and the post-1980 period during which the focus was shifted to economic growth. In the *Programs and Plans for Development of Science and Technology in China* published by the National Defense Industry Press, Yuan Guangzeng et al. (1992) offer an overview of previous plans and programs. In general, the majority of literature published on this subject has been concentrated on collecting historical data and discussing industry-specific S&T plans. Few studies have systematically analyzed China's national S&T programs and planning theories over the past 50 years.

However, there are many unsystematic studies about S&T planning in China. Academician Zou Chenglu has offered some insights into China's S&T plans. He observes that China places too much emphasis on technology but too little on science and manages science in the same way it manages technology. Policymakers must consider the essential differences between the two when making plans and ensure room for maneuver. Interpersonal relationships are central to China's current S&T system, which are not good for innovation. China needs to draw a clear line between politics and S&T, place scientists in the management positions of its S&T system,

[12]According to the note of the author, an abridged version of this article was published on the *Southern Weekend* on 27 May 2004. The full article was published in the first issue of the revamped *Science and Technology in China* on 28 May 2004.

[13]Yuan (2004).

and implement merit-based, competitive selection practices. The government may determine the overall S&T policy, but specific issues should be reviewed by experts. The larger part of China's current public R&D funding system is still based on the rule of man. R&D projects and funds are still planned by the central government. Under such a statist regime, the success of projects and fundraising depends directly on whether researchers have established a good relationship with the government. There is also a lack of standardization in the review of R&D grant applications. Academician Zou Chenglu has shed some light on deep-rooted problems in China's S&T planning system. He also points out that the formulation and implementation of S&T plans in China are two separate stages and the weakness of the Chinese scientific community has forced the government to intervene. Due to the misalignment between the Chinese government's zealous pursuit of economic growth and institutional reforms in various sectors of society, Chinese S&T plans often have functions that shouldn't be assigned to a national S&T plan.

In the *Discussion of Problems in Basic Research in China*, Li Zhenzhen, a research fellow at the Institute of Science and Technology Policy, the Chinese Academy of Sciences, maintains that, because the government plays a dominant role in R&D resource allocation, the consistency or at least non-conflicting of scientific goals with government goals is an essential condition for incremental scientific advancements in China. When the autonomy of science remains weak, this link between scientific goals and government goals was often very fragile due to a lack of resources, which was reflected in the interruptions of incremental scientific advancements in China's history. Each interruption was caused by the conflict between scientific goals and government goals. How to improve plan-based R&D resource allocation has become a key issue that demands immediate attention.

In *A Review of Previous Studies on Science and Technology Planning in China*, Hu Weijia from the Institute for the History of Natural Sciences, Chinese Academy of Sciences offers an overview of the *Science and Technology Development Program (1956–1967)*, *Science and Technology Development Program (1963–1972)*, *Science and Technology Development Program (1978–1985)*. He edited the *Collected Materials on the Science and Technology in China*, which is a collection of documents concerning S&T policies that were issued by the Chinese government after the founding of the PRC and has significant implications for future research. Weijia Hu also edited the *Researches on Programs, Plans and Policies of Science and Technology in Modern China*, which is a collection of seven research papers on S&T plans and programs. In *Technology Foresight and Science and Technology Planning*, Fan Chunliang emphasizes the importance of technology foresight to S&T planning and maintains that technology foresight is a new contract between science and society. In the *Study on Science and Technology Planning Methodology*, Xu Yaozong et al. summarize three of China's national S&T plans and put forward recommendations on how to develop S&T plans. Hu Angang, an expert in the study of national conditions, published *Some Views on the Research of China's Medium- and Long-Term Science and Technology Planning Strategies* (Hu Angang published different articles with the same title in different publications) in *Review of Economic Research*. In this article, Hu discusses S&T planning principles in the new era from an economic perspective, proposes eight guidelines for S&T planning in the new century, and suggests that China's S&T planning system should be open, people-oriented, and innovation-focused.

The Ministry of Science and Technology, Tsinghua University and other government authorities and academic and research intuitions have issued some internal reports to put forward recommendations for previous S&T plans. For example, the *Science and Technology Development Plans and Programs of the People's Republic of China (1949–2005)* were completed at the end of 2007. The staff of the Department of Development Planning, the Ministry of Science and Technology spent ten years compiling this internal report. It records the formulation, implementation, and effects of China's S&T

plans and programs over the past five decades, aiming to trace the historical evolution of the S&T planning system in China. The report also reflects the impact of political, social, and economic contexts on the content, organization, and management of R&D activities and discusses policies that reflect such impact. Overall, this report is more like a collection of relevant materials. It does not systematically discuss the intrinsic relationship between S&T planning and China's political, economic, and defense systems. The discussion of S&T planning principles in the report is also limited, but it provides a systematic review of China's S&T plans.

In a nutshell, few researchers, either domestic or foreign, have systematically studied S&T plans issued by the Chinese government after the founding of the PRC. Most of the existing literature has concentrated on superficial discussion of plans and programs and rarely delved into the theoretical discussion. Therefore, researchers should be encouraged to conduct more basic and systematic studies on S&T planning principles in the future.

This chapter of this book provides an overview of the evolution of S&T programs and planning philosophy in three periods. Chapter 2 explores the impact of the development of S&T before the founding of the PRC on the S&T programs released after the founding of the PRC. There was a debate in history about whether science can be planned. However, there is no such debate after the founding of the PRC. The S&T system of the PRC was built on the guidelines of scientific and technological work of the Communist Party of China (CPC during the Yan'an period). Wang Hao, Wu Heng, and other CPC cadres in charge of S&T during the revolutionary period continued to hold the decision-making authority in the development of a national S&T program after the founding of the PRC. This chapter also studies scientific and technological advancements and the scientific community during KMT's reign.

Chapter 3 discusses S&T programs, including *Science and Technology Development Program (1956–1967)* and *Science and Technology Development Program (1963–1972)*, and planning philosophy in the central planning era. The main thrust of these two programs was to consolidate the new political regime and lay the foundation for national science, technology, and education systems. Therefore, during this period, China made remarkable strides in national defense research and institutional establishment. S&T plans and programs, aligned with the national interests and embodied in a series of scientific and technological projects, from "Two Bombs and One Satellite" to today's manned space missions, have played a crucial role in the development of S&T in China. The assistance provided by the Soviet Union and the nationwide S&T system were central to the first exponential growth in S&T in the history of the PRC. We repudiate the view that there was no S&T plan during the Cultural Revolution period and believe there were some weak forms of S&T planning.

There were three major scientific and technological planning principles during the central planning period: putting technology at the top of the agenda, prioritizing national core interests, and strengthening political functions. Due to the termination of the Soviet aid program, self-reliance became one of the governing principles of the *Science and Technology Development Program (1963–1972)*.

Chapter 4 sum up S&T programs and planning philosophy during the transition period (1978–2000). The adoption of the reform and opening-up policy marked the beginning of the transition period. S&T programs during this period include without limitation *Science and Technology Development Program (1978–1985)*, *National Medium and Long-Term Science and Technology Development Program (1986–2000)*, *Ten-Year Program & the Eighth Five-Year Plan for the Development of Science and Technology in the People's Republic of China (1991–1995–2000)*, *Ninth Five-Year Plan for National Science and Technology Development (until 2010)*. The main thrust of S&T programs during this period is to promote economic growth and social development in an all-round way, which is clearly reflected in the planning philosophy stipulated by the Chinese government. The reform of the S&T system is another theme in this period. A great number of scientific and technological projects were completed under the guidance of the aforementioned plans and programs. Studies have shown that, despite remarkable achievements in S&T, basic research garnered little attention and lagged behind.

During the transition period, the reform and opening-up policy had a huge impact on S&T planning. With the shift of focus and deepening of reform, S&T played an increasingly central role in national economic growth and more emphasis was placed on the utility of S&T. The reform of the S&T system became an important component of the S&T planning system. "Technology first", "utility", and "economic functions" were the key words of S&T planning during the transition period.

Chapter 5 provides great details about S&T programs in the 21st century, including those prepared in the 21st century: *National Medium- and Long-Term Program for Science and Technology Development (2006–2020)*, *Tenth Five-Year Plan for National Economic and Social Development—Science, Technology and Education*, and *Eleventh Five-Year Plan for National Science and Technology Development*. The establishment of a national innovation system shows that the CPC's understanding of S&T has matured. Basic research has received more attention. A new paradigm that combines the strengths of the statist and laissez-faire regimes and engages all stakeholders, including the central government, local governments, and industries, has gradually taken shape. The introduction of the people-oriented principle shows that the Chinese policymakers have become aware of the overemphasis on the utility of science. This awareness has led to changes in planning philosophy. China's current S&T planning philosophy is that an S&T program should promote people-oriented and indigenous innovation with a focus on technological innovation.

Chapter 6 analyzes S&T planning philosophy which are influenced by national development ones. The evolution of S&T planning philosophy reflect the changes in national policies and policymakers' understanding of S&T. Greater details are provided on the utilitarianism in S&T plans during the transition period, which has a far-reaching impact on contemporary academic research. This chapter also provide a case study (Shanghai Y-10) to illustrate that an S&T program is the result of the complex effects of various factors.

Chapter 7 sums up the book and present the conclusions. Resource constraints faced by China make scientific and technological planning necessary. As we now have a better understanding of the dos and don'ts in S&T planning, this chapter will provides more details on priorities in basic research. China's current weak basic research and the lack of independence of the Chinese scientific community are related to problems in S&T planning.

Studies have shown that science and technology are more of a means to an end rather than goals in China. Therefore, S&T programs in China often have many functions which should not be vested in them. In the long run, this mindset will not be conducive to the development of S&T, especially basic research, in China. Researchers and policymakers should rationally recognize the strengths and weaknesses of China's S&T planning system, which is tilted towards the statist regime. The state has substantial control over R&D resources. In practice, science and technology are regarded by the Chinese policymakers as a tool to increase goods and improve the well-being of people. The overemphasis on economic growth has made S&T an accessory to political and economic systems. Under the current system, R&D activities are determined by the state rather than scientists.

By examining the ten medium- and long-term S&T programs and the evolution of planning philosophy since the founding of the People's Republic of China, we can see that S&T programs are affected by existing scientific and technological conditions, economic strength, international background, and long-term visions, implying the relationship between technology and basic research. China's medium- and long-term S&T programs have obvious Chinese characteristics and combine the visions of both the governing party and the scientific community. The effects of this combination depending on the environment at the time and how well the governing party understands scientific and technological development. China's S&T programs focus on technological development and reflect the country's desire to catch up with developed countries.

The evolution of S&T programs and planning philosophy during the transition period is accompanied by the establishment and reform of China's S&T system. After the reform, the system is infused with a strong utilitarian tendency, which has contributed significantly to China's fast economic growth but also has many undesirable consequences. Chinese researchers and policymakers should reflect on the problems in the S&T system and the weaknesses of the Chinese scientific community. Failure to attach importance and give freedom to basic science and research is the main reason for the lack of Nobel laureates (in the natural science category) in China. Fortunately, since the beginning of the 21st century, the CPC and the Chinese government have had a deeper understanding of the nature of S&T and human development and given more thoughts to the consequences of overemphasis on the utility of science, which has been reflected in *National Medium- and Long-Term Program for Science and Technology Development (2006-2020)*, and *Eleventh Five-Year Plan for National Science and Technology Development*. In fact, the Chinese policymakers' reflection on the consequences in this connection is also manifested in the introduction of the "Scientific Outlook on Development" philosophy.

Many lessons can be learned from the planning experience over the past five-seven decades. Will the utilitarian tendency reflected in S&T planning principles and R&D activities have undesired effects on future scientific and technological development? Whether should S&T be regarded as an end or a means to an end? What effects do the legacy of ancient times and the weak research foundation in modern times on the evolution of S&T programs? How to deal with the relationship between the autonomy of science and government intervention? There are a lot of unanswered questions, hence a long way to go in the terms of the S&T planning research.

References

Etzkowitz, H. 2005. *The Triple Helix Model*. Beijing: Oriental Press.
Fan, C. 2002. Planning and freedom of science. *Studies in Science of Science*, 1.
Fu, B. 2008. Discussion of Wu Xuezhou's science planning theory. *Studies in Science of Science*, 6.
Stokes, Donald E. 1997. *Pasteur's Quadrant: Basic Science and Technological Innovation*. Washington, D.C.: Brookings Institution Press. Chinese Edition: Stokes, Donald E. 1999. *Pasteur's Quadrant: Basic Science and Technological Innovation*. Beijing: China Science Publishing & Media Ltd.
Yuan, C. 2004. The debate over big science projects. *China Youth Daily*, 10 August 2004.

2 The Impact of Science and Technology Development on Science and Technology Planning Before the Founding of the People's Republic of China

Abstract

This chapter describes the development of science and technology in China before 1949, and its possible impact on science and technology planning after 1949. When the Communist Party of China was in the Revolutionary Base areas, it began to practice science and technology and trained a large number of senior scientific and technological cadres. The level of science, technology, and scientists in the Kuomintang controlled areas was generally higher, and some achievements with world characteristics have been made. After the founding of new China in 1949, the science and technology policymakers of the Chinese government mainly came from the science and technology cadres in the Revolutionary Base areas. They treated the Returned experts differently from those in the Kuomintang controlled areas. The effective command management in the military struggle continued to introduce science and technology planning after 1949. The policy of whether to maintain the freedom and independence of intellectuals wavered after the founding of new China, which affected the development of science and technology planning and basic research.

Keywords

Science and technology in base areas · Science and technology in kuomintang controlled areas · Chinese science and technology makers · Intellectual policy · Ideological transformation · Scientific community

2.1 Science and Technology from the Period of Revolutionary Base Area to Before the Founding of the People's Republic of China

To study the medium and long-term science and technology program and its philosophy since the founding of the People's Republic of China (PRC), we need to shift our focus on the guiding concept of China's science and technology development before the founding of the PRC. Before the founding of the PRC, especially before the victory of the anti-Japanese war, China's scientific research showed a diversified development trend due to regime separation. There were mainly three forces: the Communist Party of China (CPC), the Chinese Kuomintang, and the Japanese invaders. This article can not only examine the scientific and technological undertakings of the CPC in the field of local political power but also review technological development concepts in the Kuomintang-ruled

Z. Chen, *China's Medium and Long-Term Science and Technology Program*,
https://doi.org/10.1007/978-3-030-75146-3_2

areas so as to make comparisons. Japan had some scientific research institutions in northeast China, but most of them served the aggression by adopting inhuman methods to do experiments, so they were not designed for real scientific research. This section discusses the CPC's scientific practice before the founding of the PRC and its influence on sci-tech planning after the founding of the PRC.

I. A Brief Historical Review

If we look at the long history of China, we may see some characteristics of China's scientific and technological development. For example, after the Qin and Han dynasties, the function of attaching importance to the technology itself is far greater than the rational power of academia itself. The development of science, especially technology, is accompanied by the advancement of the maximum goal of the interests of the ruling clique. It is not the task of this book to elaborate on the history of ancient science and technology. However, judged from the open answer to "Needham's Question", the reason why China lags behind in science and technology is far more difficult to answer than why it leads. This section skims over this Question.

Emperor Kangxi of Qing Dynasty gave high priority to science and technology but did not fully accept Western sciences and the rising Western civilization. In the middle and late Qing Dynasty, the entire feudal society went to the end. The ruling clique was more concerned with maintaining and extending its regime, and its interest in science and technology became increasingly utilitarian.

Ancient science, especially ancient technology, became a tool for safeguarding the ruling interests. Beyond this goal, there will be no support. After the Opium War, the Westernization Movement, which lasted for 30 years or so, took place with the goal of "learning from foreigners and mastering skills to control them". The supremacy of technology shows that its utilitarian goal is self-evident. In the last decade of the late Qing Dynasty, there emerged the "New Deal in the late Qing Dynasty". The biggest shock was the abolition of the imperial examination system in 1905, which had been implemented for more than 1,300 years. The thousand-year-long imperial examination resulted in Chinese intellectuals had "belief in imperial examinations" but no "belief in science".[1]

In modern society after 1840, science and technology, a social factor bringing out obvious social utility, gradually became a tool in the struggle for political power. At the beginning of the twentieth century, the New Culture Movement and the May Fourth Movement held high the banner of science and democracy, achieving great victories. However, at this time, science was used as a progressive slogan representing the direction of social progress and its development did not receive much attention. Science has become a positive tool for social movements.

Dr. Sun Yat-sen had great ideals in his life. The *Nation-building Strategy* written by Dr. Sun Yat-sen has about 200,000 words, detailing his complete plan to establish China into a "country with great unity". It is divided into three parts: mindset adjustment, material construction, and social construction. Its detailed content and careful plan are incredible. Many plans have not yet been fully achieved or have just been completed. Material construction involves detailed plans covering transportation, ports, railways, water power, mining, agriculture, etc. There is no specific discussion on the development of science and technology in *Nation-building Strategy*. However, from the whole discussion, Dr. Sun Yat-sen had a full understanding of the role of technology and took science, especially technology, as a tool for the construction of social materials. "We need to use all sorts of advanced machines and well-organized manpower to vigorously develop our industries and revitalize the Chinese nation."[2] However, after Dr. Sun Yat-sen's death, the Kuomintang government did not govern and build China according to his ideal. It was not until after the founding of the PRC that the Chinese government under the leadership of the CPC began to build the country in an all-round way. Many plans are actually the same as

[1]Chen (2007).

[2]Sun (1956).

Dr. Sun Yat-sen's *Nation-building Strategy*, emphasizing the effectiveness of technology while ignoring scientific rationality. The tradition of attaching much importance to the technological efficacy and ignoring scientific rationality remains existed in the science and technology planning after the founding of the PRC, which results in each science and technology plan placing great emphasis on technology and a little emphasis on its scientific part. China's science and technology planning, called technology planning, can also be evidenced here.

II. Science and Technology during the Period of the Revolutionary Base

It is necessary to examine the concept of makers and organizations of science and technology planning when exploring the concept of science and technology planning after the founding of the PRC.

Therefore, it is obviously necessary to examine the guiding ideology of the CPC, the leadership of the Chinese government. Science and technology were not specifically covered in the first program of the CPC when it was founded in 1921. But the program reads, "The CPC recognizes the Soviet management system, organizes workers, peasants and soldiers, and takes social revolution as the main purpose of its policies. The Communist Party of China completely cut off any ties with bourgeois intellectuals desperate for fame and fortune and other similar parties."[3] It can be seen from this point that the Communist Party of China is based on the worker-peasant class and struggles against imperialism, feudalism and the big bourgeoisie. Therefore, which class intellectuals belong to is related to the division between ourselves and the enemy. Sci-tech workers are intellectuals. The CPC has repeatedly experienced a tortuous process of understanding intellectuals and dividing their class attributes. The policy of treating intellectuals holds the key to the development of science and technology. Generally speaking, intellectuals received high courtesy before the founding of the PRC. The courtesy is derived not only from intellectuals' direct contribution to the war but also from political problems that landlords and intellectuals in the middle forces of the society influence the popular support or opposition for the future of the revolution.

This part mainly describes historical facts of the development of science and technology by the CPC after gaining local power before the founding of the PRC, so as to examine the guiding principles thereafter.

1. Senior Leaders Have Gradually Attached Much Importance to Science and Technology

In its long struggle to seize power, the CPC deeply understood the importance of materials and equipment. The demand for science and technology had increased unprecedentedly under extremely difficult circumstances. Mao Zedong gave top priority to the role of science and technology in solving China's problems as early as 1926.[4] He put forward in *National Revolution and Peasant Movement* that "the peasant problem originally includes two aspects: oppression problems caused by imperialism, warlords, and landlord class and natural oppression problems such as floods, droughts, natural disasters, diseases and pests, poor technology, and production reduction. The former problem is, of course, an urgent one at present, and comrades are all focused on it. However, the latter problem is also severe. We cannot but take an active interest in it. To solve the latter problem, we need revolutionary political power and scientific methods throughout the country. It is not something that can be done immediately."[5] At the founding meeting of Yan'an Natural Science Research Association, Mao Zedong presented that "in order to gain freedom in society, people must use social science to understand and transform society and carry out a social revolution. In order to obtain freedom from nature, people must understand nature with natural science, overcome and transform nature, and get

[3]The first outline of the CPC (Translated Russian Edition) (1921) [DB/OL], xinhuanet: http://news.xinhua-net.com.

[4]Li and Junfeng (2004).

[5]Mao (1993a).

freedom from nature."[6] It can be seen that Mao Zedong, as the first-generation core leader of the Communist Party of China, had a clear understanding of the importance of science and technology. Mao Zedong also attached great importance to intellectuals who mastered science and technology before the founding of the PRC.

In 1939, Mao Zedong drafted, on behalf of the CPC Central Committee, *Recruit Large Numbers of Intellectuals*, marking the beginning of a new stage in the understanding and treatment of intellectuals by the CPC. He criticized many army cadres for not knowing the importance of intellectuals and even fearing them. It is required that "all party and Party-led armies in war zones should recruit a great deal of intellectuals to join our army, our schools, and the government."[7] In January 1941, in a letter to Mao Anying who was Mao's son, Mao Zedong said, "There is only one thing I would like to suggest to you. While you are still young, you should learn more about natural sciences and fewer politics. Politics need to be discussed, but it is advisable to study natural science with great concentration, supplemented by social science. In short, we need to pay attention to science, for science is true knowledge and will be used endlessly in the future."[8]

Comrade Zhu De presented in the celebration of the first annual meeting of the Natural Science Research Association, "Natural science is a great force. The progress of natural science, the development of workers and peasants in all walks of life, the improvement of production capacity, the development and rational utilization of natural resources, and the sound management of industry. Only by doing so can we enrich our strength, enrich the combat effectiveness of the army, and enable the people to obtain a rich life..."[9] Other top leaders also have a clear understanding of the important role of science and technology in war. Although Mao Zedong himself gave top priority to natural science,[10] the CPC's understanding of natural science also went through a process. Influenced by the reality of China at that time, the status of natural science was not as high as that of technical science.

2. The Red Army's Early Preliminary Techniques

The revolutionary struggle and bases led by the CPC after its establishment are generally divided into three stages: the Agrarian Revolution, the War of Resistance, and the War of Liberation. During the Agrarian Revolution, leaders of all revolutionary bases were aware of the extreme importance of weapons, equipment, and materials in responding to the war. Therefore, more attention would be paid to science and technology departments that were closely related to the war. In this process, related technologies such as stamping and alignment had been developed. Secondly, agricultural production also received attention. Traditionally, peasant uprisings in China were caused by agricultural bankruptcy. Therefore, agricultural production and related technologies were valued by the Red Army. Mao Zedong said in his report titled *Our Economic Policy* that "at present, it is definite that problems related to national agriculture and collective agriculture cannot be raised. However, in order to promote the development of agriculture, it is urgently needed to organize small-scale agricultural experimental fields and set up agricultural research schools and agricultural product exhibition centers."[11] Agricultural schools were set up on the bases to train agricultural cadres. In 1933, the Central Agricultural School was established in Ruijin, the then Red Capital, to carry out an initial experiment of production increase. As China is a big agricultural country, the final victory of the Chinese revolution was largely due to the support of the farmers. Thanks to the experience accumulated in the war years, the unified planning and improvement of agricultural technology were put high on the agenda of all previous scientific and technological plans

[6]Mao (1993b).

[7]Mao (1991).

[8]Mao (1983).

[9]Historical Materials of Yan'an Academy of Natural Sciences (1986).

[10]Yu Guangyuan. Mao Zedong and Dialectics of Nature. Study on Dialectics of Nature, 1993, (12).

[11]Mao (1966).

after the founding of the PRC. Agricultural technology always holds an important position in many technological plans. In addition, the medical sector is closely linked to the war. The Red Army had set up hospitals for various purposes, pharmaceutical factories, and health magazines.

The notable technological achievement in the early stage of the Red Army was the establishment of radio stations. In the winter of 1929, Li Qiang and Zhang Shenchuan successfully installed a 50-W radio transceiver in Shanghai. Radio technology emerged at the beginning of the twentieth century and was used for a short period in wars. Therefore, it is a relatively modern high technology for the Red Army. In December 1940, the CPC also set up Yan'an Xinhua Radio Station and began broadcasting.

Science and technology in the Agrarian Revolution period is still at a very preliminary stage. It is not so much science as technologies, or relatively primary technologies, that have been developed. All served the military and had no time for theoretical summary and promotion, which is far from the conscious development of science and technology in today's sense. However, the CPC began to contact and think about the direct effect of modern science, especially technology, on the armed revolution. Mao Zedong and other leaders had begun to notice the great value of science and technology. The initially trained technical cadres would further promote technological development in the future struggle.

III. Science and Technology during the Anti-Japanese War

1. Early Science and Technology Policy

During the Anti-Japanese War, the CPC's science and technology were further developed by leaps and bounds and institutionalized. This point can be seen from the Party's documents and policies at that time. In February 1940, the Shaanxi-Gansu-Ningxia Border Region's *Declaration of Natural Science Research Association* called for "strengthening the movement of natural science, and mastering and improving natural science become a fighting force in the War of Resistance Against Japan … we should use the front of natural science to smash the enemy's economic blockade and strike its cultural policy."[12] Liberation Daily, the official newspaper of the CPC Central Committee, had successively published important editorials such as *On Economic and Technological Work*, *Reward Free Research*, *Welcome Scientific and Artistic Talents*, and *Advocate Natural Science*. It stated that "In the economic construction of the border region, technological science, especially a decisive factor … natural science is the necessary step for the development of the anti-Japanese-war economic and cultural construction to achieve the goal of persisting in the long-term anti-Japanese war and improving the people's happiness … advocating natural science is to improve the development of agricultural and industrial production technology and increase material production."[13] In 1941, the Political Bureau of the CPC Central Committee approved the Administrative Program of Shaanxi-Gansu-Ningxia Border Region, clearly stating, "Respect intellectuals, advocate scientific knowledge and literary and artistic movements, and welcome scientific and artistic talents and medical talents." In May 1942, the Shaanxi-Gansu-Ningxia Border Region government issued the *Order on the Treatment Standard for Technical Cadres of Construction Department*." Later, it approved the implementation of the *1943 Favored Treatment Measures for Technical Cadres*. The two decrees provide for preferential policies to attract sci-tech talents.[14] These policies are more systematic and meaningful than those of the Agrarian Revolution and can be regarded as documents of the early scientific and technological policies of the CPC.

In January 1939, Lin Boqu, the then chairman of the Shaanxi-Gansu-Ningxia Border Region government, put forward sci-tech tasks of the Shaanxi-Gansu-Ningxia Border Region government in the new stage of the Anti-Japanese War in his Report on the Work of the First Provincial

[12]He (1990).

[13]Wu (1983: 62–63).

[14]Wu (1983: 61).

Assembly of the Government of the Shaanxi-Gansu-Ningxia Border Region. The tasks were to "set up practical scientific research institutes to develop scientific research in industry, plants, civil engineering, animals, chemistry, geology, and other fields, and to train scientific talents to meet the needs of developing the national defense and economy." In 1941, the Shanxi-Chahar-Hebei Border Region promulgated and implemented the *Regulations on Rewarding Production Technologies in the Shanxi-Chahar-Hebei Border Region*, which provides that "sci-tech personnel shall be given honor and material rewards for their technological inventions, improvements, manufacturing and discoveries in agriculture, industry, mining, forestry, animal husbandry, water conservancy, and medical treatment."[15] It can be seen that during the Anti-Japanese War, the CPC's scientific and technological policy of "serving the Anti-Japanese War and the founding of the PRC with science and technology" was gradually developed. There should be a clear guiding ideology for science and technology, which has played a reference role in the formulation of science and technology planning after the founding of the PRC.

2. Higher Science and Technology Institutions and Scientific Organizations

In May 1939, the CPC Central Committee decided to establish Yan'an Natural Science Research Institute and later renamed it Yan'an Academy of Natural Sciences (YANS) (Figs. 2.1 and 2.2). YANS joined Yan'an University in April 1943. It is the predecessor of today's Beijing Institute of Technology. It was the first institution of higher learning established by the CPC during the Anti-Japanese War to train sci-tech cadres. The school-running purpose of YANS is "to train technical cadres and specialized technical talents who serve the founding of the PRC and the Anti-Japanese War."[16] It can be seen that more attention was paid to the technical aspects that directly facilitated the Anti-Japanese War at that time. In spite of this, YANS, a very important institution of the CPC in the modern history of China, mainly focuses on natural sciences. On November 1, 1949, the Chinese Academy of Sciences was established on the basis of the former Academia Sinica, Beiping Research Institute, and Yan'an Academy of Natural Sciences.

The leaders of YANS had devoted a lot of effort to its development. Before the 359th Brigade developed Nanniwan, YANS conducted an investigation of Nanniwan and considered it suitable for reclamation, which led Zhu De to make up his mind to transfer the 359th Brigade to open up the wasteland there.[17] After the Rectification Movement began in 1942, Comrade Xu Teli proposed to hold a major discussion on the educational policy of Yan'an Academy of Natural Sciences. Through the discussion, everyone was aware that the scientific and technological education in the Anti-Japanese base must serve the Anti-Japanese War and must follow the educational policy of combining with production practice. The emphasis of everything on efficiency was the last resort when material conditions were very limited, but it also reflected some leaders' linear understanding of technology. It has a direct impact on the formulation of the concept of science and technology planning after the founding of the PRC. YANS had trained more than 500 senior and middle-level sci-tech cadres, who have played an important role in the scientific and technological undertakings after the founding of the PRC. The idea of "being useful and effective and serving one center" conceived by them in the war was brought into the science and technology policy of the PRC consciously or unconsciously, which had a subtle influence on the formulation of the science and technology planning of the PRC.

China Medical University,[18] Yan'an's second higher science and technology school, gave medical treatment to the injured to make a contribution to the victory of the Anti-Japanese War. At the same time, many sci-tech organizations

[15]Wu (1989: 45).

[16]Historical Materials of Yan'an Academy of Natural Sciences (1986).

[17]Wang (2002: 26–31).

[18]Qu (1985: 148–149).

Fig. 2.1 Experimental class of Yan'an Academy of Natural Sciences

Fig. 2.2 The site of the former Yan'an Academy of Natural Sciences

Fig. 2.3 The Yan'an conference site for celebrating May Day attended by Shen Hong in 1939[22]

had been set up in the border region. In February 1938, the Defense Science Society of the Border Region, founded by Gao Shiqi, Chen Kangbai, and Li Shijun, was the first sci-tech group in the resistance base. Its aim was to research and develop the science of national defense and increase the scientific knowledge of the public.[19] The Society had some quality of science popularization. Later, Gao Shiqi became the first director of the China Science Popularization Research Institute. In February 1940, the Natural Science Research Association of Shaanxi-Gansu-Ningxia Border Region was established in Yan'an under the sponsorship and support of Mao Zedong, Wu Yuzhang, and others, with more than 300 scientists and technicians. The Declaration of the Natural Science Research Association proposed four tasks designed "to carry out the popularization of natural sciences, to engage in the exploration of natural sciences, to solve theoretical and practical problems in natural sciences, and to use materialist dialectics to study natural sciences."[20] The activities of these scientific groups are very limited due to the influence of the war and have little effect on the popularization of science and the improvement of the quality of the people. However, these activities mean that the CPC began some attempts to organize and lead scientific communities. This had some impact on the formation and growth of the scientific communities after the founding of the PRC.

3. Achievements in Science and Technology

Science and technology laid an important material foundation for the victory of the Anti-Japanese War. Agricultural production is the foundation of the bases. The agricultural industry increased production significantly. For example, "the promotion of the Golden Queen corn increased grain production by 25–30% in the Shanxi-Hebei-Shandong-Henan Resistance Base."[21] Regarding industry, science and technology have made important contributions. In early 1938, Shen Hong moved his small machinery factory from Shanghai to Yan'an,

[19]He (1990).

[20]Ibid., 388.

[21]Zhang (1999: 83).

[22]China Society of Mechanical Engineering. Picture Display in Commemoration of the Centennial Birthday of Comrade Shen Hong: http://www.cmes.org/cmes/ftp/shenhong/shenhong/shen/toupian.htm.

bringing ten machines and seven skilled workers. The ten machines underpinned Yan'an's modern industry and made important contributions to Yan'an's machine manufacturing, especially the upgrading of military manufacturing (Fig. 2.3).

After the founding of the PRC, Shen Hong was appointed as the deputy minister of the First Machinery Industry Ministry and led the research and manufacture of China's first 12,000-ton all-welded forging hydraulic press. He advocated the working method of the consistent system including research (Fig. 2.4), testing, design and other four steps. In 1980, he was elected a member of the Chinese Academy of Sciences (Academician).[23]

During the Anti-Japanese War, the chemical industry in the border region also made great progress and was able to produce the gasoline alternative. It is also worth mentioning that the border region had its preliminary petroleum industry. China has a long history of discovering and using petroleum. As early as the Eastern Han Dynasty, the famous historian Ban Gu (A.D.32–92) recorded in the *Geography of the Han Dynasty* that "Gaonu (now Yan'an) has water that can burn." Shen Kuo (A.D.1031–1095), a famous scientist in the Northern Song Dynasty, made a meticulous investigation of Yanchang Oil in Northern Shaanxi. He wrote in "Meng Xi Bi Tan", or the encyclopedia of the Song Dynasty, that "there is petroleum in Yanchang and the old record reads that Gaonu County produces grease water or petroleum."[24] The late Qing government exploited the mine in 1905. In 1937, the Construction Department of the Border Region government took over Yanchang oilfield. In 1938, the CPC Central Committee decided to resume oil production and appointed Chen Zhenxia, an oil engineer, as the principal engineer. In the spring of 1940, the first new well began to be drilled, producing about 10 tons of crude oil per month. Technicians had also created a technology of water shutoff by the empty pipe. After the founding of the PRC, China's oil industry improved this method and it has been used to this day.[25] During the hard years of the Anti-Japanese War, Yanchang Oil Factory produced 3,155 tons of crude oil and other oils. For the anti-Japanese soldiers and civilians who were equipped with millet plus rifles at that time, the exploitation and application of oil in the War of Resistance against Japan was a major sci-tech achievement. In May 1944, Chairman Mao Zedong wrote the inscription "Hard Work" as an encouragement to Chen Zhenxia, head of the oil field, who was honored with the title of special model worker in the Border Region (Fig. 2.5).[26] After the founding of the PRC, Yanchang oil field continued to develop. By the end of 2004, the original headquarters of Yanchang Oil Field Administration had proved 588 million tons of

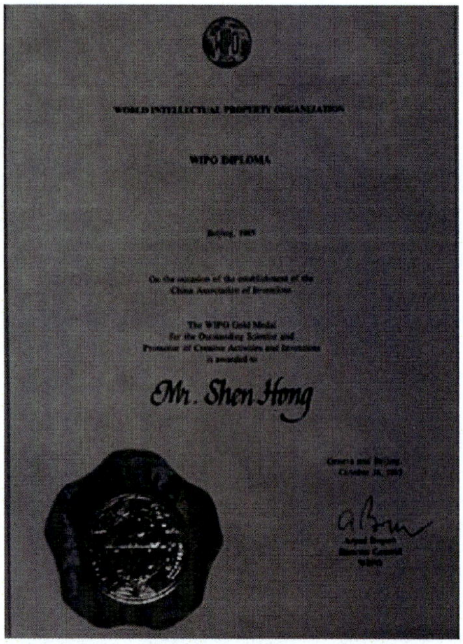

Fig. 2.4 World Intellectual Property Organization in 1985 Gold Award Certificate to Shen Hong for Outstanding Scientists and innovative inventors[27]

[23]People's Daily, May 20, 2006, 1st Section.

[24]Qu (2005: 92).

[25]Wu (1988).

[26]Shi (2006).

[27]China Society of Mechanical Engineering. Picture Display in Commemoration of the Centennial Birthday of Comrade Shen Hong: http://www.cmes.org/cmes/ftp/shenhong/shenhong/shen/toupian.htm.

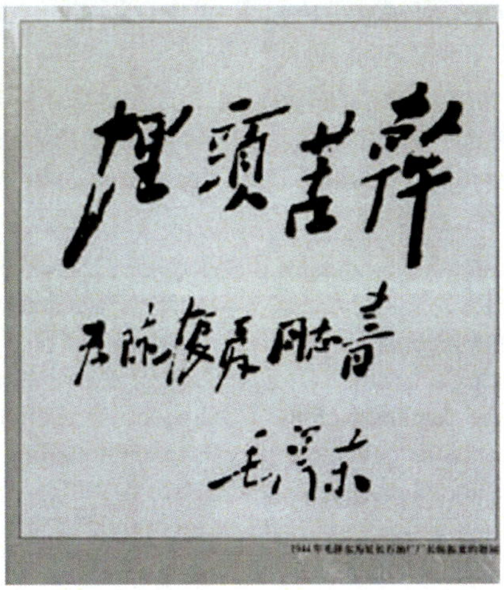

Fig. 2.5 Chairman Mao Zedong's inscription for Chen Zhenxia[28]

oil reserves, ranking 10th among China's oil fields. In 2014, the output of crude oil reached 12.77 million tons, and in 2013 the oil field entered the world's top 500 enterprises.

Through the development of science and technology, the Shaanxi-Gansu-Ningxia, Shanxi-Chahar-Hebei, and other border regions had formed their own industrial systems almost from scratch. Military technology has made rapid progress, providing necessary material support for the anti-Japanese base. Meanwhile, they had trained thousands of revolutionary sci-tech cadres, laying a solid foundation of talents for the development of sci-tech undertakings in the PRC.

Although great achievements have been made in science and technology in the anti-Japanese base areas, the main purpose is not to develop science and technology itself but to serve the anti-Japanese war. At that time, the general principle of science and technology policies was that "all scientific research, whether general or specialized, theoretical or technical, should be subordinated to the political, economic and cultural development in order to insist on the long-term War of Resistance against Japan and enhance the well-being of the people."[29] To serve the Anti-Japanese War, Zhu De said at the 7th National Congress of the CPC in 1945, "in order for the future counterattack, we must learn to master new technologies from now on. The most important thing is to learn artillery technology … Our army needs to keep learning and master all new military technologies we must grasp. It is very important."[30] The extraordinary hardships of the Anti-Japanese War enabled people to have direct perceptual knowledge of the role of science and technology, especially the role of technology. The perceptual knowledge laid a historical foundation for the sci-tech planning after the founding of the PRC.

IV. Science and Technology during the War of Liberation

The War of Liberation was shorter than the Chinese People's War of Resistance against Japanese Aggression, and the Liberated Areas went through only two stages of strategic defensive and strategic counter-offensive. Thanks to the large-scale movement of the army, science, and technology are also developing fast. However, science and technology mainly aim to serve the War of Liberation and war-related technologies continue to be developed.

To train a large number of scientific and technological cadres for the victory of the liberation war and military needs, liberated areas attached more importance to science and technology education. In January 1946, the financial meeting of the Administrative Committee of the Shanxi-Chahar-Hebei Border Area was held, pointing out that we need to cultivate agricultural and industrial technical cadres in a planned way and care for and give preferential treatment to existing technicians.[31] The resolution of the work report of the Shanxi-Hebei-Shandong-Henan Border Region government in May 1946

[28]*Photo Source* Qilu Petrochemical News, September 30, 2011.

[29]Wu (1983: 64).

[30]Zhu (1983).

[31]China Modern Economic History Team, Institute of Economics, Chinese Academy of Social Sciences (1986: 30).

provided that a large number of specialized science and technology schools should be built and we should respect sci-tech talents.[32] In 1947 and 1948, the Chinese People's Liberation Army successively set up seven aviation schools and several naval schools, as well as medical, communication, and equipment schools, training a large number of technical cadres.

Liberated areas continued to increase agricultural production. The North China Agricultural Science Institute, established in April 1949, was the first comprehensive agricultural research institution in the liberated areas. Chen Fengtong (A.D.1897–1980) served as director. Chen, a famous agronomist and agricultural expert specialized in science and technology management, was one of the main pioneers and leaders of agricultural science and technology in the PRC. It is worth mentioning that the state-owned machinery farm, established in the Northeast Liberated Area in 1947, strongly supported the Chinese People's Liberation War and set the socialist direction for the future development of individual farming.[33] Military research and production remained the focus of science and technology during China's War of Liberation. During this period, the focus of military production and research was shifted to assault weapons.

The Northeast Natural Science Research Association, established in Harbin in 1948, was aimed to "unite natural science workers and scientific workers who are willing to serve the people, promote the development of scientific theories and technologies, participate in various construction undertakings launched in northeast China and the PRC, and integrate scientific theories and technologies with the labor of the majority of the people."[34] In May 1949, the Northeast Natural Science Research Association, coupled with the Chinese Association for the Advancement of Science (CAAS), the Chinese Association of Natural Science (CANS), and the Chinese Association of Scientific Workers, proposed to hold a "National Congress of Natural Science Workers" to seek support for the scientific undertakings of the PRC.

[32]Huangfu (1989: 321).

[33]Zhu (1987: 175).

[34]Wu (1984a: 288).

V. Impact of Scientific and Technological Practice on Scientific and Technological Planning after the Founding of the People's Republic of China

Before the founding of the PRC, the CPC carried out the long and effective practice of science and technology and yielded many important results, making scientific and technological contributions to the founding of the PRC. Because the CPC became the governing party after the founding of the PRC, the scientific and technological practice of the CPC before the founding of the PRC had a great direct impact on the scientific and technological planning after the founding of the PRC.

1. Accumulating a certain sci-tech foundation

The power of the CPC is mainly composed of workers and peasants. In the revolutionary era, the cultural literacy of the worker-peasant class was low and illiteracy was common. Therefore, with little scientific knowledge, we did not know what role science and technology play at first. From the revolutionary base to the War of Resistance Against Japan and then to the War of Liberation, the scientific and technological level and achievements of the CPC have been improved with the growth of the Party. In the advancement, the CPC had accumulated a certain scientific and technological foundation, gaining a direct understanding of the role of science and technology. Not only has the basic knowledge been accumulated but also the material basis of science and technology. Military production including military hardware, industrial production, agricultural production, and medical and health care have laid the essential material foundation for the establishment of the new Chinese regime.

The revolutionary practice and sci-tech practice in the 28 years before the founding of the PRC empowered the CPC to grasp and understand the preliminary sci-tech knowledge and the interaction between modern science and society. The formulation and implementation of policies regarding science and technology accumulated valuable experience for the formulation of science and technology policies after the founding of the PRC. There was a certain perceptual knowledge of science and technology advancing the war and improving national strength. Therefore, after the founding of the PRC, the sci-tech planning, a unified national sci-tech development arrangement, was inevitable. In this sense, there is no similar debate in China about the necessity of sci-tech planning to that in Britain. China's intentions can be reflected in science and technology planning.

2. The enhanced historical position of technological tradition

It must be pointed out that, before the founding of the PRC, the CPC's perceptual knowledge of science and technology was relatively superficial and preliminary. As the function of science and technology is also based on technical efficiency, an important feature and concept of science and technology planning after the founding of the PRC are to "develop technology before planning science and technology". The CPC did not invest too much in fundamental research before the founding of New China. As such, its understanding of the role of basic sciences is not deep, even vague, and neglected.

It resulted in the neglecting of basic sciences in all previous sci-tech plans after the founding of the PRC. Science and technology planners and implementers have a tendency to filter and ignore basic sciences, which leads to the long-term weakness of basic sciences in China.

Science comes from the traditions of scholars, while technology originates from the traditions of technologies. These traditions in the two aspects, according to the development history of China's ancient science and technology, exist, but the traditions of technologies are more evident. The arduous struggle before the founding of the PRC made the position of technologies unprecedentedly improved, which were indispensable for the existence and growth of the CPC's political power and the victory of the war. Technological traditions are further powerful. The sci-tech planning after the founding of the PRC continues this long-standing technology tradition. The planning is regarded as technology planning, which originates from technology tradition. Even the basic science is managed by the method of managerial technology, such as how many concrete achievements the basic science is expected to produce in a specific period of time. After the founding of the PRC, the emergence of nationwide steel smelting, technological innovation, and nationwide science campaign is a result of the belief that science and technology are equivalent to hammer and forge technologies that anyone can practice.

Another consequence of attaching more importance to technology than science is the wavering and distrusted policies of intellectuals mastering scientific knowledge. For example, intellectuals were valued and reused during the war years, but they became targets of revolution during the "Cultural Revolution". The victory of technical traditions and the actual visible effect had enabled opinions of intellectuals inheriting scholars' traditions to be ignored when making sci-tech plans.

3. A large number of cadres were trained, and their war thinking was reflected in sci-tech planning

Thanks to the sci-tech practice before the founding of the PRC, tens of thousands of sci-tech cadres trained for the CPC became the mainstay of sci-tech undertakings in building the PRC. The batch of cadres has a high consciousness, attaches much importance to practice, and does not like empty talk, which further strengthens technical components of the well-formulated sci-tech plan in the concrete implementation process. Great achievements have been made in China's sci-tech planning, such as "atomic and hydrogen bombs, missiles, and man-made satellites," which must be attributed to these cadres cultivated in the army and the war.

Meanwhile, these sci-tech cadres have fully reflected military thinking in their planning, especially in several sci-tech plans before the reform and opening-up. For example, "concentrating superior forces to annihilating each enemy" and "large-army fighting". On the one hand, the military thinking completely guarantees the embodiment of the state's intention in science and technology planning and its implementation effect. On the other hand, it further weakens the foundation of free research and academic independence in basic science and has a long-term complex impact on China's science and technology planning.

2.2 Typical Sci-Tech Figures in the Revolutionary Period and Their Influence on Sci-Tech Planning after the Founding of the PRC

I. Typical sci-tech figures

Some sci-tech figures emerging from the revolutionary war representatively reflect the historical process of science and technology from the revolution to the founding of the PRC. They have made important contributions to the Chinese revolution, greatly influencing the formulation and implementation of science and technology planning after the founding of the PRC. The study on the figures can help deepen the understanding of science and technology planning after the founding of the PRC.

1. Wang Zheng

Radio has played an important role in China's revolutionary cause. Wang Zheng was just a representative figure. Wang Zheng (1909–1978), formerly known as Wu Renjian, came from Wujin County (now Wujin District), Jiangsu Province. In December 1930, the Red Army captured Wang Zheng and other nine Kuomintang radio personnel as well as a complete telegraph receiver. After being educated and enlightened, Wang Zheng and others volunteered to stay as members of the Red Army and became the earliest radio communication personnel in the CPC's army. The Red Army's radio unit was established. In October 1941, Mao Zedong personally wrote an inscription to signal unit, praising it as a "scientific clairvoyant" and spoking highly of Wang Zhen as "communication pioneer" of the people's army and the "forerunner of the Red Army's communication". Wang Zheng was also the technical director of Yan'an Radio Station, which was later established, making important contributions to the Station.[35]

After the founding of the PRC, Wang Zheng became a vice minister and secretary of the Party Leadership Group of Post and Telecommunications Department of the Central People's Government, and also director of the Third Bureau of the Central Military Commission. In 1955, Wang Zheng was awarded the rank of lieutenant general. In 1956, Wang Zhen was appointed the director of the Military Communications Department of the People's Liberation Army. He was also the director of the Military Electronics Institute. In 1957, Premier Zhou Enlai appointed Wang Zheng as vice president (the then president was Qian Xuesen) of the Fifth Academy and president of the Second Academy of the Ministry of Defense. Wang Zheng put into practice the idea of "independent designing after imitating". In 1963, Wang Zheng became the minister of the Fourth Machinery Industry Department, whose main task was to provide advanced electrical equipment for national defense and national economic development. Wang Zheng carried out the policy of "giving priority to the military applications and integrating military applications with civilian ones." In the 1960s, Western European industrial powers focused on the development of electron tube technology, while the United States and Japan focused on the development of emerging microelectronics technologies. In the two major development trends, Wang Zheng was aware that the future development trend of new technologies may be microelectronics technology. He proposed to establish the "National Semiconductor Device Committee," promoting the rapid development of semiconductor technology in China. In 1977 after the "Cultural Revolution", Wang Zheng, serving as deputy chief of the General Staff and the fourth minister of the People's Liberation Army, was responsible for combat intelligence matters. He died of illness in

[35]Feng (2007, (3)).

Beijing in 1978. In January 1995, the People's Daily published a commemorative article titled *Far-sighted Strategic Decision— Wang Zheng's Contribution to the Decision of the Development Policy of Electronic Industry*, giving a high evaluation of Wang Zheng's performance.[36]

Wang Zheng was a senior sci-tech cadre trained by the CPC. He has made an important foundation contribution to the development of science and technology after the founding of the PRC, especially the technological progress in radio. In the revolutionary era, the important role of radio technology in the victory of the war made it more sensitive to the progress in the field of intelligence and information technology. Therefore, Wang Zheng was capable of grasping the development of semiconductors, making important contributions to the development of electronic technology after the founding of the PRC.

Fig. 2.6 Wu Heng[38]

2. Wu Heng

Wu Heng (1914–1999) (Fig. 2.6) comes from Xuzhou, Jiangsu Province. He studied in the Department of Geography of Tsinghua University in 1934. In 1939, Wu Heng arrived in Yan'an. In the second half of 1941, Wu Heng participated in the investigation work of the Geological and Mineral Institute of the Shaanxi-Gansu-Ningxia Border Region Natural Science Association. The suggestion of building iron-works in Guanzhong was put forward. Wu discovered a large number of ancient vertebrate fossils in the red layer beneath the loess layer. The investigation confirmed his concepts of "popularizing science" and "integrating theory with practice (Fig. 2.6)".[37]

Wu Heng was transferred to the Department of Geology of Yan'an Academy of Natural Sciences in the autumn of 1942 to serve as director of Northeast Industrial Institute in 1949. The Northeast Industrial Institute, located in the former site of the "Continental Academy of Sciences" in Manchuria, became a high-level scientific research institution in the entire Northeast. He was good at making annual plans for scientific research. Obviously, this practice was related to the system that the scientific research department was subordinate to the industrial department at that time. What he was good at laid a solid foundation for his future scientific planning. In 1952, Wu Heng was appointed secretary-general of Northeast Branch of the Chinese Academy of Sciences.

Wu Heng acted as deputy secretary-general of the Academic Secretariat of the Chinese Academy of Sciences in 1954. According to Wu Heng's suggestion, the industrial sector set up its own research institutions to form an independent industrial research system by the division of labor between the Chinese Academy of Sciences and the industrial sector. In March 1956, the State Council announced the establishment of a 35-member Scientific Planning Committee, with Wu Heng serving as its deputy secretary-general. In 1958 when the National Science Planning Commission and the National Technology

[36] Lu Zhiyu et al. Biography of Wang Zheng. Beijing: Electronics Industry Press, 1998.

[37] Li Zhenzhen.Wu Heng[DB/OL].the website of the Chinese Academy of Sciences: http://www.cas.cn.

[38] Ibid.

Fig. 2.7 The World Intellectual Property Organization awards[40]

Commission merged to form the National Science and Technology Commission, Wu Heng was appointed deputy director. In 1964, Wu Heng held the position of the director of the National Invention Selection Committee. During the "Cultural Revolution", Wu Heng took part in the decision-making and organization of the Chinese Academy of Sciences in the preparation of the Long Wave Time Service Station and Navigation Station, high-energy physics, laser, and other research work.

At the end of 1978, Wu Heng participated in and presided over the formulation of the Outline of the National Science and Technology Development Plan from 1978 to 1985. In September 1977, the CPC Central Committee decided to reinstate the National Science and Technology Commission, with Wu Heng as its deputy director and deputy Party secretary. From 1979 to 1990, Wu Heng served as director of the Natural Science Award Committee. He made great efforts to raise people's awareness to promote the establishment of China's patent system. In 1980, the National Patent Office was formally established, with Wu Heng as the first director. Thanks to his efforts, the Standing Committee of the National People's Congress discussed and passed the Patent Law of the PRC on March 14, 1984. Mr. Bogsch, director general of the World Intellectual Property Organization, congratulated Wu Heng and presented him with a "gold medal for invention and creation" and a certificate of merit (Fig. 2.7). The certificate reads, "Mr. Wu Heng was specially awarded a gold medal of World Intellectual Property Organization for his achievements in establishing China's patent system and promoting international cooperation in protecting the right to the invention."[39]

In 1982, Wu Heng served as an advisor to the National Science and Technology Commission and was elected a member of the Central Advisory Committee in 1985. In 1984, the Chinese Invention Association was established, and Wu Heng was elected the president of the Association. He formed the National Committee of the Chinese Invention Association. Wu Heng devoted a lot of effort to the Invention Association.[41]

From an ordinary college student to a senior cadre in different fields of science and technology after the founding of the PRC, Wu Heng made important contributions to the foundation and development of China's science and technology programs. Especially when taking the leading position in formulating science and technology plans, he directly promoted the promulgation and

[39]Pan and Yajun (1993, (1)).

[40]National Intellectual Property Administration: http://www.sipo.gov.cn. (On August 16, 1984, Mr. Bogsch, director general of the United Nations World Intellectual Property Organization (WIPO), awarded Wu Heng the WIPO Gold Medal for Invention and Creation at the Great Hall of the People in Beijing).

[41]Yan (2004: 16–18). (Yan Dexuan served as Wu Heng's secretary from 1986 to 1990).

implementation of science and technology plans in our country. The temper of the revolutionary war made Wu Heng infinitely loyal to the Party's sci-tech undertakings.

3. Qu Bochuan

Qu Bochuan (1909–1997) was a doctor of organic chemistry studying in Germany. In 1937, he received a doctor's degree in chemical engineering from the Dresden University of Technology in Germany. Qu Bochuan first returned to Chongqing in November 1937 and was not appointed. In October 1939, Zhou Enlai introduced him to Yan'an. Qu Bochuan, the first senior intellectual in Yan'an revolutionary base, returned from abroad with a doctorate in natural sciences. In January 1940, he took part in the preparatory work of the Shaanxi-Gansu-Ningxia Border Region Natural Science Research Association, and later he participated in the preparatory work of Yan'an Academy of Natural Sciences as the education director. In September 1945, he was appointed director of the Industrial Testing Institute of the Shanxi-Chahar-Hebei Military Industry Department. In January 1947, he went to Dalian to serve as the principal of Guandong Special Industrial School. In July 1950 when the Institute of Technology of Dalian University became Dalian Institute of Technology independently (renamed the Dalian University of Technology in 1988), Qu Bochuan served as president and Party secretary.[42] In March 1979, he became the dean of Dalian Institute of Technology again. Since September 1981, he had become the honorary president of Dalian Institute of Technology (honorary president of Dalian University of Technology).

In May 1983, Qu Bochuan, together with honorary presidents Kuang Yaming, Li Shusen, and Liu Dan respectively from Nanjing University, Tianjin University, and Zhejiang University, wrote to the Secretariat of the CPC Central Committee, proposing that about 50 institutions of higher learning need to be listed as major national construction projects.[43] The main content of the proposal was to suggest that the state allocate five billion yuan, with a focus on funding 50 institutions of higher learning, so that they would reach the international advanced level in terms of instruments and equipment by the 1990s. The proposal was highly valued by Deng Xiaoping and other central leaders and gradually put into action. In 1993, the CPC Central Committee and the State Council explicitly put forward the "Project 211" in the *Outline of China's Educational Reform and Development*. The Project aims at building about 100 key institutions of higher learning and a number of key disciplines in the twenty-first century. In 1998, the Central Government put forward the "Project 985" of "building several world-class universities and high-level research universities".[44]

Qu Bochuan and other three famous educators of the older generation said the above suggestions were honored as "835 suggestions" in the higher education circle. The "835 suggestions" are the same as the "Program 863" presented by four famous older generation scientists in the PRC such as Wang Daheng and were supported by Deng Xiaoping. They are called the good news that intellectuals wrote to make suggestions and promote high-level decision-making in the 1980s.[45] In reality, the "Four Gentlemen' Statement" in the field of large aircraft in 1986 and the Knowledge Innovation Project of the Chinese Academy of Sciences in the 1990s are typical cases in which Chinese intellectuals submitted a written statement to the top leaders and achieved success. They are positive cases of the combination of science and politics in China, partly reflecting the multifaceted nature of the marriage between science and politics in China.

Qu Bochuan, as an early senior intellectual of the CPC, had made important contributions to China's scientific and technological undertakings, especially higher education.

[42] Zhang (2009).

[43] Ibid.

[44] Zhang (2009).

[45] Gong (2006: 9–12).

4. Chen Kangbai

Chen Kangbai (1898–1981), originally named Yun Huang, came from Changsha, Hunan. He graduated from Xiamen University in 1927. Afterward, he acted as a faculty member of Xiamen University and Zhejiang University as well as a research fellow of the Faculty of Science of Peking University. In 1932, he went to the research institute of Copenhagen University to study organic chemistry. He returned to Yan'an in 1937. He once served as director of the Technology Department of the Military Industry Bureau of the Central Military Commission, vice president of Yan'an natural science institute, and president of Yan'an Academy of Natural Sciences.

After the founding of the PRC, he successively served as vice minister of the Cultural Department of the Northeast People's government, president of Harbin Institute of Technology, secretary general of the Chinese Academy of Sciences, and Counselor of the State Council. He was elected vice chairman of the All-China Federation of Special Societies for Natural Sciences in 1950.

The above four are the more typical sci-tech cadres in the revolutionary period. In fact, a large number of "red experts" were born in that era. For example, Yin Xipeng, a doctor of pathology studying in Japan, Chen Qiyuan, an expert in pediatrics, and Zhang Wenqi, an expert in ophthalmology. Yan'an Academy of Natural Sciences also had well-known figures such as Xian Xinghai, Yu Guangsheng, Ge Baoquan, Qiao Guanhua, and Huang Wei. Here is a brief introduction to them. Yun Ziqiang was appointed vice president of Yan'an Academy of Natural Sciences in 1944 and taught chemistry. After the founding of the PRC, he participated in the formulation of the *Twelve-Year Science and Technology Plan* and was elected an academician in 1995. Liu Ding, an expert in military engineering, once studied in Germany, was one of the main founders and leaders of China's weapon industry. He successively served as minister of Military Industry of the Eighth Route Army Headquarters of Taihang and Deputy Director of Yan'an Military Industry Bureau, making great contributions to the upgrading of the Red Army's military technology and weapon power. Lin Hua, a chemical expert, was one of the organization leaders of China's science and technology.

What is described here is only part of red experts who are more representative and have made more achievements. Hundreds of red experts have gone through the test of revolutionary war, making great contributions to science and technology, in order to establish the PRC. Since the founding of the PRC, these experts have become leading cadres in the science and technology, realigning science and technology of the PRC with their thinking in formulating and implementing ideas. In addition to the red experts of the CPC, China's scientific and technological forces after the founding of the PRC include scientists from the former Kuomintang-ruled area, experts returning to their homeland after the founding of the PRC, and the Soviet experts assisting China. Among these four sources of experts, sci-tech cadres of the CPC held a core position, and the Soviet experts had an important influence.

II. Impact on the PRC's Science and Technology Planning and Policy

The sci-tech cadres trained before the founding of the PRC had a great influence on the formulation and implementation of sci-tech policies and plans after the founding of the PRC.

First of all, they created the atmosphere that the CPC attaches great importance to intellectuals and scientific knowledge. Without a doubt, the CPC attaches much importance to science and technology, but the significance of this emphasis varies at different levels of the Party's leadership. In the arduous war, everything served the expansion of the Party and the victory of the revolutionary cause. Therefore, it is definite for science and technology to serve as a means of service and a tool to support the war. In different periods, the Party's policies on science and technology vary, but it is a consistent policy to attach great importance to sci-tech experts in wartime. In April 1941, the Central Military Commission of the CPC issued the *Policy*

Directive on the Recruitment and Treatment of Specialists in the Army, stating that "an army cannot become a powerful organization without a large number of specialists (military strategists, engineers, technicians, doctors, etc.). All the above-mentioned talents are given full responsibilities based on their specialized knowledge, including directors of factories and hospitals."[46] These standards gave sci-tech experts high-degree trust, full freedom, generous treatment, and diligent work. Although there were some mistakes in the rescue movement in some periods, such as the late Yan'an Rectification Movement, the overall implementation was very good. Before the founding of the PRC, the correct policies of intellectuals led the CPC to recruit and train a large number of red experts and sci-tech cadres, making important contributions to the final victory of the Chinese revolution.

A large number of sci-tech talents and cadres had been trained in the revolutionary base. The vast majority of sci-tech talents have still played a huge role in the sci-tech front after the founding of the PRC. For example, Li Fuchun was the first president of the Yan'an Academy of Natural Sciences. After the founding of the PRC, he served as deputy director of the Finance and Economic Committee of the State Council, minister of Heavy Industry, deputy director of the State Planning Commission, and vice premier of the State Council. Besides Shen Hong, Lu Da, for example, was one of the leaders in the metallurgical technology in China. He successively held positions of deputy director of the Iron and Steel Bureau of the Ministry of Heavy Industry, deputy minister of Metallurgical Industry, and chief engineer. It is because of these sci-tech cadres that the CPC has accumulated rich experience in science and technology management.

Secondly, a range of policies on science and technology awards issued during the revolutionary period has provided an experience for the development of science and technology in the PRC. In August 1950, the Government Administration Council announced the *Decision on Rewarding Inventions, Technological Improvements and Rational Construction Related to Production* and the *Provisional Regulations on the Protection of Rights to Invention and Patent*. The two documents are the first-batch ones on sci-tech inventions and patents after the founding of the PRC. On August 27, 1954, the State Council also promulgated the *Interim Regulations on Rewards for Invention, Technological Transformation, and Rationalized Proposals Related to Production*. These important measures and regulations adopted at the beginning of the founding of the PRC have played an important role in the development of science and technology in China. Their policies and thoughts came from the revolutionary period. These sci-tech reward policies originated from sci-tech cadres' understanding of the importance of science and technology research before the founding of the PRC. The implementation of policies cannot be separated from the promotion of these sci-tech cadres.

Thirdly, the mechanism for sci-tech cadres to participate in national construction had been established. In August 1950, the National Congress of Science and Technology was formally convened. The conference empowered the majority of sci-tech workers to identify the direction of scientific work and decided to establish the "All-China Federation of Specialized Natural Science Societies" and the "All-China Association for Popularization of Science and Technology". In 1958, they merged to form the China Association for Science and Technology. The National Science Congress put forward a proposal to reorganize the original national scientific institutions to "establish the National Academy of Sciences," later known as the Chinese Academy of Sciences.[47] the Chinese People's Political Consultative Conference adopted the *Common Programme* with the function of a provisional constitution and formulated a policy on the development of science and technology. According to the *Common Programme*, Article 43 provides that "efforts should be made to develop natural sciences to serve the construction of industry, agriculture and national defense,

[46]Wu (1984: 22–24).

[47]Chen and Yulin et al. (1994: 6).

reward scientific inventions and discoveries, and popularize scientific knowledge." Therefore, it became the general policy of scientific and technological work in the early days of the PRC. They are consistent with the sci-tech policy of "promoting economic development with science and technology and serving the founding of the PRC and the Anti-Japanese War" presented in Yan'an during the Anti-Japanese War.[48] Sci-tech cadres organized all sci-tech workers across China to participate in the construction of the country and submitted suggestions to the government.

In addition, the good style of seeking truth from facts has been established in sci-tech circles, which set a good spiritual paradigm for the development of science and technology in China. On April 1, 1942, Xu Teli organized the Yan'an Academy of Natural Sciences to hold a nine-month discussion on educational policies during his tenure as the president of the Academy. He published in *Liberation Daily* an article titled *Rethinking How We Learn*. In this article, he put forward the idea of "seeking truth from facts" and not being "self-righteous".

In November 1950, Xu Teli wrote to Zhou Shizhao in his inscription for Hunan No. 1 Normal College, "Seeking truth from facts and not being self-conceited is the style of our leader Comrade Mao Zedong … Without it, all revolution, all construction, all work, and study will have deviations and will be in danger of going astray … Comrade Mao Zedong's style is an honest style and a thorough Marxism."[49] Cadres dedicated to the science and technology front dared to seek truth from facts before the founding of the PRC, which influence the realistic estimation of China's science and technology status when formulating the first science and technology plan after the founding of the PRC. However, after the "anti-rightist" campaign, the scientific and technological community's style of seeking truth from facts was affected.

[48]Wang Huilin. Reference for China's Modern History 1919–1949 (II). Beijing: Beijing Normal University Press, 1992: 546.

[49]Xu (1995: 224).

2.3 Science and Technology Development and Scientific Community in the Ruling Area of Kuomintang

This book mainly studies the scientific and technological planning of the PRC under the leadership of the CPC, but the science and technology in the Kuomintang-ruled area are also part of the sci-tech foundation of the PRC. It is necessary to make a brief survey of the science and technology and the contribution of overseas students in the Kuomintang-ruled area. The PRC's scientific and technological strength is composed of the following four forces:

1. Red experts under the leadership of the CPC;
2. Science and technology experts in the ruling area of Kuomintang;
3. Returnees divided into those who returned to their motherland before and after the founding of the PRC, respectively, and, before the founding of the PRC, those returnees classified into one group returning to base areas and the other group going back to the ruling area of Kuomintang;
4. Foreign experts consisting of those devoting themselves to sci-tech undertakings before and after the founding of the PRC, respectively, and, after the founding of the PRC, mainly Soviet experts.

Among the four forces, red experts were leading the scientific and technological front after the founding of the PRC. Soviet experts were in important advisory positions. However, experts from the Kuomintang areas and returnees filled actual work positions in most cases. How was science and technology in the Kuomintang areas before the founding of the PRC? This section will elaborate on this question.

I. Science and Technology Awards in the Kuomintang Areas

During the anti-Japanese war, the industry in the Kuomintang areas developed fast due to the stimulation of the war, "especially in Sichuan, Yunnan, and Guizhou. The number of private factories increased from 270 before the war to

4,764 in 1944, while the registered capital rose from 91.36 million yuan to nearly 300 million yuan (if the pre-war currency is converted into RMB)."[50] The war stimulated the development of the steel industry. "As of the first half of 1944, there were 105 private steel mills in the Kuomintang areas, of which 80 were set up after 1938."[51] In addition, other industries such as machinery, chemical industry, textile, and petroleum have been greatly developed. The fast development of industry in the Kuomintang areas has improved the overall technical level, laying a material foundation for the victory of the Anti-Japanese War.

As a government, the Kuomintang must have attached importance to science and technology that could advance national progress, social development, and people's well-being after the establishment of its political power. Therefore, there were many corresponding academic awards. It was a general introduction to the development of science and technology in the Kuomintang areas from the perspective of academic awards.[52] In 1912, the Government of the Republic of China (ROC) formulated the *Interim Regulations on Awards of Artworks* to reward scientific and technological inventions. In 1923, the ROC government extended the scope of awards to cover both new products and new methods and granted patents, which created the atmosphere of motivating people to conduct scientific and technological inventions. In 1944, the Kuomintang's government drafted China's first *Patent Law*.[53] The *Patent Law* greatly boosted the advancement of science and technology in the Kuomintang areas. The Law represented an inspiration for the tortuous understanding of patent issues after the founding of the PRC.

The Nationalist government fought in domestic warlord conflicts and the anti-Japanese war. In this context, it was more inclined to the development of science and technology producing practical effects. However, in general, a lot of promotion had been done for basic science. In 1944, the Academia Sinica presented a proposal to the Nationalist government, hoping that the government would strike a balance between all kinds of sciences so as to establish a complete and independent scientific system in China. Chiang Kai-shek also said, "If China is to stand among modern countries, pure scientific research is absolutely indispensable."[54] From 1941 to 1947, the Nationalist government issued a total of six "National Academic Awards", in which eight won the first prize in natural sciences. Many experts and scholars had made remarkable achievements in fundamental research. For example, "Wu Dayou's research on the vibrational light spectra and structures of multiple atoms and molecules, Ma Shijun's theory of atomic nuclei and cosmic rays, Zhu Ruhua's paper on molecular rearrangement and organic binding, Su Buqing's curve mapping research, Zhong Minglai's contribution to probability theory and number theory, Li Siguang's research on Nanling's geological structure, Ma Yanying's analysis on the determination of ancient climate by coral layer growth, Zhang Qinglian's research on heavy water, Zhao Jiuzhang's atmospheric vortex motion and other important academic achievements. All have produced important impacts in international academia."[55] The Nationalist government formulated a range of regulatory documents to reward inventions and technological advancements of sectors and industries.

Judged from these awards, the Kuomintang areas still attached much importance to science and technology. Thanks to rich large and medium-sized cities and vast geographical resources in China at that time, the sci-tech development in the Kuomintang-controlled area was an institutionalized development and the "national academy". A number of major achievements in natural sciences had been made, demonstrating world-class influence. These achievements and scientists who had achieved

[50]Zhu (1988, 5).

[51]Jian (1945, 10 (5,6)).

[52]Yao (2007).

[53]Han and Kefu (1996: 123).

[54]Yao (2007).

[55]Hao (1972, 59) = 123.

them are important sources of scientific strength in the PRC.

II. Scientific Community of the Kuomintang Areas

The development of science and technology in the Kuomintang areas can be verified by the development of scientific communities. Here, several representative scientific communities are selected to elaborate.

1. China-US Foundation for Education and Culture

Although the China-US Foundation for Education and Culture is not a scientific community, it plays an important role in cultivating China's early scientific communities. The Foundation's funds come from the "Boxer Indemnity". "In 1907, the U.S. Congress gave the President full power to deal with the surplus of funds."[56] In 1924, the United States returned 12 million dollars to China for the development of Chinese culture and education by setting up the China-U. S. Foundation for Education and Culture. In 1925, the first board of directors was established, with ten from China and five from the US.[57] In 1937, the working organization was as follow, "Chairman: Jiang Menglin; Vice Chairman: Weng Wenhao and Leighton Stuart; Honorary Secretary: Hu Shi; Honorary Accountant: Huo Baoshu and Hopkins; members of the Executive Committee: Weng Wenhao, Zhou Zhichun and Huo Baoshu; members of Finance Committee: Li Ming, Thomas and Zhou Zhichun; Deputy Accountant: Thomas and Ye Liangcai."[58] (Fig. 2.8). The Foundation was an independent one that had great influence on the development of science and technology in China before the founding of the PRC. During the 20 years from 1926 to 1945, (Fig. 2.8), part of the board of directors of the China-US Education Foundation discussed that the Foundation funded many academic fields, totaling nearly 100 institutions.

Fig. 2.8 A partial board meeting of China-US Foundation for Education and Culture[60]

Specifically, the foundation funded Beiping Library (the predecessor of today's China Library), Tsinghua University Library, Peking University Library, Wuchang Cultural Library, and Mingfu Library of Chinese Association for the Advancement of Science (CAAS).[59]

The China-US Foundation for Education and Culture also supports the construction of universities and individual academic research. "In 1936, the CASS and China's societies of Mathematics and Zoology jointly sponsored a scientific conference. About 60% of the 250 papers were directly or indirectly funded by the Foundation. Some of the funded scientists made the international advanced achievements at that time."[61] Before the founding of the PRC, the main work and funded objects of the China-US Foundation for Education and Culture were all in the Kuomintang-ruled area. After 1949, the Foundation moved to Taiwan. Up to now, it is still supporting Taiwan's cultural and educational undertakings, but their historical position is not as good as before.

[56]Hu (1985: 183–185).

[57]E. V. Codry. Sdence. Vol. LXV, No. 1676, P150.

[58]Zhuang (1999: 32–38).

[59]China Education Research Center of Columbia University, the website of Columbia University: http://www.tc.Columbia.edu/centers/coce/detail.asp?Id =About + Us&Info=A+Photographic+Exhibition.

[60]China Education Research Center of Columbia University, the website of Columbia University. https://www.tc.columbia.edu/coce/.

[61]Liu (2002: 97).

Fig. 2.9 Cai Yuanpei and Yang Xingfo[62]

2. Academia Sinica

The Academia Sinica is the highest scientific institution in the Kuomintang areas. After the founding of the PRC, the Chinese Academy of Sciences was established on the basis of the Academia Sinica and Beiping Research Institute in the Kuomintang areas and Yan'an Academy of Natural Sciences. The basic science is mainly from the Academia Sinica. In April 1928, the Academia Sinica was established and Cai Yuanpei became its president. Cai Yuanpei invited Yang Xingfo to be the first director-general (Fig. 2.9). The Academia Sinica adopts a decentralized system. Cai Yuanpei did not interfere in the development and research directions of all institutes, which has always been the academic tradition of Academia Sinica. Yang Xingfo was the initiator of the CAAS, China's then largest academic society. He served as the secretary-general of the CAAS and was responsible for the editorial work of the CAAS's Journal Science for seven years. Yang Xingfo used his network to recruit a lot of talents for the Academia Sinica.

Cai Yuanpei and Yang Xingfo both stressed that Academia Sinica should be developed into an academic institution free from political interference and domination. Cai Yuanpei believed that the centralized construction of the Academia Sinica was a purely academic issue. Chiang Kai-shek "argued that the centralized construction of the Academia Sinica was related to the national image, but not a political intervention."[63] A dispute arose between the two figures. However, Chiang Kai-shek did not use his authority to forcibly suppress Cai Yuanpei, so the Academia Sinica was able to develop independently and became another important academic tradition of the Academia Sinica. Before the founding of the PRC, the Academia Sinica established 13 research institutes, including meteorological institute, covering natural sciences, engineering sciences, and social humanities. The "academy—institute" management system was obviously inherited by the Chinese Academy of Sciences after the founding of the PRC. On the eve of the founding of the PRC, only a few members and organizations of the Academia Sinica followed the Kuomintang to Taiwan. After a long period of low ebb, the Academia Sinica gradually developed into a regional academy of sciences with considerable influence in the Asia–Pacific and even the world. It is worth noting that even after moving to Taiwan, successive presidents and the overwhelming majority of academicians and researchers have resisted political interference and safeguarded academic independence and democracy. The year of 2008 marked the 80th anniversary of the founding of the Academia Sinica (1928–2008). The Academia Sinica published its large-scale history—the *80th Anniversary of the Academia Sinica: the Pursuit of Excellence* and held a forum for presidents of academies of sciences across the world.[64]

When the Academia Sinica was founded, Cai Yuanpei had the idea of emulating European and

[62]The History of the Academia Sinica: the 80th Anniversary of the Academia Sinica: the Pursuit of Excellence, Volume 1: A Long Way to Go [DB/OL], 2008. The Academia Sinica: www.sinica.edu.tw.

[63]Ibid.

[64]The website of the 80th anniversary of the Academia Sinica http://apforum.sinica.edu.tw.

American countries and electing academicians. Later, President Ding Wenjiang first established the system of assessors in 1935. In October 1946, Acting President Zhu Jiahua believed that the time was ripe for the election of academicians and decided to organize the first academician election. Fu Sinian strongly advocated that "We are at loggerheads with the CPC side."[65] However, in actual elections, academic standards are still higher than political standards. The election of Guo Moruo as an academician shows that most of the assessors contended that learning surpasses politics and did not agree to evaluate academic achievements from a political standpoint.

The academicians elected for the first time include: Su Buqing, Shiing-shen Chern and Hua Luogeng in mathematics; In chemistry, Wu Xian, Zhuang Changgong and Chen Kehui; Geologically, Li Siguang, Weng Wenhao, Yang Zhongjian and Zhang Hongzhao; In meteorology, Zhu Kezhen; Academicians of other disciplines such as Hou Debang, Mao Yisheng, Fu Sinian, Chen Yinque, etc. (Figs. 2.10 and 2.11). "It is worth noting that the rate of returned students to be elected is extremely high. Of the 81 academicians, 49 had studied in the United States, 23 in Europe and 5 in Japan. At that time, there were only six people who had no experience in studying abroad, namely, Dong Zuobin, Yu Jiaxi, Zhang Yuanji, Gu Jiegang, Liu Yizheng and Chen Yuan, whose knowledge was related to traditional historiography and classics."[66]

As far as basic science is concerned, the Academia Sinica in the Kuomintang areas at that time represented China's scientific and technological level in the 1930s and 1940s. It was also a full member of the International Council for Scientific Unions (ICSU). After the founding of the PRC, a lot of basic work in science and

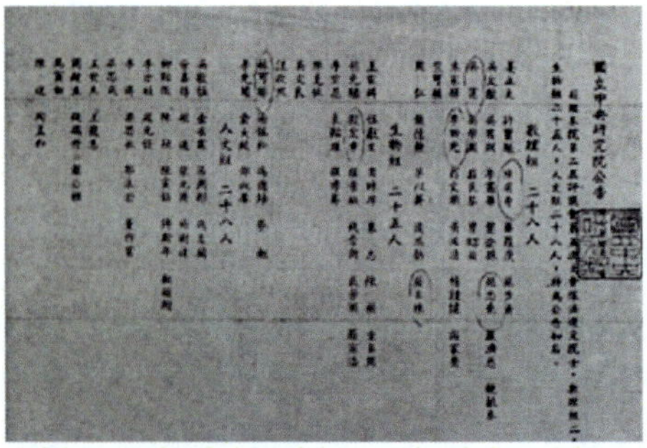

Fig. 2.10 The list of the first-batch 81 academicians of the Academia Sinica

Fig. 2.11 The seal of an academician[67]

technology planning was developed from the foundation of the Academia Sinica. We must see this. After the founding of the PRC, we received most equipment and personnel of the Academia Sinica, but we did not inherit its system and spirit well. Scientists left behind by the Academia Sinica continued to engage in scientific research after the founding of the PRC, but the proportion of scholars holding leading positions in sci-tech undertakings in the PRC was not high. Their influence in scientific and technological planning after the founding of the PRC was smaller than that of sci-tech cadres from the base and Yan'an.

[65]The History of the Academia Sinica: the *80th Anniversary of the Academia Sinica: the Pursuit of Excellence*, Volume 1: A Long Way to Go [DB/OL], 2008. The Academia Sinica: www. sinica. edu. tw.

[66]Ibid.

[67]The Academia Sinica: www.sinica.edu.tw.

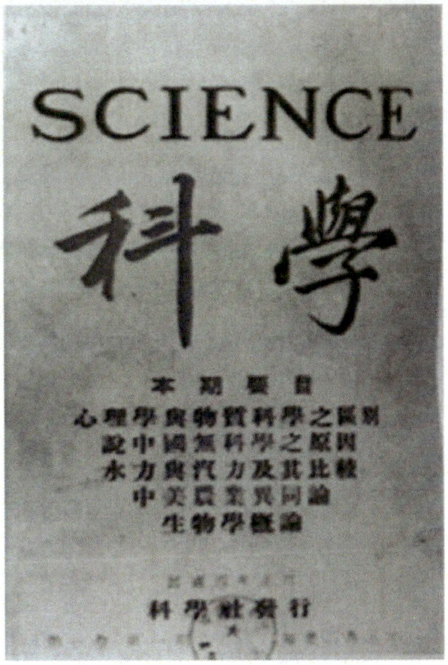

Fig. 2.12 The inaugural issue of Science in 1915[68]

3. Chinese Association for the Advancement of Science (CAAS)

This is the community of folk science that had the greatest influence before the founding of the PRC. If the mainland's scientific and technological representative after 1949 were the Chinese Academy of Sciences and the former was the Academia Sinica, it was the CAAS that represented Chinese sciences before the birth of the Academia Sinica in 1928. The CAAS, established at Cornell University in 1914, was founded by Hu Mingfu and Ren Hongjuan. The initial core work was the publication of *Science* magazine (Fig. 2.12). The mission was to serve the country by improving Chinese people's science literacy. In April 1915, it was reorganized into a society with the aim of "connecting comrades, studying academia, and making common efforts to develop science and technology in China."

Fig. 2.13 The logo of the CAAS[69]

The members of the first Board of Directors of the CAASare: Ren Hongjuan, Zhao Yuanren, Hu Mingfu, Bing Zhi and Zhou Ren. Figure 2.12 is the first issue of *Science* in 1915. Ren Hongjuan served as president and Yang Xingfo editorial director. In 1918, the headquarters of the CAAS moved from the United States to China and set up an office. The CAAS is the first non-governmental organization for comprehensive natural science in modern China (Fig. 2.13). It is also the academic organization with the widest coverage and greatest influence in China before 1949. The number of members had grown to nearly 4,000 in 1949, with members specializing in almost all fields of science and technology. Before the founding of the Academia Sinica in 1928, the CAAS participated in international scientific activities as a representative institution for foreign exchanges among Chinese scientific circles.

The CAAS held annual scientific conferences to promote academic exchanges. Members of the CAAS represented the mainstream of Chinese sciences at that time. Cai Yuanpei was the first special member (with special contributions). Zhang Jian, a big industrialist, was elected honorary member at the second annual meeting of the CAAS. Famous members include Ren Hongjuan, Zhu Kezhen, Ding Wenjiang, Weng

[68] "The 90th Anniversary of CAAS and *Science*" [DB/OL]. CAAS: http://www.kexuemag.com/90nian/kexue90zl.htm.

[69] Ibid.

Fig. 2.14 The cover of Wissen und Wissenschaft, Volume 3 and Issue 1[70]

Wenhao, Qian Baocong, Li Siguang, Mao Yisheng, and Shiing-shen Chern.[71] Many of them were well-known masters in China's history of sciences at that time or later.

The special status of the CAAS in China won many prominent sponsorships from the political and military circles. According to the "Membership Record of the CAAS" published in 1930, the sponsoring members include Xu Shichang, Li Yuanhong, Xiong Kewu, Chiang Kai-shek, Yen Hsi-shan, etc.[72] It can be seen from the lists that the social impact of the CAAS was great.

In the 1920s and 1930s, specialized societies were established one after another, such as the Geological Society of China, China Astronomical Society, and China Meteorological Society. The CAAS played a leading role in connecting various specialized scientific communities.

Holding the beautiful ideal of "saving the nation through science and establishing a scientific China," the CAAS had done a lot of hard work and made indelible contributions to the nurturing and development of modern science in modern China. In January 1923, Volume 8, Issue 1 of *Science* published Ren Hongjuan's article the *Introduction to the Outline of Science*, which introduced the Outline of Science compiled by British Professor J. A. Thomson. At that time, young Mao Zedong, who was determined to "try to supplement basic scientific knowledge," benefited from the Outline impressing him. When Mao Zedong met with Tsung-Dao Lee in 1974, he talked about the *Outline of Science*. Tsung-Dao Lee recalled, "Chairman Mao said that he didn't have much time to study science when he was young, and most of his ideas about science came from a set of *Outline of Science* written by Thomson. Chairman Mao also gave Lee a set of *Outline of Science* in English."[73] It can be seen that the influence of the CAAS is multifaceted.

After the founding of the PRC, the CAAS held "six symposiums to learn how to better build the PRC" in more than half a year from the end of 1949 to June 1950.[74] At the beginning of liberation, the Central Government hoped to nationalize science from folk science. In December 1950, the National Federation of Science and Technology intended to take over *Science* as an organ-related publication. In 1951, *Science* ceased publication. In July 1957, *Science* resumed publication as a quarterly magazine. Only four volumes and 12 issues were published after the resumption of publication. The publication was suspended in 1960 and resumed in 1985. Some scholars think that there are the following reasons for the suspension of *Science*. "First, the PRC should break up all the old things, including the old science, and the title of

[70]*Overview of the Chinese Scholarship Society* was compiled by the General Office of the Society in July 1936. The author copied it from the Institute for the History of Natural Sciences, Chinese Academy of Sciences.

[71]Fan Tiequan. CAAS and China's Science and Culture. Doctoral Thesis of Nankai University, 2003: 67.

[72]Fan Tiequan. CAAS and China's Science and Culture. Doctoral Thesis of Nankai University, 2003: 67.

[73]"The 90th Anniversary of CAAS and *Science*" [DB/OL]. CAAS: kexuemag.com/90nian/kexue90zl.htm.

[74]Science, vol. 32, issue 1–8, 1951.

the magazine cannot be used. Second, it is believed that only by unifying scientific journals and merging various scientific organizations can scientific unity be realized. Third, the CAAS itself has high expectations of the PRC and regards it as the Party's decision to implement it. Fourth, there are indeed many difficulties in running the journal. Both human and financial resources are relatively in short supply."[75] This analysis is reasonable and reflects the revolutionary characteristics of understanding the original scientific basis after the founding of the PRC. The CAAS's other publications and institutions were also gradually integrated into the scientific and cultural undertakings of the PRC. In 1954, the CAAS transferred its Institute of Biology to the Institute of Biology of the Chinese Academy of Sciences. Mingfu Library was donated to Shanghai Municipal People's Government in 1956 and later became part of Shanghai Municipal Library. "At the meeting of the Council of CAAS held in the autumn of 1959, it was decided to donate all the financial resources of the CAAS to the country so as to play a better role in socialist construction. With the consent of all members, the CAAS donated all its existing houses, property (including bank deposits, bonds, and cash, totaling 83,542.79 yuan), books, and equipment to the Central People's Government."[76] On May 4, 1960, the CAAS and Shanghai Association for Science & Technology completed all handover matters. The CAAS withdrew from the Chinese science arena.[77]

The CAAS is a non-governmental scientific organization that played an important historical role in introducing and developing science and technology in China before the founding of the PRC. After the founding of the PRC, the "unified" scientific system deprived the CAAS of the necessity and environment for its existence. However, after the reform of the sci-tech system, especially after entering the new century, China's non-governmental scientific organizations have gradually emerged and increased. Still, none of them has reached the social influence and historical position of the CAAS before the founding of the PRC. If the CAAS had not been disbanded but had continued to develop, this kind of "folk science" must have been a good supplement to the government-led "government-run science". All previous sci-tech plans after the founding of the PRC rarely took into consideration the problems of non-governmental scientific organizations. The weak non-governmental academic organizations are somewhat related to the weak independent spirit of China's scientific communities.

4. Chinese Scholarship Society

The Chinese Scholarship Society (CSS), a famous academic community, experienced the same period together with the CAAS. It was initiated by Chinese students studying in Japan. If the CAAS focuses on the research of natural sciences, the CSS lays more stress on the popularization and research of humanities and social sciences. "The CSS aims to study truth, advocate civilization, exchange wisdom and knowledge, and promote culture. In the fifth year of the Republic of China (1916), 47 students, including Chen Qixiu, Wang Zhaorong, and Wu Yongquan, from Imperial University of Tokyo, Waseda University, Higher Institute of Technology and Higher Normal School, initiated the organization and named it Bing-Chen Association."[78]

On June 22, 1922, the Bing-Chen Association was renamed the Chinese Scholarship Society (CSS) that established its Head Office in Shanghai. Zheng Zhenwen and Zhou Changshou were elected respectively as the first director-general and deputy director-general. In September 1925, the Xueyi University established by the

[75]Xu (1992, 3).

[76]"The 90th Anniversary of CAAS and *Science*" [DB/OL]. CAAS: http://www.kexuemag.com/90nian/kexue90zl.html.

[77]Fan Tiequan. CAAS and China's Science and Culture. 47–49, Doctoral Thesis of Nankai University, 2003: 4.

[78]Overview of Chinese Scholarship Society compiled by the General Office of the Chinese Scholarship Societyin July 1936. The author copied it from the Institute for History of Natural Sciences of Chinese Academy of Sciences.

CSS opened in Shanghai, but it provided that "students must be prohibited to participate in any party membership or any political or social activities ... By 1936 when the CSS celebrated its 20th anniversary, its membership had already exceeded 1643."[79]

Publishing was the most important business of the CSS. The Journal of *Wissen und Wissenschaft* and *A Collection of Ancient Books*. *Wissen und Wissenschaft* (Fig. 2.14) was founded in April 1917 and ceased publication in April 1958. *A Collection of Ancient Books* is the crystallization of the cooperation between the CSS and the Commercial Press. Guo Moruo was also a member of the CSS, which had a large pool of talented people. "After joining the CSS, Guo Moruo successively published the opening words of the historical play Su Wu and Li Ling, the Flood Era, the Essence of Literature, and the novel Mid-Lake Pavilion."[80] Among them, Li Yan and Qian Baoqiong made pioneering contributions to the study of the history of ancient Chinese mathematics under extremely difficult circumstances.

After 1956, Xu Junyuan, Wu Zaoxi, Li Zhidao, and Ge Suicheng, 4 of the 22 directors, were classified as rightists. On February 4, 1958, the Publicity Department of Shanghai Municipal Party Committee approved the voluntary suspension applied by the CSS. The Journal of *Wissen und Wissenschaft* ceased publication in April 1958. The library of the CSS was received by the Municipal Bureau of Culture, while the headquarters of the CSS was incorporated into the Compiler of Ci Hai, a Chinese encyclopedia, Zhonghua Book Company. The CSS was announced to be disbanded in August 1958.

Different from the CAAS, the CSS paid more attention to the integration of natural sciences and humanities and social sciences. Fan Shoukang, the father of the famous contemporary scholar Fan Dainian, is an important member of the CSS. "According to the historical experience of the CSS, it is possible to establish an academic group with the participation of natural scientists, applied science and technology experts, humanities experts, philosophy and social scientists, and literature and art experts, and to edit and publish a comprehensive academic publication ... which may play a positive role in promoting the communication between arts and sciences and facilitating the integration of humanities and science."[81] This shows that the CSS has made important contributions to boosting the development of China's social sciences and humanities and the integration of sciences.

5. The Chinese Association of Natural Science

The CAAS and the CSS were both founded abroad, while the Chinese Association of Natural Science (CANS) is an indigenous organization. The Natural Science Society of West China, the predecessor of the CANS, was founded by Zhao Zongao, Li Xiufeng, and other two people at the former Central University of Nanjing in September 1927, with the aim of researching and developing natural science.[82] When the Natural Science Society of West China held its first annual meeting in Nanjing in July 1928, it was renamed the Chinese Association of Natural Science (CANS). In 1934, the CANS recruited Xin Shuzhi, Han Zukang, Ye Xiufeng, Chen Lifu, and other five sponsoring members, and in 1941, Zhu Kezhen, Xiong Ji, Zhu Jiahua, and other six sponsoring members.[83]

In 1944, the CANS promoted international academic exchanges by publishing two publications, *China Science* and *Collection of Scientific Literature*, through its International Cooperation Committee on Scientific Work. In 1943, Dr. Joseph Needham, a British scientist, came to Chongqing on behalf of the British scientific community and brought a letter from the British Association of Scientific Workers to the CANS, "expressing admiration for Chinese scientists' hard work in the War of Resistance against Japan, and hoping to maintain close ties and strengthen cooperation with the CANS. The CANS recruited Dr. Joseph Needham as a

[79]Ouyang (2004: 39).
[80]Deng (1979: 74).
[81]Fan (2004: 68–85).
[82]Li (2008: 179–186).
[83]Shen and Langming (1982: 58–73).

consultant for *Collection of Scientific Literature*."[84] In December 1946, the CANS and the CAAS formed the China Association for the Promotion of Science to seek the great unity of scientific workers across the country.

In July 1949, under the leadership of the CPC, the CANS, together with the Northeast Natural Science Research Association, the China Association of Scientific Workers, and the CAAS to organize the preparatory meeting for the National Congress of Natural Science Workers of China. After one year's preparation, the first National Congress of Natural Science Workers of China was held in August 1950 under the leadership of the CPC. The All-China Federation of Special Societies of Natural Sciences and the All-China Association of Science Popularization were established to undertake the promotion of national science and science popularization respectively and to contribute collective strength to the national construction. The CANS issued a statement on April 10, 1951, to end its social affairs.

The CANS, which was born locally, is relatively pragmatic and its research is closer to reality. Also, the CANS have a better understanding of the reality of Chinese sciences. It has played an important role in the unification of the Chinese scientific community after the founding of the PRC.

6. The Tension Between Planned Science and Free Research

Judged from these scientific communities in the Kuomintang-controlled area, most of them held the good wish of saving the nation by science, improving the scientific literacy of the people and pushing forward Chinese learning, and made great achievements, laying a solid foundation for the scientific cause of the PRC and especially in natural sciences, greatly promoting the scientific level of China before the founding of the PRC. However, science as a social program, the instability of the country, and the lack of independence of the nation have, to a great extent, affected and hindered the development of science in China. It is in this sense that the transformation of folk science into national science, after the founding of the PRC, has become a trend. However, by taking a comprehensive view of these representative academic organizations in the Kuomintang-ruled area, they naturally resisted political interference and insisted on academic autonomy and research freedom, which has provided a good reference for the development of the academic community after the founding of the PRC.

Before the founding of the PRC, among the scientific communities, the rest were non-governmental academic organizations, except the official Academia Sinica. Free research and academic autonomy are their common goals. They did not conduct research under a unified institutional framework. The existence of the Kuomintang government enabled them to serve the "national learning" with science and related organizations. However, the semi-official scholars, represented by Fu Sinian, director of the Institute of History and Linguistics of the Academia Sinica, were not entirely dependent on the Kuomintang government. Their political tendencies are very complicated. "On the one hand, nationalism was strong, firmly opposed to imperialism, especially Japan; On the other hand, the scholars supported Chiang Kai-shek, having loyalty to the king, but they have always adhered to a liberal stance and opposed governmental corruption and interference."[85] The lukewarm relationship with the Kuomintang government is a characteristic of semi-official scholars. However, other completely non-governmental scientific communities are far from the Kuomintang's power center and the Kuomintang government has no intention of forcibly integrating them. On the whole, the scientific research situation in the Kuomintang-ruled area was relatively free and scattered before the founding of the PRC.

This very messy and unplanned research is generally at the stage of small science. It does not conform to the era background of big science. After the founding of the PRC, under the leadership of the CPC, the research has moved

[84]Ri and Jin (2007).

[85]Chen (2008).

towards a systematic national science system. On the one hand, the scientific community concentrated various scientific forces to form advantages and made many great achievements that could not be done before the founding of the PRC. On the other hand, the scientific community became a political force favored by the CPC. As a result, the unified scientific community after the founding of the PRC has both scientific and political nature. In some time and space, political nature is much higher than scientific nature. This is an important difference between the scientific community before and after the founding of the PRC.

A necessary condition for the emergence and implementation of science and technology planning after the founding of the PRC is to have a relatively centralized group of scientists who can be dispatched in a unified way. Therefore, the unification of scientific communities is inevitable after the founding of the PRC. From a broader perspective, the unity and disorder of scientific communities should maintain a necessary tension. The scientific communities, completely scattered, without a centralized goal, under a unified state power, or without a unified national will, obviously do not conform to the national interests in the era of big science and the interests of scientists themselves. However, excessive unification would restrict the development of the spirit of free creation and independent judgment to a certain extent. The two sides should keep a reasonable space distance. This tension is reflected in the science and technology planning. In other words, it needs to be kept a balance between the unified research to achieving the national interests and the free research to achieving the innovative spirit in science and technology planning.

After entering the twenty-first century, the sci-tech planning of China has undergone rational reflection. The formulation and implementation of sci-tech plans serve the construction of the national innovation system. The way of building the national innovation system and maintaining the necessary free research, and the free research of the scientific community before the founding of the PRC can give us some enlightenment.

III. Contribution of Returnees

A large number of scientists in the Kuomintang-ruled areas served the War of Resistance against Japanese Aggression. Here, only the contributions made by scientists and technicians who had returned from abroad to fight against Japan are introduced. Despite harsh conditions, the eight-year war of resistance was an important period in the development history of science and technology, culture, and education in China. During this period, returned students made a great contribution to science and technology, which should be remembered by history.

As early as before the July 7th Incident (Lugou Bridge Incident) of 1937, the patriotic anti-Japanese activities of Chinese returned students had already begun. After the September 18th Incident in 1931, more than 2000 students returned from Japan to carry out the large-scale anti-Japanese national salvation movement. After the July 7th Incident broke out, China's students studying in Japan responded to the National Government's call to "withdraw from the enemy country and return to the homeland for the Anti-Japanese War" in large numbers. In just over two months from the Lugou Bridge Incident to September 1937, more than 4,000 students returned to China from Japan.[86] Meanwhile, Chinese students studying in Europe and the United States also returned home after the full-scale outbreak of the Anti-Japanese War, forming the second large-scale return climax in the history of studying abroad after the Revolution of 1911. Overseas students used their sci-tech knowledge to serve the War of Resistance against Japan. Returning from Germany, Gong Zutong developed China's first military telescope and machine gun sight in 1938.[87] Zhao Zongao, who returned from Germany, designed and built a synthetic oil substitute factory to solve the problem of a wartime gasoline shortage. Zhao Tian, who returned from studying in Britain, invented the Zhao's Antimony Refining Method, which has been used for decades. Zhang

[86]Returned students in the War of Resistance Against Japan. Xinhuanet: http://news.Xinhuanet.com.

[87]Lai (1990).

Changshao, who returned from Britain, compiled books such as *Penicillin and Streptomycin* and *Sulfasamide*, and introduced these new drugs into China. Li Wenping, who returned from studying in the US, paved the way with asphalt for the first time in China, making a great contribution to the completion of "Stilwell Road".

Returned students brought fresh strength to China's higher education during the Anti-Japanese War. In 1941, of 179 professors and associate professors of Southwest Associated University, 87% were returned, students. 26 heads of department, with the exception of the Department of Chinese Literature, are all returned, students. For example, Dr. Chen Xingshen of Hamburg University returned to China in 1937, Hua Luogeng and Zhao Jiuzhang in 1938, Doctoral Physicist Zhang Wenyu of Cambridge University in 1938, and Doctoral Physicist Ma Dayou of Harvard University in 1940. The return of Chinese overseas students had enabled the Chinese academic community, almost isolated from the world during the war, to keep up with the international trend. Apart from guns and other weapons, the War of Resistance against Japan was also a battle of science and technology, in which returned students became the main force to boost sci-tech progress.

Dr. Joseph Needham, a well-known British scholar, wrote, "Chinese scientists from big cities in east China are as surprised as we are about the original situation in west China and their poor technical conditions. In Jiading, Sichuan, someone discusses nuclear physics in an ancestral hall with a remote view of Tibetan peaks. In Dali, a local tribe assists a plankton biologist in pushing a boat into the water by Erhai Lake under Wuhua Tower."[88] Although the conditions are simple and the fighting is incessant, China's sci-tech workers demonstrate their lofty mission of serving the country and their patriotic enthusiasm.

Obviously, science and technology and corresponding planning after the founding of the PRC are based on the foundation before the founding. It is a problem for the PRC to inherit and develop these foundations.

[88]Xu and Xiaoqing (1981: 650.46).

2.4 Review and Sublimation

This chapter describes the development of China's science and technology before the founding of the PRC and its possible impact on science and technology planning after the founding of the PRC. The science and technology system after the founding of the PRC is quite different from that before the founding, but the foundation comes from that before the founding of the PRC, including the science and technology foundation and the source of scientists. The research in this chapter lays a solid foundation for the entire book.

The CPC had already started the practice of science and technology on a revolutionary basis. Although the practice is a preliminary and basic work compared with the present, the PRC had accumulated experience in the formulation of scientific and technological policies and related management after the founding of the PRC. The practice of science and technology serves the revolutionary practice, so the science and technology work of the CPC was revolutionary at the beginning. Emphasizing actual effects, the function of technology, and sci-tech personnel, and bringing military management into science and technology have become the characteristics of sci-tech practice before the founding of the PRC. The characteristics are reflected in the sci-tech planning of the PRC, thanks to the new government established by the CPC.

A large number of senior sci-tech cadres, such as Wang Zheng and Wu Heng, trained by the CPC during the war years, have played an important leading role in the formulation and implementation of sci-tech plans after the founding of the PRC. A large batch of sci-tech cadres from the base before the founding of the PRC and in the Yan'an period served different fields and links of the sci-tech system of the PRC. They are one of the backbone forces for the smooth implementation of science and technology programs. In the war years, the military style and management methods were introduced into the sci-tech planning, empowering China to succeed in carrying out its first sci-tech plan.

The level of science and technology and the quality of scientists in the Kuomintang-ruled areas were generally higher. It is also possible to conduct research on self-selected topics and achieve some results of a global nature. The science and technology and scientists had contributed to the victory of the Anti-Japanese War. The vast majority of scientists had beautiful expectations and ideals for the PRC and were willing to make contributes to the PRC. In the formulation and implementation of science and technology plans, most of these scientists undertook actual work. Although they seldom served as senior leaders, the scientists had played an important supporting role in the implementation of science and technology planning.

Before the founding of the PRC, China's scientific and technological development was scattered and multi-regional, which could be regarded as the "multi-regional development in diverse ways". After the founding of the PRC, science and technology came from both the bases and the Kuomintang-ruled areas, including the enemy-occupied areas and other military separatist regions. Integrating these resources to form a unified research model is the natural result of the reunification of the state power. The CPC had an edge in the sci-tech foundation of the base and its sci-tech cadres had taken up a high proportion in the integration of resources. Therefore, the research on the sci-tech planning after the founding of the PRC needs to be extended to the sci-tech practice of the CPC during the military struggle before the founding of the PRC because this practice has had a far-reaching impact on the future thanks to the establishment of the people's political power.

The sci-tech workers in the PRC mainly come from red experts under the leadership of the CPC, scientists and technicians in the Kuomintang-controlled areas, returned students (divided into those who returned before and after the founding of the PRC and those who returned to the bases and the Kuomintang-controlled areas before the founding of the PRC), and foreign experts (classified into those who came to China before and after the founding of the PRC, and mainly the Soviet experts after the founding of the PRC). Among the four forces, red experts under the CPC were leading the scientific and technological front after the founding of the PRC, while the Soviet experts served important advisory positions. However, experts from the Kuomintang-ruled areas and returnees held actual positions in most cases. At that time, the economic conditions and scholarship foundation of the Kuomintang-controlled areas were better than those of the bases, yielding some achievements made in the development of science and technology. There emerged scientific achievements and scientists who occupied a certain position in the world. As scientists and scientific communities in the Kuomintang areas are generally separated from the political center, their research is relatively free and scattered. After the founding of the PRC, this case underwent a proletarian transformation, and the original scientific community became a stratum of unity, education, and transformation in the new political power. These intellectuals often became one target of social movements.

The swing and twists of the CPC's policy towards intellectuals before the reform have had some negative effects on China's science and technology development and the formulation and implementation of sci-tech plans, resulting in the shortage of scholar temperament in independent research and academic autonomy. The development of scientific communities is not perfect, which, in the long run, will bring adverse effects on China's science and technology, especially basic science. Therefore, after entering the twenty-first century, the CPC has a fuller understanding of science and technology, examining itself. Reviewing the development of science and technology and scientific communities in the former Kuomintang-ruled areas before the founding of the PRC will help the rethinking and deepen the understanding of China's science and technology development laws and the complex relations between science and society.

References

Chen, Hongbo. 2008. The practice of the institute of history and linguistics and the rise of Chinese scientific archaeology (1928–1949). Doctoral Thesis of Fudan University.

Chen, Jianxin, Zhao Yulin, et al. 1994. *Development History of Science and Technology in Contemporary China*. Wuhan: Hubei Education Press.

Chen, Zhenghong. 2007. Scientific belief and "imperial examination belief". *Journal of Huainan Normal College*. (6).

China Modern Economic History Team, Institute of Economics, Chinese Academy of Social Sciences. 1986. Selected economic historical materials of revolutionary bases (Part II). Nanchang: Jiangxi People's Publishing House.

Collected Works of Mao Zedong (Volume 1). 1993. Beijing: People's Publishing House.

Collected Works of Mao Zedong (Volume 2). 1993. Beijing: People's Publishing House.

Deng, Niudun. 1979. Guo Moruo and "Chinese Scholarship Society". Fudan University (social science edition), (4).

Fan, Dainian. 2004. The rise and fall of Chinese scholarship society and scholarship committed to blending humanities and science. *Science and Culture Review*, (3).

Feng, Xiaowei. 2007. Wang Zheng: From a captured soldier to a general (part I). Literature for Party History, Wen Hui.

Gong, Fang. 2006. Characteristics of Kuang Yaming's strategic decision-making: Enterprising spirit, big-picture view, and creativities, Seen from " 835 Suggestions". *Journal of Nanjing University* (Philosophy, Humanities, and Social Sciences Edition), (4).

Han, Xinfu, and Jiang Kefu. 1996. *Memorials of the Republic of China*, vol. 5. Beijing: China Literature and History Publishing House.

Hao, Jingsheng. 1972. Science in seven years of the anti-Japanese war. *Revolutionary Literature*, (59).

He, Zhiping. 1990. *China's Science and Technology Organizations*. Shanghai: Shanghai Science Popularization Press.

Historical Materials of Yan'an Academy of Natural Sciences. 1986. Beijing: History of Chinese Communist Party Publishing House.

Hu, Shi. 1985. US returns boxer indemnity// Jichen, Hu. Unofficial history in the Qing Dynasty. Changsha: Yuelu Publishing House.

Huangfu, Shuyu et al. 1989. Educational record in Chinese revolutionary base 1927.8–1949.9. Beijing: Educational Science Publishing House.

Lai, Chunxia. 1990. Pioneer of applied optics in China—Gong Zutong. *Knowledge of Modern Physics* (5).

Jian, Zi. 1945. Private iron and steel industry growing up in the anti-Japanese war. *Southwest Industrial Communication* 10.

Li, Xuetong. 2008. Overview of the Chinese association of natural science. *Journal of Chinese History of Science and Technology* (2).

Li, Zhihong, and Ma Junfeng. 2004. *Mao Zedong's Thought on Science and Technology*. Changsha: Hunan University Press.

Liu, Xiaoyun. 2002. On the China culture and education fund and the modernization of science and education in China. *Journal of Luoyang Normal College* (1).

Ouyang, Liang. 2004. Research of Chinese scholarship society. Master Thesis of East China Normal University.

Pan, Yuntang, and Ji Yajun. 1993. Revolutionary career and scientific process—Record of Scientist Wu Heng. *Historical Materials of Chinese Science and Technology* (1).

Qu, Chunhai. 2005. A brief introduction to the establishment of Yanchang oil field in the late Qing Dynasty. *Historical Archives* (2)

Qu, Shipei. 1985. *Higher Education in Liberated Area during the Anti-Japanese War*. Beijing: Peking University Press.

Ri, Yue, and Zhu Jin. 2007. *Letters from Zhu Shuping*. Beijing: Ocean Press.

Selected Works of Dr. Sun Yat-sen. 1956. Hong Kong: Zhonghua Book Company, reprinted in 1974, 186.

Selected Works of Mao Zedong.1991. Beijing: People's Publishing House.

Selected Letters of Mao Zedong. 1983. Beijing: People's Publishing House.

Selected Works of Mao Zedong (Volume 1). 1966. Beijing: People's Publishing House.

Selected Works of Zhu De. 1983. Beijing: People's Publishing House.

Shi, Baoheng. 2006. Early exploration and exploitation of Yanchang oil field (1905–1949). *Journal of Chinese History of Science and Technology* (1).

Shen, Qiyi, and Yang Langming. 1982. A brief history of the Chinese association of natural science. *Journal of Chinese History of Science and Technology* (2).

Wang, Guangrong. 2002. Xu Teli and Yan'an academy of natural sciences. Xiang Chao (4).

Wu, Heng. 1983. *Historical Materials on the Development of Science and Technology in People's Liberated Area during the Anti-Japanese War (Part 1)*. Beijing: China Academic Press.

Wu, Heng. 1984a. *Information on the Development History of Science and Technology in Northeast China (During China's War of Liberation and the Early Days of the Founding of the PRC) (Comprehensive Volume)*. Beijing: China Academic Press.

Wu, Heng. 1984b. *Historical Materials on the Development of Science and Technology in People's Liberated Area during the Anti-Japanese War (Part 2)*. Beijing: China Academic Press.

Wu, Heng. 1988. Metallurgy, fuel and chemical industry technology in the Shaanxi-Gansu-Ningxia border region. *Historical Materials of Chinese Science and Technology* (2).

Wu, Heng. 1989. *Historical Materials on the Development of Science and Technology in People's Liberated Area during the Anti-Japanese War (Part 8)*. Beijing: China Academic Press.

Wang, Huilin. 1993. Reference for China's modern history 1919–1949 (II). Beijing: Beijing Normal University Press, Yu Guangyuan. Mao Zedong and Dialectics of Nature. *Study on Dialectics of Nature* (12).

Xu, Liyan, and Ye Xiaoqing. 1981. Dr. Joseph Needham's scientific activities in China during the anti-Japanese war. *Journal of Nature* (9).

Xu, Teli. 1995. *On School Spirit Collected Works of Xu Teli (Volume 4)*. Guangzhou: Guangdong Education Press.

Xu, Weimin. 1992. The two suspensions and resumes of science. *Newsletter for Dialectics of Nature* (3).

Yan, Dexuan. 2004. Remembering comrade Wu Heng for ever. *Invention and Innovation* (03).

Yao, Kunlun. 2007. Research on China's award system for science and technology. Doctoral Thesis of University of Science and Technology of China.

Zhang, Dianqing. 2006. Financial assistance of the board of directors of China culture and education fund to modern Chinese libraries. *Journal of University Libraries* (2).

Zhang, Kewei. 1999. *A Brief Introduction to Chinese Science and Technology Experts (Agronomy)*. Beijing: China Agriculture Press.

Zhang, Xiuping. 2009. Research on Qu Bochuan's strategic thought of higher education. *Journal of Dalian University of Technology* (Social Sciences Edition) (2).

Zhu, Jianhua. 1987. *The Script of Financial and Economic History of Northeast Liberated Area*. Harbin: Heilongjiang People's Publishing House.

Zhu, Xiuqin. 1988. On the development of national industry in the Kuomintang-ruled area during the war of resistance against Japan. *Nankai Economic Research* (5).

Zhuang, Zhiling. 1999. Historical materials of China-US foundation for education and culture. *Archives and Historiography* (1).

3 An Analysis of the History and Philosophy of Science and Technology Program in the System of Planned Economy

Abstract

This chapter expounds the historical evolution and philosophy analysis of science and technology planning before China's Reform and Opening-up, which including *the Science and Technology Development Program (1956–1967), the Science and Technology Development Program (1963–1972)*, as well as the difficult development of science and technology in the Cultural Revolution. During this period, science and technology planning embodied the following philosophy: "technology comes first in S&T planning", "national core interests come first", "enhanced political functions", and "self-reliance based on domestic conditions".

Keywords

The 12-year S&T program · Program philosophy · Planned economy · Enhanced political functions · Fourteen science opinions

3.1 Creating the Tradition: Science and Technology Program 1956–1967

I. Political background of the *12-year Science and Technology Program*

When the People's Republic of China was founded, the country had a thousand things to do on its agenda. The fledgling government was busy liberating the southwest, eliminating bandits, restoring the economic order, and helping North Korea to fight American troops. Therefore, there was no official science and technology (S&T) program from October 1949 to 1956, but China's scientific and cultural undertakings were gradually restored in that period. The first plenary session of the Chinese People's Political Consultative Conference (CPPCC) passed the *Common Program of Chinese People's Political Consultative Conference*, in which Article 43 stated that "(we will) energetically develop natural sciences to serve industrial, agricultural and national defense development. We will reward scientific findings and inventions and popularize scientific knowledge." From August 18 to 24, 1950, the first national meeting of natural science workers was held in Beijing, which was attended by more than 500 representatives from around the country. An editorial on *People's Daily* pointed out, "with the government's assistance, science workers can thus divide duties and

Z. Chen, *China's Medium and Long-Term Science and Technology Program*,
https://doi.org/10.1007/978-3-030-75146-3_3

cooperate in an organized and planned way and conduct researches in different fields according to the nation's and the people's needs. They can concentrate all their wisdom and energy on the great undertakings of the people."[1] This meant New China would carry out its S&T work in a planned manner, and making S&T plans was natural.

The international environment had both pros and cons for New China when it was founded. The Cold War landscape forced the newly established government of the Communist Party of China (CPC) to put consolidating political power and safeguarding people's interests first. Mao Zedong put forth three diplomatic policies for the new government: (1) Starting anew. It didn't recognize any diplomatic relation formed by the KMT government with other countries, but re-opened negotiations of diplomatic ties; (2) Cleaning the house before inviting guests. It cleared out the imperialists' privileges in China and re-established diplomatic ties at proper times; (3) Taking sides, which was the most important diplomatic policy. The new China took sides with the socialist camp led by the Soviet Union.[2] In Mao Zedong's opinion, "taking sides was the outcome of Sun Yat-sen's 40-year experience and the CPC's 28-year experience. There is no third way…It's wrong to think that we can succeed without international assistance."[3] By 1950, 17 countries had formed diplomatic ties with China, including the Soviet Union, Bulgaria, India, and Switzerland.

On February 14, 1950, China and the Soviet Union signed the *Sino-Soviet Treaty of Friendship, Alliance and Mutual Assistance* (for 30 years), *Agreement on China's Changchun Railway, Lvshunkou and Dalian* and *Agreement on Providing Loans to the People's Republic of China*. The first treaty replaced the *Sino-Soviet Treaty of Friendship and Alliance* signed by the Soviet Union and the KMT government in 1945, and the new treaty was basically one of equality and mutual benefit.[4] The Soviet Union would provide USD300 million of loans to New China within five years with the annual interest rate of 1%. Later the two sides signed many economic and trade agreements and Sino-Soviet trade was predominant in China's foreign trade.[5]

Mao Zedong had a clear idea about "taking sides with" and learning from the Soviet Union. He said, "We must carefully study the Soviet Union's advanced experience. Members and non-members of CPC, old and new cadres, technicians, intellectuals, workers and farmers—we should all sincerely study from the Soviet Union, not only the theories of Marx, Engels, Lenin and Stalin but also advanced S&T."[6] Learning from the Soviet Union is a traditional practice of the CPC since its formation. Although it conscientiously maintained a critical attitude toward and a certain distance from it during the revolutionary period, it still hoped to draw on the successful Soviet model to save New China twists and turns.

A surge of learning from the Soviet Union was waged across the nation. In the academic circle, the Chinese Academy of Sciences (CAS) which was responsible for planning and guiding the national scientific undertakings, made overall arrangements for "studying from the Soviet Union". In October 1952, CAS held an extended presidential meeting to discuss strengthening the study and introduction of advanced science from the Soviet Union, and a resolution was adopted for that purpose. CAS took steps to intensify the introduction and study of Soviet scientific results, submitted suggestions to relevant administrative authorities on adopting those results in work, and planned to send a delegation to visit the Soviet Academy of Sciences.[7] One of the main outcomes of this Soviet

[1] Carry out Science Work for the People in an Organized and Planned Way, *People's Daily*, August, 27, 1950.

[2] Qizhi and Xiancai (2002, 16).

[3] Selected Works of Mao Zedong (1966).

[4] It was not a completely equal treaty since the Soviet side had predominant interests, but it was more equal than the one signed with the KMT government. Refer to Baichun et al. (2004, 14–19).

[5] Baichun et al. (2004, 14–15).

[6] Zedong's (1990, 45–46).

[7] "CAS extended presidential meeting decides to strengthen the study and introduction of Soviet's advanced sciences," *Chinese Science Bulletin 10* (1952): 655.

study campaign in the academic circle was making five-year plans and S&T plans. Bernal thought highly of the Soviet Union's planned S&T development model and the achievements it made. At that time, this model was very effective, and it was natural for China to make its own S&T plans in imitation of this model.

II. Economic background when New China was founded

When New China was founded, domestic economy was in extreme difficulty, even on the verge of collapse. In 1949, the per capita output of main industrial and agricultural products ranked almost last in the world. The industrial sector had a weak basis and backward technology, and a complete industrial system hadn't been established. The new China confiscated the bureaucratic capital of the "big four" families but allowed them to keep on the production, while the confiscated capital became the main source of state-owned enterprises and later the backbone of national economy. As soon as it was formed, New China confiscated all assets in the country that belonged to the Fascist countries of Japan, Germany and Italy. "We allow the existence of British and American enterprises, but give them no prerogatives."[8] After the Korean War broke out, China controlled about USD190 million of American assets in the country, and very few foreign enterprises were left in 1954. In 1950, the nation began to unify the national fiscal and economic work. Price of goods fell and we finally made ends meet, putting an end to the long history of inflation and price hike during KMT's rule. Mao Zedong described this victory as "no less significant than the victory of Huaihai Campaign."[9] The enormous economic achievements and fast economic improvement made by the new CPC government refuted the imperialists' fallacy that the CPC had zero economic knowledge.

In June 1950, the 3rd plenary session of the 7th CPC Central Committee was held, planning to launch planned economic construction. In August that year, the CPC Central Committee held the 1st national meeting on planned work and discussed formulating the plan for 1951 and three-year objectives.[10] In August 1952, Mao Zedong believed "after two and a half years' hard work, the national economy has recovered and begun planned construction."[11] In September that year, CAS convened a meeting in Beijing, at which heads of a number of departments, including the Ministry of Heavy Industry, Ministry of Fuel, Ministry of Textile Industry, Ministry of Agriculture, Ministry of Forestry, Ministry of Transportation, Ministry of Health, Ministry of Water Resources, Ministry of Education and Meteorological Bureau of the Military Commission, reported their work and proposed requirements for CAS' research. Ding Zan, on behalf of CAS Planning Bureau, made a report titled *Several Opinions Regarding the Formulation of Science Work Plan*, in which he expounded on the objectives and principles of the plan, analyzed the basic situation of national science work, and offered suggestions for reference.[12] The National Planning Commission began to study the 15-year program in May 1953 and asked CAS for assistance. In late August 1954, the Commission issued the *Reference Materials for Formulating the 15-year Program* to all central departments, asking them to form their own 15-year programs in reference to it based on their specific situations.

On May 29, 1954, CAS decided to help the industrial, agricultural and transportation sectors to form the 15-year plan, and set about preparing its own plan as well. In June, it invited experts, both inside and outside CAS, to a series of symposiums targeting math, physics, chemistry, biology, geology, science, technology, water conservancy, electric appliances and power machinery respectively. On August 3, CAS submitted the *Report on the Long-term Plan for*

[8]Qizhi and Xiancai (2002, 19).

[9]Ibid.

[10]Huanfen (1995, 38–43).

[11]Selected Works of Mao Zedong (1977, 69).

[12]History of Chinese Academy of Sciences, 1952, published on CAS website: http://www.cas.ac.cn. Before the book was published, the materials on CAS website were authoritative, so the book directly used some of them.

National Economic Construction to the National Planning Commission and Cultural Heritages Committee, along with a dozen special reports, including that for the iron and steel industry. In the general report, CAS gave specific opinions and suggestions regarding a range of industries and issues, covering iron and steel, nonferrous metal, dynamics, organic/inorganic chemistry, electronics, freshwater and oceanic fishery, handicraft industry, as well as economic zoning, utilization and reclamation of different parts of soil, and treatment of locust plague.[13] A weak foundation and starting from scratch—this was the background then; consolidating the government and improving people's livelihood—this was the general requirement then. They determined the founding and fundamental nature of the first S&T program.

III. The formulation process and contents of the *12-year S&T Program*

(I) CAS' 15-year development program

1. Establishment of CAS' leading position in academic circles

On October 19, 1949, the third meeting of the central people's government committee appointed Guo Muoruo president of CAS; on November 1 that year, CAS began to work in Beijing, marking its official establishment. The CAS was formed on the basis of restructuring and reorganizing the 22 units that used to belong to the Central Research Institute and Beiping Research Institute. In early 1956, it had 44 subsidiary institutes with more than 2,500 researchers. Later, departments of the central government, especially the industrial departments such as geology, metallurgy, machinery, railway, transport, fuel, electronics, construction, textile, and agriculture, all formed their own research institutes in succession, while administrative regions, provinces, municipalities directly under the central government and autonomous regions also formed new research institutes. In 1955, a system of S&T research organizations primarily took shape in new China, which comprised four parts: CAS, colleges and universities, industrial sectors (and their mines and plants), and research institutes affiliated to local governments. Statistics show that there were 410 independent research institutes across the country in 1956, increasing more than ten times from the nearly 40 when New China was just founded. Those institutes had 64,000 employees, including 19,600 researchers, increasing about 33 times from no more than 600 researchers in the same period. Among them, CAS had 66 research institutes and 5,100 researchers.[14]

At the highest state meeting held on January 25, 1956, Mao Zedong said "Chinese people should have an ambitious plan. Within a span of several decades, we will strive to change our backward status in economy, science and culture, and quickly reach an advanced level in the world."[15] In February 1956, the central government issued the "instructions on the *Report of the Work Group of Securing Overseas Students on Calling Back Chinese Students still in Capitalist Countries.*" According to the instructions, we would spend three years attracting the 7,000 Chinese in capitalist countries to come back to China, and the guideline was "attracting all overseas students with the focus on those in America."[16] There were about 7,000 Chinese students studying abroad at that time, scattered in universities in the United States, Britain, France, and Japan. From 1950 to early 1957, lots of overseas students overcame countless difficulties and came back to the homeland, waging a surge of overseas returnees. Among them was Hua Luogeng, Li Siguang, chemical expert Hou Xianglin, physicist Xie Xide, chemist Tang Aoqing, nuclear physicist Zhang Wenyu, metallographist Shi Changxu, meteorologist Gu Zhenchao, as well as Zou Chenglu, Wu Guanzhong, Liang Sili, Hu Jimin, and Lu Zhaojun. The 23 founding fathers that made incomparable contributions to China's development of the atomic and hydrogen bombs and man-made

[13] History of Chinese Academy of Sciences, 1954, published on CAS website: http://www.cas.ac.cn.

[14] Overview of S&T in China for 50 Years, published at http://www.sts.org.cn/.

[15] Zedong (1999, 2).

[16] Weijia (2006, 168).

satellite all came back—Yu Min, Wang Daheng, Wang Xiji, Zhu Guangya, Sun Jiadong, Ren Xinmin, Wu Ziliang, Chen Fangyun, Chen Nengkuan, Yang Jiachi, Zhou Guangzhao, Qian Xuesen, Tu Shoue, Huang Weilu, Cheng Kaijia, Peng Huanwu, Wang Ganchang, Deng Jiaxian, Zhao Jiuzhang, Yao Tongbin, Qian Ji, Qian Sanqiang, and Guo Yonghuai. From 1949 to the spring of 1957, about 3,000 overseas students came back, but this surge subsided after 1957 due to the international tension and the Anti-Rightist Movement in China.

By the end of 1956, China, with the Soviet Union as an example, primarily put in place a new S&T system centered on CAS and assisted by the research institutes of various industrial departments and enterprises under the State Council, those affiliated to the People's Liberation Army (PLA), those of provinces, municipalities, autonomous regions and local departments below the provincial level, and those in colleges and universities. In this system, CAS was the center of academic leadership, research institutes of industrial departments and colleges were two main supporting forces, and local research institutes were indispensable help. They had clear regional and industrial boundaries and were under collective management based on administrative affiliation. When New China was formed, CAS was responsible for guiding the national scientific undertakings, the same as the Central Research Institute. But later they both became independent academic organizations and weren't in charge of national academic affairs.[17] This led to a question: was it possible and necessary to have an organization responsible for academic affairs on behalf of the nation?

In March 1951, Premier Zhou Enlai signed on the following instructions: "strengthen the connection between CAS and industrial, agricultural, health, educational and national defense departments in order to plan and guide the national scientific research undertakings."[18] In March 1954, the CPC Central Committee roughly divided the duties of CAS, colleges and universities, and research institutes under productive sectors: "CAS focuses on the research on basic scientific theories and solving key scientific issues of great significance for national economy; research institutes under productive sectors mainly solve practical problems in production; and colleges and universities carry out research on basic scientific theories or scientific problems in production based on their conditions." This division of duty was consistent with the Soviet model.

2. Study of the Soviet Union's planned science development model

On February 24, 1953, a CAS delegation led by Qian Sanqiang, which included 26 scientists of 19 disciplines, left Beijing for a 3-month visit to

[17]The 1st session of the National People's Congress (NPC) was held from September 15 to 28, 1954. The *Law of the Organization of the State Council of People's Republic of China* was adopted on September 21, which said that CAS was no longer a government department. See http://www.cas.ac.cn/html/cas50/bns/1954.html.

[18]The specific methods are as follows: "(1) When the departments hold professional meetings related with scientific research, they should invite CAS to attend and inform it of the main contents at an early date, so it would have time to study them and put forth opinions at the meeting. The meeting materials should be sent to CAS. (2) Scientific research institutes under those departments should contact CAS when formulating research plans and regularly send a copy of the research report to it so that CAS can have a comprehensive idea of the national situation of scientific research. CAS should give those institutes technical and procedural guidance and assistance. (3) CAS should make a point of systematically communicating the scientific research results in China and abroad, either advise the productive and educational sectors to selectively use them or advise relevant departments to organize the joint further study of them. All departments should actively work with CAS in adopting and promoting new scientific achievements and changing unscientific old habits. (4) CAS should systematically investigate the productive sectors' needs for scientific research, and try to adapt its work plans and the work plans of scientific researchers nationwide to those needs. For that purpose, CAS should convene meetings of national scientific researchers when necessary, announce work assignments and ask relevant departments for assistance." Refer to "Instructions of the State Council on Strengthening the Connection Between CAS and Industrial, Agricultural, Health, Educational and National Defense Departments," Party Literature Research Center of CPC Central Committee (1992, 100–101).

the Soviet Union.[19] Its main mission was to understand and learn how the Soviet Union organized and led scientific research work, especially how it developed science from the old basis after the October Revolution, and the current status and direction of its science development. The two sides exchanged views on science cooperation. Wu Heng believed that in the Soviet Union, science served the socialist and communist construction and the further development of national economy. The distinct objectives and correct guiding thought made its scientific work highly planned. At that time, it was believed that planned social management and development was the feature of socialism, including planning for S&T and management of the scientific circle. Science planning already yielded marvelous results in the Soviet Union, so it was considered an important aspect of building New China.

In the *Report to CPC Central Committee on CAS' Current Work and Future Tasks* dated November 19, 1953, the Party group of CAS said, "(we will) continue the efforts to have a comprehensive picture of the current scientific foundation and forces in China, and work out feasible work plans and development plans accordingly. That is the priority and most fundamental item in CAS' current agenda."[20] In late August 1954, the National Planning Commission issued the *Reference Materials for Formulating the 15-year Plan* to all central departments, kicking off the work on formulating long-term plans for science development.

Soviet experience was highly important for the first S&T planning. In January 1955, CAS advisor Ковда Виктор Абромович[21] (1904–1991) proposed *Some Methods of Planning and Organizing the National Scientific Research Work in People's Republic of China*, in which he suggested creating and developing certain subjects, making reasonable layout of research institutes, conducting general studies of China's productive forces and land resources, and setting up academic titles and ranks. Most importantly, he advised China to plan for national scientific research work and formulate the 15-year science development program, in a bid to solve the most important S&T issues raised in the 15-year plan for national economic development.[22] КобДа's suggestions received close attention from the CAS. On February 12 and April 7, 1955, CAS Party Group and its president Guo Muoruo reported to Premier Zhou Enlai and Vice Premier Chen Yi about КобДа's suggestions in succession. As a lot of work he suggested involved a wide range of fields and required the joint efforts of relevant government departments, they made the following proposal: the National Planning Commission, CAS, Ministry of Higher Education and other related departments formed the "National Planning Commission for Scientific Research Work" to work out the five-year and long-term programs for scientific research work; the National Planning Commission set up a Planning Bureau of Scientific Research Work as the unit in charge of science planning management.

In April 1955, the Political Bureau of CPC Central Committee discussed the report submitted by CAS Party Group. Liu Shaoqi said in his summary, "КобДа's suggestions are very

[19]History of Chinese Academy of Sciences, 1953, published on CAS website.

[20]Reply and instructions from the CPC Central Committee to CAS Party Group regarding the *Report to CPC Central Committee on CAS' Current Work and Future Tasks*. Party Literature Research Center of CPC Central Committee (1993, 164–184).

[21]КобДа was a pedologist with remarkable attainments in the formation and modification of solonchak and solonetz. He won the Soviet National Award in 1951 and 1953 and was selected as corresponding academician of Soviet Academy of Sciences in 1953. He was vice chairman of the Committee of Assistance for Communist Construction under Soviet Academy of Sciences, and had rich experience in planning and organizing scientific work. After coming to China, he visited CAS' branches in Beijing, research institutes in east and south China and read a great amount of materials. On that basis, he suggested making national plans for science development. He also surveyed the Yellow River Basin. His wife came to China with him but returned to Soviet Union in April 1955 to treat her worsening heart condition. КобДа didn't accompany her because of work, and his wife died in Moscow on June 26. After he left Beijing for Moscow on June 27, КобДа never came to China again for health reasons. See CAS website: http://www.cas.ac.cn/html/cas50/bns/1954.html.

[22]Zitong et al. (2004, 59–63).

important. We should attach importance to them. The National Planning Commission, CAS and relevant departments should put forth ideas as to how to put those suggestions into practice and submit them to the CPC Central Committee for discussion and decision."[23] Making the long-term plan for China's science development was a central topic at the inauguration conference of Academic Divisions of the Chinese Academy of Sciences (CASAD) in June 1955. In his report at the conference, CAS President Guo Muoruo announced the ten key tasks that CAS planned to carry out during the first 5-year plan, including research on peaceful utilization of atomic energy, research on oil and earthquake, research on natural and economic zoning in China, research on various basic theories about national construction in the transition period, and research on China's contemporary and modern history as well as contemporary and modern ideological history.[24] As CAS was the national academic leader then, it set about developing its own 15-year S&T program.

On September 15, 1955, the executive meeting of CAS Council discussed and passed the *Instructions on Formulating CAS' 15-year Development Plan*, and decided to begin the discussions and development of the plan in October 1955. The *Instructions* even stipulated the procedures of making the long-term plan and emphasized that making CAS' long-term plan for science development was an important measure for national science development and an important part of planning for national scientific undertakings.[25] In March 1956, based on the long-term plans submitted by each department and overall and balanced considerations, and under the guidance of the President's advisor professor Лазаренко (1910–1979),[26] the Academic Secretariat of CAS put forth the *Major Scientific Research Projects CAS Should Carry out in the Next 12 Years* (natural science and technical science). There were 53 projects in total, some of which concerned scientific research for national defense and sophisticated technology, including atomic energy, semiconductor, radio electronics, computer, automation system, rocket, precision machinery and instrument, and new materials. CAS' arrangements for making the 15-year plan were good preparations for making the national 12-year program for S&T development later, and most tasks it proposed were included in the latter.[27] CAS played a leading role in formulating the *12-year S&T Program*. In the formulation of later S&T plans, CAS also made its own plans, but none of them was as important for the national plan as to when the *12-year S&T Program* was developed. The Ministry of Science and Technology wasn't formed then, so CAS, in a sense, performed some of its functions.

(II) Formulation of the national *12-year S&T Program*

On January 31, 1956, the State Council held a mobilization meeting. It was chaired by Zhou Enlai, Chen Yi and Li Fuchun and attended by leaders of CAS, departments of the State Council and colleges and universities as well as technicians. The meeting announced that National Planning Commission would take the lead in working with relevant departments to develop the *12-year S&T Program*. Premier Zhou Enlai

[23]History of Chinese Academy of Sciences, 1955, published on CAS website: http://www.cas.ac.cn.

[24]"Research on high polymeric compound" was added to CAS' outline of the 1st 5-year scientific work plan reported on July 24, 1955, totaling 11 key tasks.

[25]"Instructions on Formulating CAS' 15-year Development Plan," *Chinese Science Bulletin 11* (1955): 16–18.

[26]Ъ.Р.Лазаренко arrived in Beijing in November 1955 to succeed КобДа as the advisor of CAS president. He invented electrical discharge machining (EDM). While he stayed in China, he participated in the formulation of CAS' 15-year development plan; joined the formulation of China's 12-year science development plan in 1956 and followed it through and personally wrote the part about electromachining and new applications of electric energy under task 43. He urged China to set up a branch of combined electromachining in CAS Institute of Electrical Engineering, and cultivated the first-generation electromachining scientists for China. Лазаренко completed his mission and went back to Soviet Union in February 1958. See http://www.cas.ac.cn/html/cas50/bns/1955.html; Chunyi (1983, 56).

[27]History of Chinese Academy of Sciences, 1955, published on CAS website: http://www.cas.ac.cn.

was in overall charge, and Vice Premier Chen Yi and Li Fuchun, the former being in charge of scientific work and the latter also being director of National Planning Commission, were specifically leading the project. The meeting also announced the decision to form a ten-member scientific planning group in the State Council, with Fan Changjiang being the group leader and the members included Zhang Jinfu, Liu Jie, Zhou Guangchun, Zhang Guojian, Li Dengying, Xue Muqiao, Liu Aifeng, Yu Guangyuan and Wu Heng (mentioned earlier). This planning group had more politicians than scientists, not to mention famous ones, which in a way indicated that the formulation of the *12-year S&T Program* was a top-down process.

The Political Bureau and Secretariat of CPC Central Committee held multiple meetings to study the formulation of the 12-year plan. After the formulation began, to obtain a better idea of how far China lagged behind world advanced level in S&T development, a group of CPC and state leaders, including Mao Zedong, Liu Shaoqi, Zhou Enlai and Chen Yun, gathered at Huairen Hall of Zhongnanhai on the afternoon of January 21, 1956 and listened to the academic reports made by a number of CAS scientists, such as Wu Youxun, Zhu Kezhen, Yan Jici and Pan Zinian (Fig. 3.1).[28]

More than 1,000 provincial, municipal, ministerial and commission leaders who were attending the meeting of matters concerning intellectuals also listened to the reports. Zhu Kezhen wrote in his diary that day that "It was a solemn meeting today. I didn't expect the people's government to attach so much importance to science." After the meeting, CPC Central Committee forwarded the reports to the CPC Committee of all provinces (municipalities), autonomous regions, institutions directly under CPC Central Committee, state organs, the military commission and the general political department, to give the Party cadres a general idea of science development in China and the world and the main tasks of China's scientific work in the current stage.[29] Thanks to the close attention of Mao Zedong and other central leaders, the formulation of the 1st S&T plan quickly unfolded. For a new government, S&T plan could quickly consolidate the government and raise the nation's status, so China's S&T plan had some inherent political functions. On February 24, 1956, a meeting of the Political Bureau of CPC Central Committee officially approved the restructuring of the ten-member science planning group into the Science Planning Committee of the State Council, which was officially formed on March 14. Vice Premier Chen Yi was director of the Committee (succeeded by Vice Premier Nie Rongzhen in November 1956); Vice Premier Li Fuchun, Director of National Construction Committee Bo Yibo, CAS President Guo Muoruo and Vice President Li Siguang were deputy directors; Zhang Jinfu was secretary general and the other nine members of the ten-member science planning group were deputy secretary generals. Members of the Science Planning Committee were experts from CAS, colleges and universities and various industrial sectors.

Based on intensive investigations and studies and under Zhou Enlai's leadership, the ten-member science planning group drafted 11 special reports,[30] including the *Instructions of CPC Central Committee about Matters Concerning Intellectuals*, *Conditions of Scientists' Research Work and Opinions*, and *Opinions about the Treatment of Senior Intellectuals*. When the meeting was held on January 14, 1956, Zhou Enlai gave the *Report on Matters Concerning Intellectuals* on behalf of CPC Central Committee. The report described intellectuals as "part of the working class," defined science as "a decisive factor concerning our national defense, economic and cultural development," and called on all people to "work hard on modern science".[31]

[28]Formulating China's 1st S&T plan, published on the website of China Association for S&T: http://new.cast.org.cn.

[29]History of Chinese Academy of Sciences, 1956, published on CAS website: http://www.cas.ac.cn.

[30]Huayuan (1991, 240–247).

[31]History of Chinese Academy of Sciences, 1956, published on CAS website: http://www.cas.ac.cn.

Fig. 3.1 Mao Zedong and other central leaders meet with scientists formulating the 12-year S&T Program

The issue of intellectuals was considered a barometer of China's S&T policy by Richard P. Suttmeier, a famous expert on China's S&T policies. Before the reform, Chinese policies on intellectuals had changed several times, but the S&T plan wasn't shocked as much as the intellectuals. Why was that? That was because S&T plan had fairly stable political functions and was closely linked with the CPC and the national government. The CPC's ruling position and a stable national government ensured the steady formulation and implementation of S&T plan, which in turn was able to stabilize and consolidate the government.

In January 1956, Mao Zedong convened the highest state meeting to discuss the draft outline of national agricultural development (1956–1967) put forth by CPC Central Committee. He said at the meeting that to achieve the grand goal of reaching world advanced level in several decades, the decisive factor was to have cadres and sufficient excellent S&T experts.[32] Doubtlessly, the formulation of the *Draft Outline of 12-year Plan for National Agricultural Development* set an example for developing the outline of S&T plans.

With the official establishment of the Science Planning Committee in the State Council, the formulation of the 12-year plan unfolded on a large scale. In addition to CASAD members specializing in natural sciences, CPC Central Committee also selected nearly 800 renowned scholars in different fields from 23 units across the nation. There should be a clear guiding thought for formulating the plan. When debriefing the planning work, Zhou Enlai stressed that we should adopt world advanced technology as much as possible, keep our eyes on emerging sciences and technologies at the time, and seize the right moment to "catch up". Meanwhile, we should bear in mind the reality of limited resources and forces in the newly established PRC, concentrate resources on key areas, and avoid the scattering of forces and the prolongation of work. The guiding principle for the *12-year S&T Program* was thus determined— "developing with focus, catching up". Under this principle, the experts and scholars studied the draft long-term plan for national economy developed by National Planning Commission and the long-term plans for production and S&T developed by each department. They had to figure out what specific requirements the national economy had for S&T. On that basis, they had to make reports on subject development, discuss the

[32]"Abstract of Chairman Mao Zedong's speech at the highest state meeting," *Xinhua Bimonthly 4* (1956): 1.

status quo of S&T development in China, and come up with catch-up methods and conditions. It was their thorough discussions that ensured that the *12-year S&T Program* was scientific. Лазаренко exclaimed, "such a comprehensive science plan, such democratic way of discussion—the work itself is of advanced level. Even the Soviet Union didn't do it this way."[33]

While the plan was in process, Лазаренко advised the Science Planning Committee to invite prestigious Soviet scholars in relevant fields to join the formulation. The State Council and Science Planning Committee accepted his advice after careful considerations, and invited famous Soviet scientists specializing in various subjects to give lectures in China on a short term. From March 26 to April 5, 1956, an expert team comprising 18 academicians of the Soviet Academy of Sciences helped draw up and review the science plan. Their assistance enabled Chinese scientists to timely know about the world S&T level and developing trend, and made the formulation more targeted. The Soviet expert team also suggested some emerging subjects that had to be prioritized, such as semiconductor and computer. In accordance with their suggestions, China designated several key areas for special development, which played an important role in future development.

When the *12-year S&T Program* was being developed, there was a debate over whether to focus on subjects or tasks. Scientists mostly put subject development first, but the State Council hoped to focus on scientific research tasks as it urgently needed to fortify national defense and enhance the economic strength. The debate was rather intense, and the final decision, which took into account the opinions of all sides, was to promote subject development by carrying out tasks.[34] As regards the situation that some people ignored basic sciences, Zhou Enlai said we must intensify basic theoretical research while strongly developing applied sciences. Without theoretical research as foundation, fundamental progress and revolution in technology was impossible. Under the direct guidance of Premier Zhou Enlai, the *12-year S&T Program* added the *Study of Several Basic Theoretical Issues in Modern Natural Sciences*, and a basic scientific research plan was specifically made later on that basis.[35] Premier Zhou Enlai went to France on a work-study program when he was young, so he knew the global trend of science development and the importance of basic sciences. In the first-generation leading group, he was one of those on top that had a deep understanding of basic sciences.

The different focus—either on subject or on task—a difference in the planning philosophy. As mentioned earlier, China's S&T plan had inherent political functions and safeguarding the people's government and national interests were its paramount purpose. As New China was just founded, if we focused on scientific research tasks, we could better concentrate the advantageous forces to accomplish national tasks and meet political needs. In that process, subjects would develop too, thus ensuring national interests and scientists' interests to the largest extent.

(III) Contents of the *12-year S&T Program*

Hundreds of experts and scholars worked hard for nearly 5 months and basically completed the *12-year S&T Program* in August 1956. The *Long-term Program for S&T Development 1956–1967*, which totaled more than 6 million characters, included an *Outline of the Plan* and four appendixes: *Task Descriptions and Descriptions of Central Matters*, *Descriptions of the Plan for Basic Scientific Subjects*, *List of Tasks and Central Matters*, and *Emergency Measures for 1956 and Key Points of Research Plan for 1957*.[36]

Based on elaborate and thorough investigations and research, the *12-year S&T Program* listed 57 important S&T tasks and 616 central matters. Among them, the 12 key tasks were the

[33]Wenli and Meng (2007).

[34]Dictated by He Zuoxiu, organized by Zhang Boling, "A review of the *12-year S&T Plan*," *S&T in China* 2 (2006): 53.

[35]Heng (2006).

[36]*Selected Articles of Nie Rongzhen on Science and Technology* (Beijing: National Defense Industry Press, 1999), 10.

peaceful utilization of atomic energy; jet technology; semiconductor and computer (electronics); remote control; production automation and precision machinery, apparatus, and instruments; prospecting of important resources like petroleum; building China's own alloy system and new smelting technology; comprehensive utilization of important resources; new-type power machinery and large machinery; overall development of the Yangtze River and Yellow River; agricultural mechanization and electrification and chemical fertilizer; prevention and treatment of several key diseases; and research on several important basic theories.[37] Making task-centered S&T plans gradually became a tradition, and later plans all had terms like major tasks, including the major special projects in the 21st century. The demand for subject development was satisfied through other methods than the S&T plan, such as special educational plans and development plans like "985" and "211". The principle that S&T plan should serve the nation's central needs and highest interests was implemented persistently.

The experts discussed and decided on "several emergency measures" aimed to accelerate science development, namely atomic energy, jet and rocket (or missile), computer, semiconductor, electronics, and automation, the first two being confidential national defense tasks. They were key measures for developing sophisticated S&T. Qian Xuesen played a crucial role. He said that it was easier to make breakthroughs in missile technology than in plane technology, and "radio technology has extensive applications while confidentiality can be solved through other means."[38] Besides, Zhou Enlai paid close attention to the work of S&T intelligence. At his suggestion, the *12-year S&T Program* made the creation of S&T intelligence work an important task. At the 8th CPC National Congress, Zhou Enlai pointed out, "to develop scientific research in China, we should also timely resolve a series of problems, including necessary books, materials, apparatus and test fields…and collect and exchange Chinese and foreign S&T materials." Therefore, in May 1958, the State Council approved the *Plan for Carrying out S&T Intelligence Work* proposed by the former Science Planning Committee and Technology Committee, determined the tasks, management systems, and organizational setup of this work, and set the principles for building a domestic network of S&T intelligence. In the course of a year and more, various departments set up more than 50 professional intelligence units, 7 regional intelligence agencies were set up nationwide, and 457 kinds of intelligence publications, 28 kinds of science and technology digests and over 20 kinds of newsletters were published.[39]

Chen Yi, Li Fuchun and Nie Rongzhen gave an overview of the plan formulation in the *Report to CPC Central Committee on Science Planning* on October 29, 1956. "The plan listed 57 important S&T tasks and 616 central matters concerning national construction, and pointed out the direction of various subjects. This is the main achievement of the planning work and a grand initiative in Chinese science history. The plan was completed based on the joint efforts of hundreds of Chinese scientists and nearly 100 Soviet experts over more than half a year. They made the greatest efforts in current conditions. The plan was developed according to the important S&T tasks needed by the nation. This approach made sure the plan was closely linked with the national needs for socialist construction. Meanwhile, great importance was attached to theoretical studies."[40] The report suggested keeping the current science planning committee and making it a standing high-level coordinating organization that could supervise and coordinate the implementation of the science plan. This comment was pertinent and started China's tradition of making S&T plans through concerted efforts of multiple parties. When S&T plans were

[37]Medium/long-term S&T Program (1956–1967), published on the website of the Ministry of S&T: http://www.most.gov.cn/kjgh/lskjgh.

[38]Zuoxiu (2006, 55).

[39]Heng (2006).

[40]Party Literature Research Center of CPC Central Committee (1994, 425–435).

made later, the first step was convening a large group of experts and scholars to carry out collective studies, tackle difficulties together, form consensuses and identify central tasks.

To implement the plan, the State S&T Commission set up several taskforces to perform specific tasks. The head of some taskforces was a vice premier, the deputy head was a person responsible for scientific research and production, and the members included scientific research, production, and design personnel. The taskforces' work was disturbed by political movements, especially the "Great Leap Forward", later but they still supervised and guaranteed the implementation of the plan.

On May 10, 1957, a new Science Planning Committee was formed in the State Council, with Nie Rongzhen being the director, Guo Muoruo, Lin Feng, Li Siguang and Huang Jin being deputy directors, and Fan Changjiang being secretary general. The new committee had seven tasks: "supervising the implementation of the long-term plan, especially the key research tasks; formulating the long-term plan and annual plan for scientific research as part of the general national plan, etc."[41] The committee held the 4th extended meeting in mid-June 1957, at which it discussed the annual plan of the *12-year S&T Program*—national S&T research plan of 1957. The annual plan included the system of scientific research work and conditions such as books, materials, apparatus, and chemical reagents.

The meeting arrived at a rough consensus regarding the scientific work system that had been argued for a long time, and members of the committee agreed to what Nie Rongzhen proposed in his speech. "The unified scientific research system in China consists of four parts: CAS, colleges and universities, research institutes under central industrial departments, and local research institutes. In this system, CAS is the academic leader and key research center, while the other three parts constitute the vast base of our scientific research."[42]

The 12-year S&T development program was completed, the leading institution and research work system were primarily formed for its implementation, efforts were made to actively create the conditions for research work, and the system of implementing the long-term plan through annual plans began to emerge.[43] The 1956–1967 S&T program initiated the fine tradition of formulating and implementing S&T programs in China.

IV. Implementation of the first S&T program and results

To better supervise and urge the implementation of the *12-year S&T Program*, the State S&T Commission in 1959 formed 35 specialized taskforces, covering overall survey, ocean, meteorology, surveying and mapping, geogrpahy & mineral, metallurgy, petroleum & geology, coal, construction, agriculture, forestry, aquatic products, medicine, railway, traffic, transportation, communication & broadcast, radio technology and manufacturing, semiconductor and measurement. It then formed the technical science taskforce in 1962. In their own scope, the taskforces made suggestions on the guideline, policies, and major measures of S&T development, understood and examined plan implementation, and conducted coordination and organization for it.[44] While the State S&T Commission made the annual plan, the competent department also examined the plan implementation annually, and had two comprehensive inspections in 1958 and 1960. In 1962, the Commission organized S&T experts to comprehensively check the implementation of the *12-year S&T Program*, and the results showed that 50 of the 57 important tasks had basically reached the original target. Five tasks were not finished, namely "rational utilization of national energies and study of power technology," "study

[41]News release by Xinhua News Agency, May 12, 1957.

[42]4th extended meeting of the Science Planning Committee of the State Council, *Xinhua Bimonthly 13* (1957): 166–167.

[43]Weijia (2007).

[44]Development Planning Department of the Ministry of S&T, *S&T Development Plans of the People's Republic of China (1949–2005)*, 2008.

of the production process of heavy organic chemical products and high molecular compound and expansion of their application scope," "production mechanization and automation," "study of seismic activity in China and disaster prevention and defense," and "agricultural mechanization and electrification & agricultural machinery manufacturing". They were not finished because they were beyond the national conditions and scientific strength then. Besides, the "overall survey of Tibetan plateau and Kangding-Yunnan Hengduan Mountains and research on their development plan" and "titanium metallurgy and its alloy" were not finished either due to changed conditions.[45] Therefore, the *12-year S&T Program* was achieved five years in advance. After 7 years of strenuous efforts, China's S&T undertakings presented conspicuous changes in every respect. During the 7 years, the number of research institutes of all levels across the nation increased from 381 in 1956 to more than 1,000 at the end of 1962, and the number of research personnel increased from 62,000 to nearly 200,000. We had a primarily clear idea of our land resources, laying the foundation for economic development. In the industrial perspective, Chinese S&T personnel was able to design and build iron and steel enterprise with the annual output of 1.5 million tons, oil refinery with the annual output of 1 million tons, electric railways and other important projects of economic construction. China basically realized self-sufficiency in usual raw materials and oil products. It began scientific research for petrochemical industry and rare metal industry and made quite many results. Non-agricultural raw materials took a rising proportion in light textile production.

Chinese scientists were able to produce synchrotron and other experimental devices of nuclear physical research; radioisotope was used in industry, agriculture and medical research and testing; China independently designed and manufactured electronic digital computer and other types of computers and set up the computer center; substantial progress was made in automation technology, semiconductor technology, and radio electronics. In the medical field, China approached or reached world advanced level in many aspects, and traditional Chinese medicine (TCM) was sorted through and improved. Basic sciences were strengthened, and new scientific subjects were created, such as computational mathematics, differential equation, biophysics, element-organic chemistry, macromolecular physics, chemical physics, plasma physics, hydroacoustics, solid-state physics, pneumatics, and engineering thermophysics.

All in all, thanks to the formulation and implementation of the *12-year S&T Program*, especially the efforts from 1956 to 1962, S&T, particularly basic technical categories and industrial sectors, developed rapidly in China. This laid a solid foundation for consolidating the fledgling government, improving people's livelihood, and for our self-reliance after the Soviet Union cut off its assistance in the 1960s.

V. Significance of the *12-year S&T Program*

The *12-year S&T Program* had an extremely important position in China's contemporary S&T history and left profound and far-reaching scientific, technological and historical heritages. It was formulated and implemented when China and the Soviet Union were in a "honeymoon", and many of its historical conditions might never reappear, so it was a very special plan in unique historical conditions.

First of all, it intensified the integration of national S&T forces and laid the scientific, technological and educational foundation. As far as its most direct utilitarian function—consolidating the government and developing production—was concerned, the plan fully performed its duties with extra high standards, with the four emergency measures being a highlight. The Science Planning Committee made its judgment about world S&T development and considered China's needs for economic and national defense development. On that basis, to develop radio electronics, automation, semiconductor and computing technology—four new subjects that were critical in modern S&T development, and

[45] Ibid.

elevate them to the world level in a short term, the Science Planning Committee put forth *Emergency Measures for Developing Computing Technology, Semiconductor Technology, Radio Electronics, Automation and Remote Operation.* In July 1956, CAS decided to form a preparatory committee for the establishment of computing technology institute, institute of automation and remote operation, and electronics institute, as well as a research group of semiconductor physics. Apart from allocating CAS' researchers to these units, CAS also transferred relevant researchers from industrial sectors and colleges and universities. The return of the first group of overseas students to Europe and America after New China was founded, including Qian Xuesen and Ye Duzheng, was of critical importance for the formulation and implementation of the 12-year plan as they brought back the most advanced S&T results and developing trend from the west. China assigned people to study in the Soviet Union for one to two years and added new subjects for them, who made tremendous contributions to the development of atomic bomb, hydrogen bomb, and artificial satellite. The 12-year plan officially established the postgraduate system, and CAS and universities all set up graduate schools to cultivate a new generation of scientists living in the transition period. The first group of postgraduates in the People's Republic of China laid the talent foundation for our future S&T development. The implementation of the four emergency measures laid the foundation for our success in the "two bombs and one satellite", which we often talked about with great relish. China's development of nuclear weapon was very timely as the *Comprehensive Nuclear Test Ban Treaty* wasn't signed then and an anti-nuclear-weapon atmosphere hadn't been formed in the international community yet. From 1960 to 1970, we had a string of successful first launches—the first liquid sounding rocket, the first independently developed middle-short-range missile, the first atomic bomb, the first extended middle-short-range missile with a nuclear warhead, the first hydrogen bomb, and the first man-made earth satellite—the Dongfanghong No. 1. These launches largely raised China's international standing, became a major cornerstone for the formation of the China-US-Soviet strategic triangle, and reserved long-term resources for our S&T development. The "two bombs and one satellite" were the most significant S&T achievements since the founding of new China, but we should know that they were technical achievements rather than substantial breakthroughs in basic sciences. The manned space program represented by the Shenzhou series around the turn into the 21st century also stemmed from "two bombs and one satellite", which fully demonstrated their historical significance.

Second, the 12-year plan formed the China-style S&T system of "big science", which can be called the "nationwide system". In the 1920s–60s, communism was in a stage of rise and steady development in the world. The Soviet Union set the best example for new China with its immense accomplishments in national construction and S&T. The planned S&T system that was formed because of the all-round study of the Soviet Union gave a strong boost to China's overall S&T strength and lifted China up a level in S&T development, more in the technical sense.

For instance, we basically figured out the characteristics of natural conditions and the quantity, quality, and distribution pattern of natural resources in the border areas, and proposed an overall development plan and long-term blueprint for the exploitation of natural resources and for the layout of productive forces. We created the system of research institutes and the system of research management organizations with complete categories and reasonable resource allocation, and fostered a large batch of research personnel. A research institute was set up in almost all main subjects and technical fields. The industrial foundation was laid. Large oilfields such as Daqing, Dagang, and Shengli were discovered successively, and iron and steel bases such as Ansteel, Wusteel, and Baosteel were built, with the Panzhihua base just starting construction. Regarding machinery, more than ten basic industries had taken shape, including equipment manufacturing for metallurgy, mining, power generation and petrochemical and the manufacturing of plane, automobile, and engineering machinery. They were the outcomes of the "big science" system and in return reinforced that system.

At last, the 12-year plan created the precedent of CPC top leaders personally caring about S&T. Globally speaking, S&T was more closely combined with politics since the advent of the "big science" age, but China might be the first country whose leaders proactively pursued the utilitarian function of science. The *12-year S&T Program* was developed under the organization of one premier and two vice premiers of the State Council, which was rare not just in China, but in the whole world. Chairman Mao Zedong was the first to say that we should carry out a technological revolution of the CPC, and he called on all Party cadres to learn science and be determined in developing our own atomic bomb, nuclear submarine, and artificial satellite.

To keep away the interference of "leftish" force, the CPC Central Committee and Chairman Mao Zedong in 1961 approved the *14 Opinions about the Current Work of Natural Science Research Institutes* (known as the "14 science opinions") jointly compiled by the Party Group of State S&T Commission and the Party Group of CAS. As this report especially elaborated on CPC's policies on intellectuals and S&T, it was called the "S&T constitution" by S&T personnel. Its formulation and implementation was an important guarantee for the implementation of the S&T plan.

When revising the *Outline of the Study and Publicity of CPC's General Line in the Transition Period* in 1953, Mao Zedong re-emphasized the necessity to "wage a revolution in technology". At the meeting about intellectuals held by CPC Central Committee in 1956, he officially put forth the concept of "technological revolution" and called on the whole Party and nation to "march toward science". In January 1958, he made the further request to "wage a technological revolution" and "put the Party's work focus on this revolution". In 1969, he clearly defined "technological revolution" in his instructions on a document. The close attention paid by Mao Zedong and other top leaders ensured the successful formulation and implementation of the plan. Such a tradition was carried on by later S&T plans, almost all of which were cared about by top CPC leaders. Usually, if the top leaders paid close attention, the plan formulation and implementation would yield great results, which again attested to the inherent political function of China's S&T plans.

VI. **Limitations of the first S&T programs**

The *12-year S&T Program* was not without limitations despite its historic success. First, politics and science were combined too closely. Analysis of the ancient S&T history in China showed that ancient emperors urgently needed science, especially technology, to serve the interests of the ruling class and the emperors themselves. After the People's Republic of China was founded, the close attention paid by top state leaders brought this historical tradition to the peak. The leaders' foresight and vision were indispensable for scientific development, which, being unpredictable, had to be put in a predictable, potential framework, thus causing limitations. The future S&T plans looked weak without the top leaders' push and affirmation. S&T tended to take the initiative to seek political support. The combination of politics and science cannot be simply generalized as good or bad, but generally speaking, keeping them apart and independent from each other is good for their development. Maintaining a necessary tension between science and politics is conducive to thorough scientific development. The *12-year S&T Program* came into being in a special historical period. Its science-politics combination was more special and it set an example for future plans of the same kind.

Second, the lack of understanding of basic sciences affected their future development in China. Although Mao Zedong and other leaders, especially Zhou Enlai, attached great importance to basic sciences, their perception was quite limited because they didn't truly penetrate the nature of S&T, and their attitude toward Chinese intellectuals, the people who grasped S&T, was unstable. They didn't have a sufficient understanding of the infinite complexity and connections of S&T development. The central research institute and high-caliber oversea returnees represented the world trend of scientific development and the highest scientific level in China at that time. Sometimes the right opinions of some famous scientists couldn't be understood and adopted. As a result, basic sciences in China were quite weak compared with technological achievements.

In the government work report of 1957, Premier Zhou Enlai made the following remarks about the necessity of science planning. "Some people doubt science planning. In their opinion, it's impossible to make national, long-term plans for the scientific undertakings, which can only be carried out by scientists in a spontaneous and scattered manner...But it's absolutely necessary. If we don't properly organize the existing scientific forces in China, align them closely with the needs of socialist construction and make comprehensive and long-term plans, our scientific development will not have a direction and we won't be able to achieve what we aim for... We cannot deny the necessity of the plan because without the primary blueprint today, there won't be increasingly better scientific work plans in the future. **The socialist economy is a planned economy, so scientific undertakings that serve socialist economic construction should also be planned**. Apart from the national plan, there is no doubt that we should allow scientists to carry out some researches they are good at in order to give full play to their potential. But objection to the planning for scientific research work, which would leave the scientific cause in anarchy, is detrimental to both the socialist and scientific undertakings."[46]

This shows that at that time, top CPC leaders believed planned scientific development was one of the characteristics of socialism, so scientific development had to be planned. However, the insufficient or extremely insufficient understanding of the uncertainties of basic sciences remained and continued.

If CPC top leaders more or less had some knowledge about natural sciences, fewer people in the middle level that communicated policies and instructions from above to the lower level and in the grassroots executive level had a right perception of basic sciences. Having won the military and power struggles that had lasted long years, many S&T cadres in the executive level held that S&T development had the same principles as military struggle. They didn't think there was essential difference between "brain work" and "manual work". Moreover, the executive level followed the principle of "minimal administrative" cost, which meant minimizing the cost within the fixed administrative resources or funds for individuals, so as to maximize individual administrative profits. Due to this economic and managerial principle, many grassroots S&T cadres didn't approve and even criticized the intellectuals' free research as the administrative cost of free research far exceeded the cost of standard research.

Third, the *12-year S&T Program* was too detailed in certain aspects, so much so that the S&T institution wasn't flexible enough, which made its reform more difficult after the state of reform and opening up. Given China's centralized system, traditions, once formed, were hard to change. When an institution was first established, another system about its reform wasn't instituted at the same time—this must have happened in every country. Whenever two presidential candidates ran horse and horse and a president couldn't be produced, the Americans would wonder why the founding fathers, when they established the national institution, didn't put in place a system to reform it. After China launched the reform and opening-up, reforms never stopped, including the reform of the S&T institution, which cost much more than before.

The first *12-year S&T Program* created some traditions and planning philosophy. Later we will analyze three concepts—"technology comes first in S&T planning," "national core interests come first," and "enhanced political functions". These three concepts were never written down in the S&T plan, nor could they be found in any government document. They were just an academic viewpoint based on analysis and research. They were inherited by all S&T plans later, and the

[46]Enlai (1957).

essential spirit was carried on although with variations in different periods.

Besides, thanks to the centralized planning system, the scientific community that was scattered before the founding of new China was incorporated into the unified national system. This, to some extent, curbed the initiative of free research and exerted mixed effects on the implementation of S&T plans.

3.2 S&T Program 1963–1972

I. Historical background: "Great Leap Forward" and deterioration of Sino-Soviet ties

1. "Great Leap Forward" in the S&T world

From late 1950s to early 1960s, the People's Republic of China was making primary explorations for socialist construction, and practices and ideas based on previous experiences naturally existed in state governance and economic construction. What with the successful completion of the first five-year plan, with the accomplishment of Socialist Transformation in advance, and with the victory in the anti-rightist movement and the war to resist U.S. aggression and aid Korea, Mao Zedong's confidence in socialism and China's CPC government was significantly boosted, which led to the "Great Leap Forward". At the meeting of Communist Parties and Workers Parties of 12 socialist countries held in November 1957, he told Khrushchev, "you (the Soviet Union) spent 10 years surpassing the United States. We will spend 15 year surpassing the United Kingdom."[47] A lot of Chinese didn't even know what the US and the UK were like, but they were excited about the possibility of China surpassing the UK and took annual steel output as an indicator. In December that year, Liu Shaoqi announced that China would surpass the UK in the output of steel and other heavy industrial products in 15 years. During the "Great Leap Forward", setting excessively high goals, giving arbitrary and impractical orders, exaggeration and "Communist Wind" prevailed. In agriculture, "grain was the key link," "high yields" were promoted, and unit grain output reportedly kept increased; in industry, the wrong target of producing 10.7 million ton steel a year was set, which waged an "all for steelmaking campaign" with "steel as the key link," and this boosted the "Great Leap Forward" in other industries. The "Great Leap Forward" disrupted the national economic order, wasted multitudes of manpower and materials, and resulted in a serious industry-agriculture disproportion. The three-year movement was followed by three years of dire difficulties and left deep-going impacts on the S&T circle. In November 1958, the CPC Central Committee began to correct the mistakes during the "Great Leap Forward", and the movement came to an end in the winter of 1960.

Scholars generally held the following views about the "Great Leap Forward". On one hand, against the Cold War background, Chinese Communists felt a sense of urgency and were eager to change China's poor and backward status. Its economic construction needed a "Great Leap Forward" that broke the routine. On the other hand, the abnormal relations that gradually appeared between China and the Soviet Union injected the "Great Leap Forward" that was aimed to surpass the UK and the US with an implicit political motive of "surpassing the Soviet Union". The movement exaggerated the effect of human initiative and the impetus that socialist institution could give to productive forces, and impractically copied the successful experience in the days of revolutionary warfare.

The "Great Leap Forward" was considered a good working method back then that suited all fields, including the S&T. Since the scientific community in contemporary China had limited space for independent growth, it wasn't independent enough to fend off external interferences, especially the effect of political power, so the scientific circle was soon embroiled in the "Great Leap Forward". In February 13–15, 1958, CAS held a meeting of heads of all Beijing-based research institutes to make arrangements for a "Great Leap Forward" in the S&T circle.

[47]Baichun et al. (2004, 57).

President Guo Muoruo gave a report titled "Mobilization of the Scientific World" and called on "all scientific personnel to go all out for a great leap forward in science."[48] A few scientists reminded the S&T circle to remain calm. For example, Qian Xuesen said in his speech that to better achieve scientific progress, we must adopt a long-term view and see the national needs and its S&T situation ten years later or even longer.[49] But the circle was so infected by the nationwide passion and enthusiasm that "Great Leap Forward" soon swept it. Scientists took turns to take a stand. Seventeen scientists in Shanghai advocated that "(we must) be leftish and socialist-minded...In the next five years, leftish and mid-leftish forces should account for more than 60%, and we should transform ourselves through the 'Great Leap Forward' of socialist scientific and cultural causes."[50] Nineteen scientists at the Institute of Geophysics of CAS wrote a joint statement, vowing to "double our efforts for the 'Great Leap Forward' of China's scientific undertakings, which is premised on the 'Great Leap Forward' in the minds of scientific personnel. We have to accomplish the 2nd five-year plan for four key research projects—earthquake, physical prospecting, cloud and fog physics and numerical weather prediction—one or two years in advance." Among them were prestigious scientists such as Wang Yinglai, Luo Zongluo, and Zhao Jiuzhang, and there was also Ye Duzheng, who later won the highest S&T award of the nation. The "Great Leap Forward" in the scientific circle showed that at that time, the Chinese scientific community had a weak tradition of independent thinking and could barely resist external interference and impact. Meanwhile, the scientists took part in the construction of new China and the "Great Leap Forward" movement with the best wishes. As no other country had ever had a similar social movement, they kindly believed that it might be a socialist-exclusive and effective way of scientific research.

Against such a background, the medium/long-term S&T plan of 1963–1972 reflected the "leaping" thought about S&T development.

2. Worsening of Sino-Soviet ties

Another main background for the formulation of the *10-year S&T Program 1963–1972* was the escalating divergences between China and the Soviet Union till their relations completely deteriorated. The Sino-Soviet ties went down drastically in the late 1950s and early 1960s, exerting profound impacts on the world landscape, including the two countries themselves, especially China. Many scholars conducted special researches on this field.[51] This paper will only present how this happened and how it affected China's S&T development and the *10-year S&T Program*.

China and the Soviet Union were on best terms during Stalin's rule although not without problems, but they were far from causing tension and rupture. After Khrushchev came in power, he made a point of maintaining the good relationship with CPC at first as he needed our support,[52] but the bilateral relations began to change after 1958 when Khrushchev's position in the Soviet Union was secured. At the 20th national congress of the Communist Party of the Soviet Union (CPSU) in 1956, Khrushchev made a secret report titled *About Personal Worship and its Consequences*, which criticized Stalin's mistakes. Arguments appeared between China and the Soviet Union. In April 1958, the Sovict Union hoped to set up long-wave radio stations in China and later asked to form a joint fleet, but both requests were rejected by Mao Zedong. On June 20, 1959, the Soviet army unilaterally broke the new technical agreement for national defense and refused to provide samples and technical materials for atomic bomb production.[53] On September 9 that year, the Soviet Union published a pro-India statement during the Sino-India border dispute. In April 1960, CPC

[48]Chengsi (1993, 173).
[49]Ibid.
[50]Ibid.

[51]Yanxia (2003).
[52]Baichun et al. (2004, 15).
[53]Ibid.

published three articles, including *Long Live Leninism*, which criticized Khrushchev without calling names. In late June that year, Khrushchev, at the meeting of Party representatives in Bucharest, suddenly condemned the CPC and criticized its domestic and foreign policies. On July 16, 1960, the Soviet Union unilaterally broke the treaty, called back all Soviet experts in China and went back on more than 600 agreements and contracts.

The Sino-Soviet ties improved a little after the conference of Communist Parties and Workers Parties of 81 countries in November 1960, but after the 22nd national congress of CPSU, the Soviet Union cut off its diplomatic ties with Albania and accused CPC of supporting the Albanian Labor Party, aggravating the disputes between the two parties.[54] On April 16, 1962, employees at the Soviet consulate in Xinjiang, based on long-term planning and arrangements, cajoled and coerced about 60,000 Chinese residents in Tacheng, Yumin, and Huocheng to flee to the Soviet Union, and created riots in Yili on May 25.[55] During the Caribbean Crisis in 1962, China opposed America's provocation of war but also criticized the Soviet Union's adventurism and capitulationism. From November 1962, leaders of CPSU and Soviet newspapers and periodicals published a large number of anti-China speeches and articles. Directed by CPSU, more than 40 parties released resolutions, statements, and articles against CPC and the Albanian Party and caused the division of many parties. Under such circumstances, the CPC was forced to publish a series of articles in self-defense. An open argument broke out between China and the Soviet Union, and their relations further plummeted. Then the Li Wanlin Event took place in 1963.[56]

On July 5, 1963, a CPC delegation went to Moscow for talks with its Soviet counterpart, but no agreement was reached and the meeting broke up in discord. During the talks, the Central Committee of CPSU published an open letter on July 14 to Soviet Party organizations of all levels and all Party members, refuting our suggestions on the general line across the board. From September 6, 1963 to July 1964, a massive argument was on during the International Communist Movement.

After Khrushchev stepped down and Brezhnev came in power in October 1964, the relation between China and the Soviet Union and their communist parties went worse and arguments escalated into hostility. In early March 1965, the Soviet Union held a meeting in Moscow on the split of the International Communist Movement, which marked the division of the Movement and the socialist camp. CPC didn't send any representative to the 23rd national congress of CPSU in March 1966, thus severing their relations. In August that year, the Soviet Union expelled all Chinese students and China called back all its workers. Soviet Union also tore up the "pork for timber" contract with China, and bilateral trade decreased sharply. On December 13, 1966, CPSU held a plenary session of the Central Committee, at which Brezhnev made an anti-China report, adopted a resolution and criticized Chairman Mao Zedong directly. In March 1969, the Soviet Union caused the Zhenbao Island Event, provoked border conflicts, deployed millions of troops on the Sino-Soviet border and also sent troops to the Mongolian People's Republic. All these seriously threatened China's security and pushed bilateral ties to the verge of war.

The worsening relation between China and the Soviet Union, from argument to confrontation all the way to the verge of war, had immense

[54]Albania played an important role in restoring China's legal position in the United Nations. See http://news.xinhuanet.com/world/2006-10/25/content-5245069.htm.

[55]Yili riot, http://zh.wikipedia.org/zh/yitaevent.

[56]Li Wanlin was assigned by the Chinese Defense Ministry to study missile design at Soviet Military Academy. He knew China's missile development. Soviet Union wanted to instigate him to betray his country, and he was killed after refusing. See http://news.china.com/zh-cn/history/all/11025807/20050420/12255403-2.html.

effects on China's S&T development. Besides, the US began to penetrate Vietnam in 1961, and China secretly aided northern Vietnam during the Vietnam War, once again standing in political confrontation against the US after the Korean War.[57] The consequence of the Soviet Union tearing up the contracts and withdrawing its experts was that in the 1960s-1970s, "China was basically in confrontation against the Soviet Union, the United States and their allies, and was almost unable to obtain any advanced science and technology."[58] This book will not discuss the losses and gains of the argument or the pros and cons for China's overall industrial strength and S&T development if the Sino-Soviet rupture could be postponed, but they did directly affect the formulation and philosophy of the medium/long-term S&T program for 1963–1972.

3. Economic situation

After it was founded, New China carried out national construction in reference to the Soviet Union's "five-year plans". The period from 1949 to the end of 1952 was a time of national economic recovery and the period from 1963 to 1963 was a time of national economic adjustment. Other than that, China began to implement the five-year plan in 1953.

The "1st five-year plan" formulated under the leadership of Zhou Enlai and Chen Yun aimed at the economic development from 1953 to 1957. According to the general line for the transition period, this plan set the following basic tasks: concentrating the forces on industrial construction, which comprised 694 large and medium-scale projects, with the 156 ones designed with the Soviet Union's assistance in the center, and laying the primary foundation for socialist industrialization.

The "1st five-year plan" was successfully completed at the end of 1957. It laid the primary foundation for China's socialist industrialization and basically established the socialist relation of production. A large group of important projects were completed. A total of RMB55 billion capital construction investment was made in the five years, and there were RMB46.05 billion new fixed assets, 1.9 times as much as the total fixed assets in China at the end of 1952. Altogether 595 large/medium-scale projects were put into operation, initially forming the framework of China's industrial landscape.

The 8th CPC National Congress held in September 1956 officially passed the *Report of Suggestions on the 2nd Five-year Plan for National Economic Development* formulated under Zhou Enlai's leadership. The report explicitly set the basic tasks of the 2nd five-year plan: China would continue industrial construction centered on heavy industry, push the technical transformation of the national economy, and lay a solid foundation for socialist industrialization; China would continue the socialist transformation, cement and expand the collective ownership and ownership by the whole people; China would further develop industry, agriculture, and handicraft and develop transportation and business appropriately; China would work hard to foster construction talents and strengthen scientific research, so as to meet the needs of socialist economic and cultural development; China would enhance national defense and improve people's material and cultural life, etc. The report also set some indicators for the five years from 1958 to 1962: industrial output value would roughly double; agricultural output value would increase by 35%; steel output would reach 10.6–12 million tons in 1962; and the ratio of capital construction investment in total fiscal revenue would rise from 35% during the 1st five-year period to about 40%. However, the "Great Leap Forward" movement and "anti-rightist" movement since 1958 resulted in disproportion among main economic sectors, the national finance was in red for years in a row and people were living a hard life. Economic construction in China could no longer continue according to the 2nd five-year plan, and the nation decided to adjust the economy. In September 1960, CPC Central Committee put forth the economic guideline of "adjustment, enrichment, consolidation and improvement," which was approved

[57]Vietnam War, http://zh.wikipedia.org/wiki/VietnamWar.
[58]Baichun et al. (2004, 26).

by the 9th plenary session of the 8th CPC Central Committee held in January 1961.

Formulated and implemented against such a political and economic background, the *10-year S&T Program* was obligated to continue fulfilling the mission of consolidating the government and developing livelihood, but a new planning concept was created then—self-reliance.

II. **Policies on intellectuals and the formulation of the *10-year S&T Program***

After the *12-year S&T Program* was completed in 1962 ahead of schedule, a new S&T plan was put on the agenda.[59] Due to changes in the international situation, the new S&T plan must reflect the thought of "self-centrism" and invest more resources in nuclear weapon, missile, and other high-tech weaponry. In February 1962, the national S&T work conference was held in Guangzhou, at which the attendees discussed the formulation of the 2nd medium-to-long-term S&T plan. Here it is necessary to give an account of the Anti-Rightist Movement in the intellectual circle and the "14 Science Opinions" as policies on intellectuals always affected the formulation and implementation of the *12-year S&T program* and the *10-year S&T Program*.

1. Anti-Rightist Movement in the intellectual circle

At the extended meeting of the Political Bureau of CPC Central Committee held on April 28, 1956, Mao Zedong proposed the working guideline of "all flowers bloom together, all schools of thoughts prosper" for literary, artistic, and academic development. "The floor is open for all kinds of academic schools. We shouldn't overwhelm other schools with one school. As long as you are on the same side with truth, more people will believe in your school."[60] Apart from responding to the call of the academic circle, the proposition of "all flowers bloom together, all schools of thoughts prosper" was perhaps more a reflection on the "Lysenko event" in the Soviet Union and a disguised defiance against the Soviet Union's "dominant say". Later Mao Zedong made the following remarks at the 7th meeting of the highest state meeting.

"We talked about 'all flowers bloom together, all schools of thoughts prosper' at the meeting of secretaries of provincial and municipal Party Committees convened by CPC Central Committee. Spring is here. We should let all kinds of flower blossom instead of letting a few flowers bloom but suppressing the others. 'All schools of thoughts prosper' was about a story in the period of Spring and Autumn and Warring States more than 2,000 years ago. At that time there were a lot of schools of thoughts and people were free to have arguments. That's what we need now… Within the scope of the *Constitution of the People's Republic of China*, we should give space to all schools of thoughts, right or wrong, and not interfere in them. We don't know whether someone is a second Lysenko. There are so many theories, so many schools of natural sciences, and even social sciences have different schools. Just let them argue, let them express their opinions freely on publications and newspapers."[61]

The guideline proposed by Mao Zedong was warmly welcomed by the literary and art circle. Guo Muoruo even wrote 101 poems titled "All Flowers Bloom Together" to support it. However, while explaining this guideline at the highest state meeting, Mao Zedong also said, "But is it proper to let John Dewey join the argument? No. Is it proper to let Hu Shi join the

[59]The S&T plans made when the new China was just founded had different terms from the plans now because the newly established government was too busy securing its footing in 1949–1953. The 1st 5-year plan was formulated in 1953 and completed in 1956 ahead of schedule, so the S&T program for 1956–1967 coincided with the 2nd and 3rd 5-year plans. After the *12-year S&T Program* was against completed in advance in 1963, the S&T program for 1963–1972 was formulated to match the 4th 5-year economic development plan, which, however, was disrupted by the Cultural Revolution. After that, national 5-year plans and S&T plans adopted the 5-year or 10-year term, the term we are familiar with today.

[60]Hu Weijia, *Selected Materials about China's Science & Technology Policies (I)* (Jinan: Shandong Education Press, 2006), 184.

[61]Hu Weijia, *Selected Materials about China's Science & Technology Policies (I)* (Jinan: Shandong Education Press, 2006), 185.

argument? No. Is it ok for Hu Shi to come back? Of course he can come back if he wants to, but that's after we criticized him. Then the criticism is in the past. But we cannot allow the expression of counterrevolutionary remarks. That's what people's democratic dictatorship does. Should we permit the publication of Hong Kong-based and Taiwan-based newspapers in Beijing? No. Prohibition of their publication is helpful."[62]

This means that the guideline proposed by Mao Zedong did not allow unlimited freedom, but had restrictions. He had an ideal at heart. Logically, it was impossible to give floor to all ideas and remarks. For instance, it was unlikely to publicly discuss how to kill people, how to break ethics and violate the bottom line of humanity. Therefore, "all flowers bloom together, all schools of thoughts prosper" implied a framework, which could be interpreted in different ways regarding what it was like and how big it was. That was why the intellectual circle and Mao Zedong had a different understanding of this guideline. In the west, a word with the suffix of "ism" would represent a thought or viewpoint with no difference in position or influence, and it was subject to free argument and would run its own course. In China, however, it was impossible to treat all academic viewpoints equally at that time.

Soon after the guideline was put forth, the Rectification Movement began. On August 30, 1956, Mao Zedong planned to strongly rectify the style of subjectivism, bureaucracy, and sectarianism at the 8th CPC National Congress. On March 8, 1957, he mentioned that the Party was going to rectify the three styles when meeting with literary and artistic representatives. On April 30, he held talks with the heads of democratic parties and democrats without party affiliation to solicit their opinions on the rectification. During the talks, Mao, with great sincerity, explained to the non-CPC members the purpose and significance of the Party carrying out the rectification movement, and warmly welcomed their criticisms and opinions. On May 1, 1957, *People's Daily* published the "Instructions of the CPC Central Committee on the Rectification Movement," which lifted the curtain on the movement across the Party.

At the onset of the movement, Mao Zedong remonstrated the whole Party to look at criticism with the right attitude. According to the deployments of CPC Central Committee and Mao Zedong, the United Front Work Department of CPC Central Committee held 13 symposiums of representatives of democratic parties and democrats with no party affiliation from May 8 to June 3, at which more than 70 people gave speeches. At the symposiums, the attendees put forth a lot of criticisms and suggestions on the Party's and the government's work. As far as the S&T community was concerned, a famous example was the *Several Opinions Concerning China's Science System* submitted by the Central Committee of China Democratic League to the Science Planning Committee of the State Council. Jointly compiled by Zeng Zhaolun (chemist), Qian Weichang (dynamicist), Hua Luogeng (mathematician), Tong Dizhou (biologist), and Qian Jiaju (economist), the *Opinions* enumerated many prominent issues in scientific research then, including the protection of scientists. It said "efforts must be made to make China's most urgently needed scientific sectors approach the world-leading level in 12 years. We must protect scientists. This means we must adopt measures to make sure that scientists, especially accomplished ones, have full conditions to work on science, and to change the current situation that scientists are not engaged in science."[63] Regarding the over-rigorous confidentiality of science at that time, the *Opinions* said "the confidentiality system is so rigorous and rigid that it has become one of the main obstacles to scientific research. It wastes manpower and time, causes academic monopoly and private occupation of academic results, and prevents exchanges. Some researches are repeated and cause the waste of equipment. We suggest that materials owned by scientists and professors not be kept confidential unless they concern military and

[62]Ibid.

[63]Several Opinions Concerning China's Science System, *Guangming Daily*, June 9, 1957.

diplomatic matters and new inventions."[64] With regard to social sciences, it stated that "to develop social sciences, we should first change the attitude toward old social sciences…After New China was founded, some disciplines were abolished or no longer stood as independent disciplines. Many people that used to specialize in sociology, political science and law changed fields. Some courses were canceled because the Soviet Union didn't have them. Another misorientation about social sciences was that policy measures or government decrees were usually taken as objective laws…As to academic leadership and the 'locomotive' of scientific research, we believe it must come forth during practices. We shouldn't subjectively appoint a leader or locomotive. In the past, political conditions were over-emphasized in enrollment, assessment for promotion, and the selection of graduate students and overseas students. We think academic competence and political conditions should receive equal emphasis in the future."[65] The *Opinions* also passed comments on other issues.

Seen on the present day, those opinions were good for S&T development, but they were soon labeled as "anti-socialist guiding principle on science". At the 4th session of the 1st National People's Congress (NPC), Guo Muoruo condemned the *Opinions*, saying that "*Several Opinions Concerning China's Science System*, which was proposed and published in the name of the temporary group of the Central Committee of China Democratic League on science planning under the instigation of Zhang Bojun and others, is a deposition of the vicious attempt of bourgeois rightists."[66] Guo Muoruo's criticism was very long. He said, "the *Opinions* submitted by the temporary group of the Central Committee of China Democratic League on science planning is a downright anti-CPC and anti-socialist program on science work…Scientific research needs unified academic leadership and needs a locomotive to take it forward, but the *Opinions* said such unified leadership and locomotive is not needed…At the bottom, the so-called bourgeois social sciences are unscientific. The assertion to restore bourgeoir social sciences is a step back and an objection to studying social sciences with Marxism…This is not a stand-alone document, but part of the plot schemed by Zhang Bojun and other rightists to seize the national leadership, with the leadership of literary and educational work being the first target…It seems that the *Several Opinions Concerning China's Science System* is the first batch of goods 'designed' by Zhang Bojun's 'political institute'."[67] Later Nie Rongzhen also criticized the *Opinions* at the extended meeting of All-China Federation of Natural Science Societies.[68] But the *Opinions* and relevant figures were wrongly criticized. On May 8, 1980, the United Front Work Department of CPC Central Committee wrote in the *Report on the Reexamination of Relevant Patriots*: when we criticized the "Zhang-Luo alliance", we once called the *Several Opinions Concerning China's Science System* and *Our Suggestions on the Leadership System of Colleges and Universities* submitted by the Central Committee of China Democratic League reactionary programs of the "Zhang-Luo alliance". But in fact, the two documents were not initiated by Zhang and Luo, and their contents were basically acceptable and were not anti-CPC or anti-socialist. The wrong done to the *Opinions* and relevant figures was finally redressed.

In the meantime, the United Front Work Department of CPC Central Committee and the 8th Office of the State Council jointly held 25 symposiums of the business circle and 108 people gave speeches. The editorial department of *Guangming Daily* held symposiums in nine big cities, including Shanghai, and invited democrats and senior intellectuals to express their opinions about the Party. CPC Committees of the departments of the State Council, provincial and municipal CPC Committees and those of colleges and universities all held symposiums of

[64] Ibid.

[65] Ibid.

[66] Refuting an anti-socialist guiding principle on science, *People's Daily*, July 6, 1957.

[67] Chengsi (1993, 106–118).

[68] *People's Daily*, July 18, 1957.

non-Party members successively, asking them to help rectify the Party's style.

As the Rectification Movement was carried out on a large scale and high profile, a small number of people misestimated the international and domestic situation then and spread remarks against the CPC's leadership and the socialism system. Mao Zedong kept a close eye on the developments and changes in domestic situation. On May 15, 1957, he wrote an article titled *Changes are Taking Place.*[69] Mao Zedong described the battle with a tiny group of rightists as a "large-scale ideological and political battle," and believed that "this is a massive warfare (the battlefield is both inside and outside the Party). Socialism will not be accomplished if we cannot win this war, and there will even be the danger of another 'Hungarian Event'." He worked out the anti-rightist strategy of "luring the snake out of its hole". On June 8, *People's Daily*'s publication of the editorial *Why Is This* marked the beginning of the Anti-Rightist Movement. The Rectification Movement and Anti-Rightist Movement gradually spread to all regions, circles and sectors of the nation, and the mistake of extending the anti-rightist campaign quickly spread too. On October 15 that year, CPC Central Committee issued the *Standards for Identifying Rightists*, which listed six standards for the rightists, and the most important ones were objecting the socialist system, the proletarian dictatorship, and the CPC's leadership. The article provided that the chief planners and backbone members of rightist activities should be labeled as ultra-rightists.[70]

The wrong extension of the anti-rightist campaign was not curbed and it caused far-reaching adverse effects on CPC's cause. A large number of intellectuals, revolutionary cadres, and democrats were wrongly classified as rightists and subjected to unjust treatment and undeserved persecution, disabling them from exerting their intelligence and wisdom in socialist construction for a long time. During the movement, more than 20 leaders, either in the central government, state organs, or of provincial, municipal and autonomous-region level, were wrongly labeled as "rightist anti-Party bloc", "local anti-Party bloc" and "local nationalist anti-Party bloc," and more than 550,000 people were classified as rightists nationwide. It wasn't until after 1978 that CPC Central Committee decided to redress the wrong classification of rightists. It announced to clear the names of the rightists whose wrong had not been undone yet, and assigned proper jobs for them. Nevertheless, five "central level rightists", namely Zhang Bojun, Luo Longji, Peng Wenying, Chu Anping, and Chen Renbing, did not see the injustice done to them officially redressed before they died.[71]

When the CPC was still fighting to control the national government and before New China was founded, democrats offered many good opinions, including criticisms, and they were accepted with pleasure. The change later didn't totally rest with the opinions themselves or the way of giving them but was because there was no longer the political need to "secure the neutral intellectuals to help take over the government," and the top priority became how to better consolidate the government. As mentioned earlier, China's policies on intellectuals were unstable before the reform because they had political implications. Everything had to serve the central political purpose, including S&T and S&T plan. The CPC and the central government didn't want to criticize the patriotic intellectuals. The movement happened due to a range of reasons: the enthusiasm and ideal for building the newly established nation stood no tarnishing; they didn't fully grasp the infinite complexity of things; their skills and ability to draw a conclusion on something after obtaining all the information was poor. Against such a politics-centered background, the "leftist" way of thinking and work became a habit and path dependence, and it reached the peak when the "Cultural Revolution" broke out. After China launched the reform and opening-up, especially in the 21st century, we

[69]Selected Works of Mao Zedong (1977, 393).

[70]Hu Weijia, *Selected Materials about China's Science & Technology Policies (I)* (Jinan: Shandong Education Press, 2006), 327–330.

[71]Anti-Rightist Movement: http://zh.wikipedia.org/wiki/anti-rightist.

could understand things better and would do things differently. This whole process was the "centennial background" of China's S&T planning.

The Anti-rightist Movement led to the cancelation of the Supervision Ministry and the patent system. In 1959, the Ministry of Supervision and Ministry of Justice were abolished at the same time. The former's functions were taken over by the Supervision Committee of CPC Central Committee and the lawyer system was terminated. The Anti-rightist Movement also resulted in the comeback of the political theory centered on class struggle. At the 3rd plenary session of the 8th CPC Central Committee held in October 1957, Mao Zedong announced when summing up the meeting that the conflict between the proletarian class and the bourgeoisie and that between the socialist path and the capitalist path were the main conflicts in China.[72] The patent system was abolished too. Chapter 1 mentioned that the patent system was quite mature in the period of the Republic of China and it played an important role in promoting knowledge production. But no patent right or invention right had been approved again after 1957. In November 1963, the State Council issued a decree to specifically revoke the *Provisional Regulation on Protecting Invention Right and Patent Right.* It wasn't until March 12, 1984 that the *Patent Law of the People's Republic of China*, the first patent law of new China, was adopted at the 4th plenary session of the 6th NPC standing committee.[73] The twists and turns of the patent system reflected the winding journey of our perception of knowledge production. The Anti-rightist Movement led to the substantial fall in the intellectuals' social position, which was detrimental for the formulation and implementation of China's S&T plans. It also further increased the technical elements and political functions of S&T plan.

Seeing how the anti-rightist campaign persecuted intellectuals, many S&T leaders suggested protecting them. When the campaign entered a climax in the summer of 1957, and based on the discussions of the CAS Party Group, Zhang Jinfu, Secretary General of the Party Group and Vice President of CAS, made the following suggestion to Mao Zedong: China had to rely on scientists to march toward science, and measures had to be adopted to protect the old scientists because they were "national treasures" that were "scarce and valuable". Having obtained Mao Zedong's approval, Zhang Jinfu reported to Deng Xiaoping, who then instructed CAS Party Group to draw up a document on behalf of the central government. The CAS Party Group assigned Du Runsheng to draft the document. On September 8, CPC Central Committee issued the *Instructions on Anti-rightist Movement Concerning Natural Sciences*, which stated in the very beginning that more detailed policies must be made regarding the anti-rightist battle in the field of natural sciences, and different situations should be dealt with differently. In particular, natural scientists and technical workers with substantial achievements should be resolutely protected except certain individuals that had made serious mistakes. The *Instructions* drew some lines. "Some senior natural scientists made anti-Party and anti-socialist remarks, but they are generally dedicated to scientific research and have made high scientific achievements. We need them to perform scientific work in the future, and should not easily classify them as rightists. Some natural scientists with outstanding achievements are highly reputed at home and abroad. They don't care much about politics even though some should be labeled as rightists because of their remarks. Their supervisors should talk with them and point out their mistakes, but these scientists should not be denounced in public. For scientists who tried very hard to come back to China after the Geneva Conference in 1954, don't put them in any political team. We should observe them and patiently educate and transform them during

[72] Zedong (1977, 475).

[73] *Patent History of China*: http://www.sipo.gov.cn.

work in the long term, so as to attract more overseas students to come back and join the socialist construction."[74] The protection of "rightists" engaged in natural sciences again proved that the CPC and the Chinese government had no intention of oppressing intellectuals. The Anti-rightist Movement was the outcome of path dependence and political inertia formed in the atmosphere that "everything centered on politics".

2. "Fourteen Science Opinions"

The Great Leap Forward, Anti-Rightist Movement and natural disasters resulted in the "three-year famine". In face of the reality, Mao Zedong and other top leaders began to realize the problems with previous practices. At the 9th plenary session of the 8th CPC Central Committee in 1961, Mao Zedong broached the issue of investigation and research, and designated 1961 as the "Year of Seeking Truth from Facts" and the "Year of Investigation and Research". In March that year, CPC Central Committee issued *A Letter to the Party Committees of Central Departments, All Provinces, Cities and Autonomous Regions on Careful Investigation*, and also printed and distributed Mao Zedong's article *On Investigations* written in May 1930.

In the winter of 1960, Nie Rongzhen began to organize an extensive investigation and research of the S&T line. The investigation started with the missile institute and expanded later. When the investigation team came to CAS, the scientists there complained most about the fact that the time for scientific research couldn't be guaranteed. In the spring of 1961, in response to Mao Zedong's advocacy for diligent investigation and research, Nie Rongzhen asked relevant departments to hold symposiums, or "Shen Xian Hui"[75] in Shanghai and Beijing, and invited scientists to fully and freely express their opinions about the policies on intellectuals and S&T. On the basis of extensive investigation and research, the leaders found that the intellectuals' policies indeed had many problems. After rounds of consultation, the *14 Opinions about the Current Work of Natural Science Research Institutes* (*14 Science Opinions* in short) was drafted under Nie Rongzhen's lead. Nie directed Han Guang and Zhang Jinfu to hold meetings of CAS branch leaders and Party members in Beijing and the northeast, and the draft was discussed at the iron and steel, geological, agricultural and medical institutes. The central leading group of science work, State S&T Commission and CAS Party Group also discussed the draft multiple times. Based on repeated discussions, practices and revisions, Nie Rongzhen believed the *14 Science Opinions* were mature. On June 20, he submitted the *Report on Several Policy Issues Regarding Current Natural Science Work* to CPC Central Committee and Mao Zedong, with the *14 Opinions about the Current Work of Natural Science Research Institutes* (draft) attached as an appendix, written in the name of the Party Group of State S&T Commission and CAS Party Group.[76]

Half a month after Nie Rongzhen submitted the report, the Political Bureau of CPC Central Committee held a meeting to discuss it. Deng Xiaoping called it a good report and found it necessary to put it into practice on a trial basis. Additions could be made according to the trial implementation to turn it into the "constitution" on science work. The meeting passed Nie Rongzhen's report and the *14 Science Opinions* unanimously, which were then submitted to Mao Zedong and approved by him. In July 1961, CPC Central Committee distributed the report and the *14 Science Opinions* to local areas as central documents.

Around 1961, different departments usually had several guiding rules as their most important guidelines. For instance, in August 11–14, 1961, the Secretariat of CPC Central Committee held a

[74]History of Chinese Academy of Sciences, 1957, published on CAS website: http://www.cas.ac.cn.

[75]"Shen Xian Hui" refers to meeting where the attendees can express their opinions fully and freely and will not be persecuted or criticized or labeled in any way because of what they say. As the attendees have nothing to worry about at this kind of meeting and are as free as celestials, it is generally called "Shen Xian Hui", meaning the meeting of celestials literally.

[76]*Selected Articles of Nie Rongzhen on Science and Technology* (Beijing: National Defense Industry Press, 1999), 248–290.

meeting in Beidaihe chaired by Deng Xiaoping, at which they discussed and summarized 70 methods of industrial construction, which were called the *70 Industrial Opinions*.[77] On March 15, 1961, CPC Central Committee held a central work conference in Guangzhou, at which it drew up the *Work Regulations on Rural People's Communes*, known as the *60 Agricultural Opinions*. In 1961, the *Eight Opinions on Artistic Development* was drafted under Zhou Yang's lead, and the *60 Opinions on Higher Education* were drafted by the Department of Publicity of CPC Central Committee and the Ministry of Education under Lu Dingyi's lead.[78] In this context, the *14 Science Opinions* served as a science constitution to protect scientists and ensure normal S&T work.

The 14 opinions were as follows:

(1) Delivering scientific findings and cultivating researchers are the fundamental task of research institutes;
(2) Scientific research work should be kept relatively stable;
(3) The principle of "combining theories with practices" must be correctly implemented;
(4) Plans must be developed and checked based on reality and should suit the characteristics of scientific work;
(5) Scientific workers should dare to think, speak and act and should be serious, rigorous, and prudent at work;
(6) The time for scientific research must be guaranteed (calculated on the 5/6 basis);
(7) A systematic cadre cultivation system must be established;
(8) Coordination and exchanges should be strengthened;
(9) Carry out scientific work in a frugal way;
(10) Let all kinds of flowers bloom together, all schools of thoughts prosper;
(11) Unite with, educate and transform the intellectuals;
(12) Reinforce ideological and political work;
(13) Carry out extensive investigations and research;
(14) Perfect the leadership system.[79]

Given the historical conditions then, the three most critical and sensitive issues of the *14 Science Opinions* were as follows: clarifying the fundamental task of research institutes; respecting and protecting scientists; and differentiating political issues from academic ones, ideological issues from action ones.[80] The 14 opinions ensured on the most fundamental, institutional level that scientific research wouldn't be affected by external interferences and scientists and scientific work would be protected. It was an important policy guarantee for the implementation of the *12-year S&T Program* and the formulation of the *10-year S&T Program*. After the *14 Science Opinions* was distributed as a central document, CPC Central Committee thought it was not only suitable for science institutes but also for all places with intellectuals, so it was distributed in a wider scope and was reputed as the "constitution of scientific research institutes".

3. Meeting of intellectuals

After promulgating the *14 Science Opinions*, the CPC Central Committee continued to "de-label and recognize" the intellectuals (de-labeling them as bourgeoisie and recognizing them as part of the working class). After New China was founded, the nation's policies about and attitude toward intellectuals was like a barometer of its S&T development. Zhou Enlai was one of the top CPC leaders that had the most thorough understanding of the intellectuals' issue. On September 19, 1951, upon the invitation of President Ma Yinchu and some professors of the Peking University, Zhou Enlai made the *Report on the Transformation of intellectuals*.[81] He talked about the change in intellectuals' class

[77]Party Literature Research Center of CPC Central Committee (2004).

[78]Yuzhi (2006).

[79]*Selected Articles of Nie Rongzhen on Science and Technology* (Beijing: National Defense Industry Press, 1999), 271–290.

[80]History of Chinese Academy of Sciences, 1961, published on CAS website: http://www.cas.ac.cn.

[81]Selected Works of Zhou Enlai (1984, 59–71).

stance. "(Intellectuals) moved from the national stance to the people's stance and then to the working class' stance. That's the general process. We must promote this process, urge the intellectuals' progress, prevent any possible deviation in this process and gradually solve the stance problem."[82] On November 22, 1955, Zhou Enlai reported to Mao Zedong the issue of intellectuals and gave his own opinions. On November 23, Mao convened a meeting comprising all members of the Secretariat of the CPC Central Committee—Liu Shaoqi, Zhou Enlai, Zhu De and Chen Yu—as well as heads in the central government. They held discussions and decided to host a large-scale conference in January 1956 to comprehensively solve the intellectuals' issue. Moreover, the meeting decided to form a ten-member leading group of intellectuals' issue in the CPC Central Committee. Zhou Enlai was the group leader and the members included Peng Zhen, Chen Yi, Li Weihan, Xu Bing, Zhang Jichun, An Ziwen, Zhou Yang, Hu Qiaomu and Qian Junrui. The leading group had an office to prepare for the upcoming conference.

From January 14 to 20, 1956, the conference of intellectuals was hosted in Beijing. Altogether 1,279 people participated in it, including members and alternate members of CPC Central Committee, heads of Party Committees of provinces, municipalities, autonomous regions and municipalities directly under the provincial government, heads of central departments, heads of Party groups of state organs, heads of Party groups of national mass groups, and heads of Party committees of key colleges and universities, research institutes, factories, mines, design institutes, hospitals, art groups and military organs.[83] It was a large-scale conference and the CPC Central Committee attached great importance to it. Zhou Enlai gave a long report on *The Issue of Intellectuals*, in which he fully affirmed the important role played by intellectuals in socialist construction and elaborated on the CPC's policies on intellectuals and other issues. Regarding the intellectuals' class nature, he said "The Chinese intellectual circle has seen fundamental changes in the past six years…A majority of intellectuals have become national workers and part of the working class that serve the socialist cause."[84] However, he spent more time expounding the continued transformation of intellectuals. Zhou Enlai's report was verdantly welcomed by the intellectuals.

On February 14, 1956, CPC Central Committee issued the *Instructions on the Issue of Intellectuals*, which further confirmed the analysis of the issue and relevant policies in Zhou Enlai's report. Later CPC Central Committee and the State Council issued a string of instructions and policies concerning how to solve the intellectuals' issue. For instance, on April 16, 1956, the State Council issued a notice on improving the working conditions for senior intellectuals.[85]

In early 1962, the national S&T work conference was held in Guangzhou with the aim of developing the next national S&T plan after the *12-year S&T Program*. Zhou Enlai gave a speech at the meeting. He said intellectuals were not an independent class, but a social stratum comprised of brain workers. Generally speaking, most people in this stratum, in certain social conditions, were affiliated to the ruling class and served it. After Zhou Enlai's speech on March 2, Chen Yi made the famous speech about "de-labeling and recognizing" the intellectuals at the S&T conference in Guangzhou on March 5 and 6. He said with deep sincerity, "You are scientists of the people, socialist scientists, proletarian scientists and revolutionary intellectuals. The label of bourgeoisie intellectuals should be taken off. Today I shall take my hat off for you."[86] Chen Yi's speech was warmly welcomed by the attending intellectuals. It was clear that the intellectuals needed a correct political position in order to serve the nation without being disturbed.

[82]*Selected Works of Zhou Enlai (Vol. 2)* (Beijing: People's Publishing House, 1984), 59–71.

[83]*People's Daily*, January 30, 1956.

[84]Hu Weijia, *Selected Materials about China's Science & Technology Policies (I)* (Jinan: Shandong Education Press, 2006), 147.

[85]Collection of Laws and Regulations of the People's Republic of China (1956, 388–389).

[86]Qingquan (2007).

The speeches given by Zhou Enlai and Chen Yi about the issue of intellectuals aroused the intellectuals' enthusiasm but triggered heated arguments within the CPC. When Party members kept criticizing their speeches in Guangzhou, Zhou Enlai asked Mao Zedong in face to declare his stand on the class nature of intellectuals, but Mao didn't say a word. On August 9, 1962, Mao Zedong said, "some of the bourgeoisie intellectuals have adopted the right appearances but old problems remain, while some don't even have the right appearances yet." On August 13, he said again, "from the ideological point of view, bourgeoisie intellectuals still exist."[87] Why didn't Mao Zedong want to solve the issue of intellectuals? There were complicated reasons. Some people believed that he never really trusted the intellectuals, not just because he was born in a rural family but also because intellectuals seemed to have brought bad luck to him all his life. For example, when he studied at Peking University, he went to a lecture by Hu Shi, but Hu Shi didn't answer his questions; and the "left-leaning", red intellectual Wang Ming inflicted tremendous damages on the Chinese revolution and Mao Zedong himself. It is hard to guess what Mao Zedong really thought about intellectuals, and it isn't a focus of this book either. But one thing is certain—Mao Zedong's insufficient trust in intellectuals undermined the formulation and implementation of S&T plans in China and was especially bad for forming a free environment of scientific research.

At that time, there were serious divergences within the CPC about intellectuals and the dominant stand was continuing to classify them as bourgeoisie. Later the issue of intellectuals still wasn't thoroughly solved, and intellectuals suffered devastating miseries during the "Cultural Revolution". It wasn't until the 3rd plenary session of the 11th CPC Central Committee that the scientific assertion was reaffirmed that "intellectuals are part of the working class." On June 27, 1981, the 6th plenary session of the 11th CPC Central Committee passed the *Resolution on Several Historical Issues of the CPC Since the Founding of the New China*, which re-emphasized, "we must resolutely eradicate the completely wrong idea of despising education, science and culture and discriminating intellectuals that has existed for a long time and reached a peak during the 'Cultural Revolution'. We should uplift the position and effect of education, science and culture in the modernization drive, and expressly recognize the intellectuals as a force equal to workers and farmers that the socialist cause relies on. It is impossible to develop socialism without culture and the intellectuals." In 1982, the *Constitution of the People's Republic of China* was adopted at the 5th session of the 5th NPC, which stated, "the socialist undertakings must rely on workers, farmers and intellectuals, and should unite with all forces that can be united." The inclusion of intellectuals' social status and role in the Constitution was the final conclusion drawn by the CPC through long S&T policy practices. The development history of intellectual policies after the founding of new China was a reflection of its S&T development history.

4. Formulation of the *10-year S&T Program*

As most tasks set in the *12-year S&T Program* were accomplished around 1962 ahead of schedule, a new S&T plan was in the making in 1962. According to Nie Rongzhen's report to CPC Central Committee at the end of November 1961, it was proposed that a 7-year S&T plan for 1963–1969 be developed first as part of the 7-year national plan and also a supplement and continuation to the *12-year S&T Program*. "On the one hand, this plan should deliver S&T results to serve economic and national defense development in the 7-year national plan; on the other hand, it should also take a longer view and make S&T preparations for achieving long-term goals."[88] But the 7-year S&T plan wasn't formulated in the end, instead an S&T plan for 1963–1972 was formed.

[87] Yibo (1991).

[88] "Report on Several Measures to Promote China's Science & Technology Level More Quickly," *Selected Articles of Nie Rongzhen on Science and Technology* (Beijing: National Defense Industry Press, 1999), 310–314.

Around 1962, CPC Central Committee decided to draw up a 10-year plan for economic development; to support it a 10-year S&T plan would be formulated at the same time. The initial idea was that S&T should adapt to the development of national economy and be adjusted, consolidated, enriched and enhanced. "We should make full use of the existing human resources, materials and capital. Do not run around setting up many new institutes. Do not disperse our energy on new research work that we are unable to do right now."[89] Following the guideline of "adjust, consolidate, enrich and enhance," the scientific community was adjusted after the Great Leap Forward. Take the CAS for example. It had 97 independent organizations and 17,294 employees at the end of 1957, which bloated to 111 and 57,976 at the end of 1960. After the adjustment, the CAS had 86 independent organizations and 40,957 employees at the end of 1961. CAS branches were adjusted too in that period. By the end of 1962, the branches in all provinces, municipalities and autonomous regions except Xinjiang were all revoked, but five branches in the central south, east China, northwest, southwest and northeast and a representative office in north China were established.[90] The 10-year economic plan was tasked to lay the foundation for four modernizations in 10 years and realize self-reliance in economic and national defense development. This affected the determination of the objectives of the 10-year S&T plan. Some people proposed to catch up with world-advanced level, but Nie Rongzhen kept a clear mind and said, "Let's approach the world-advanced level first. That's the goal in the next 10 years. If, based on our efforts, we reach the world-advanced level in some aspects, that's better and we should strive for that. Meanwhile, we should be aware that S&T develops fast in the world and we have a weak base in China, so it won't be easy to approach world-advanced level in the next 10 years."[91] The Soviet Union had stopped its assistance to China then, and "self-reliance based on domestic conditions" became a new planning concept.

The *10-year S&T Program* 1963–1972 was formulated under the organization of the State S&T Commission, aimed to catch up with world-leading S&T level of the 1960s. It vowed to modernize agriculture and industry, establish an advanced industrial system in China, and fill the blanks in key basic disciplines and technologies. CAS organized scientists to participate in the planning by the State S&T Commission and the National Defense Science Committee respectively, while the national development plan for basic sciences and technologies was jointly formulated by more than 900 scientists in a period of over 5 months under the lead of CAS and the Ministry of Education. In December 1962, the *National Development Program for Basic Sciences and Technologies 1963–1972 (draft)* was primarily completed and submitted to State S&T Commission. In May 1963, CAS submitted the *Outline of 10-year Program 1963–1972 of Unit 04 (draft)* and *1964–1972 Program of Unit 04 (draft)* to the National Defense Science Committee.[92] More than 10,000 S&T experts directly participated in the formulation and discussion of the plan, so did a lot of S&T personnel in research institutes, colleges and universities and state-owned enterprises. The plan was finalized in June 1963. After Mao Zedong heard the report in December that year, he said "the plan must be carried out carefully. Relevant conditions should be in place. The S&T battle has to be fought and won. We cannot improve the productive forces if we don't develop S&T."[93] At the end of that year, the *10-year S&T Program* was approved by CPC Central Committee and the State Council and officially put into effect.

[89]"Closing remarks at the national science and technology work meeting in Guangzhou," *Selected Articles of Nie Rongzhen on Science and Technology* (Beijing: National Defense Industry Press, 1999), 333.

[90]History of Chinese Academy of Sciences, 1961, published on CAS website: http://www.cas.ac.cn.

[91]"Closing remarks at the national science and technology work meeting in Guangzhou," *Selected Articles of Nie Rongzhen on Science and Technology* (Beijing: National Defense Industry Press, 1999), 333.

[92]History of Chinese Academy of Sciences, 1962, published on CAS website: http://www.cas.ac.cn.

[93]Selected Works of Mao Zedong (1999, 351).

III. The planning philosophy of "self-reliance based on domestic conditions"

When the 1963–1972 S&T program was being formulated, the needs of economic and national defense development then were fully taken into account and the new movements in world S&T development were considered too. Due to drastic changes in the overall environment, the *10-year S&T Program* not only continued to uphold such planning concepts as "technology comes first in S&T planning," "national core interests come first," and "enhanced political functions". More importantly, it highlighted the concept of "self-reliance based on domestic conditions." The *10-year S&T Program* made the following general requirement: "mobilize and organize all S&T forces nationwide to solve key S&T problems in our socialist construction independently; quickly strengthen and enlarge a socialist-minded and professionally competent S&T team; grasp S&T of the 1960s in important and urgent areas, and strive to make big strides on the way to approaching and catching up with world-leading S&T level."[94] The *10-year S&T Program* consisted of six parts: outline of the plan; plan for key projects; plan for public institutions; special plans for agriculture, industry, resource survey and health care; plan for technical sciences; and plan for basic sciences. It determined 374 key research and experimental projects (including 333 projects directly serving the economic and national defense needs and 41 basic research projects). In the first phase, 32 key projects, 3,205 central matters and 15,000 subjects would be carried out.[95]

The objectives of the *10-year S&T Program* could be summarized as follows:

(1) "Provide S&T results in all aspects for increasing agricultural output and systematically solve S&T problems in the agricultural technological reform;
(2) Prioritize the grasp of industrial S&T of the 1960s and provide S&T results for establishing a complete modern industrial system, developing emerging industries and raising the technical level of the current industrial sector;
(3) Ensure the primary establishment of cutting-edge technologies for national defense;
(4) Strengthen the general survey of resources across China and the research of their protection and overall utilization, and provide necessary resource data for national construction;
(5) Make remarkable achievements in important S&T concerning the protection and improvement of people's health, prevention and treatment of main diseases, family planning, etc.;
(6) Develop basic sciences and technical sciences more quickly, reserve more scientific theories, strengthen scientific investigations and the accumulation of experimental materials, and establish and reinforce important and weak sectors;
(7) Energetically foster talents, obtain more modern experimental devices, set up research centers in important S&T fields, and foster a socialist-minded and professionally competent S&T team that is able to independently solve the S&T problems during national construction."[96]

These objectives had two highlights—agriculture and new technology, the latter including new industrial technologies and cutting-edge technologies for national defense. It was clear that the *10-year S&T Program* upheld a guideline of national economic development with agriculture as the basis and industry as a leading sector.[97] The part about agriculture took up 1/5 of the outline of the *10-year S&T Program*. This

[94] Outline of Science and Technology Development Program 1963–1972, published on the website of the Ministry of Science and Technology: http://www.most.gov.cn/kjgh/lskjgh.

[95] Jun (2001, 223).

[96] Speech on the work of organs of State S&T Commission, *Selected Articles of Nie Rongzhen on Science and Technology* (Beijing: National Defense Industry Press, 1999), 489.

[97] Speech at the national work meeting of agricultural science and technology (1999, 377).

plan used the "big agriculture" concept that encompassed agriculture, forestry, animal husbandry, sideline production and fishery, and vowed to spend 20–25 years on basically realizing the technical reform of the agricultural sector. The part on agriculture in the *10-year S&T Program* could be generalized as below.

Section 1—Agriculture: (1) Carry out rational crop rotation and make agriculture and animal husbandry promote each other; (2) Select and plant fine crop breeds; (3) Step up plant protection and prevent farming and warehousing losses caused by main diseases, insects and weeds; (4) Strengthen research on how to resist meteorological disaster; (5) Reclaim low-yield soil; (6) Conserve water and soil; (7) Develop gardening and local specialties.

Section 2—Animal husbandry: (1) Reasonably utilize and modify the grassland and mountainous and hilly areas in the south; (2) Increase feedstuff output and utilize it reasonably, and develop the feedstuff industry; (3) Improve the production capacity of workstock, poultry and fur animals; (4) Control and reduce main diseases of livestock and poultry; (5) Intensify the research on animal husbandry machinery.

Section 3—Forestry: (1) Well operate and protect the existing forests and increase their growth; (2) Plant fine tree types and improve the technology to increase the yield of fast-growing timber forest and woody grain and oil non-timber forest; (3) Carry out barren hill afforestation at a faster pace and perfect the planting technology of soil and water conservation forest, sand-fixation forest and farmland shelter forest; (4) Raise the mechanized level of forest culture and management as well as timber felling and transport.

Section 4—Aquaculture: (1) Survey and protect fishing resources; (2) Improve the breeding technology and increase the types and areas of aquaculture; (3) Study the technical reform of aquaculture.

Section 5—Tropical crops: (1) Comprehensively develop and reasonably utilize tropical plants; (2) Study the breeding, cultivating and processing technology of rubber trees and related basic theories; (3) Study the introduction, selection, cultivation and processing of tropical economic crops, mainly tropical woody oil product, hard fiber and medicinal plant; (4) Strengthen the research on basic subjects such as tropical biology and geoscience.

Section 6—Agricultural mechanization and electrification, agricultural water conservancy, and the use of chemical fertilizers.

Section 7—Research on basic theories of agricultural sciences and application of new technologies in agriculture. (1) Study the laws governing the genetic variation in animals and plants; (2) Study the laws governing crop growth and find new ways of increasing the yield; (3) Study the basic laws governing animal physiology and biochemistry; (4) Study plant pathogenic pathology and the pharmacology and resistance of pesticides; (5) Study the theoretical basis of veterinary technology; (6) Study the biological characteristics of forest; (7) Investigate and accumulate basic materials; (8) Study the agricultural application of new technologies.[98]

In light of the contents of the plan, the top leadership in the CPC Central Committee attached great importance to agriculture and the promotion of it by virtue of S&T. "If, in a not-too-long period of time, China accomplishes the great task of agricultural technological reform, achieves satisfactory agricultural development and builds a prosperous and strong socialist society, how tremendous its political influence will be in the world!"[99] Another highlight was the development of new technologies. At that time, China was besieged by the two powers of the United States and the Soviet Union. It saw through the former blackmailing with sophisticated technologies and the latter stamping on China's principles in the same approach. The new technologies mentioned in the *10-year S&T Program* referred to those of the 1960s, including the extensive application of various new materials and processes and the popularization of

[98]Outline of Science and Technology Development Program 1963–1972, published on the website of the Ministry of Science and Technology: http://www.most.gov.cn/kjgh/lskjgh.

[99]Speech at the national work meeting of agricultural science and technology (1999, 375–377).

machinery, automation and electric apparatus. The establishment of new industrial sectors was high on the agenda, including organic synthetic chemistry, petrochemistry, new materials, electronics industry, instruments and apparatus, precision manufacturing, atomic energy and jet technology. After the implementation of the 1st S&T plan and thanks to the Soviet Union's industrial assistance, China's general technical level was from the 1940s to the 1950s. Therefore, the *10-year S&T Program* planned to spend 10 years lifting the general S&T strength to the level of 1960s, ready to transition to the 1970s.[100] This denoted that the *10-year S&T Program* had a basically similar historical mission as the *12-year S&T Program*, both being tasked to lay a solid foundation while making big strides to catch up. The two highlights actually reflected the planning philosophy of "self-reliance based on domestic conditions."

Unlike the S&T program for 1956–1967, the *10-year S&T Program* followed the general guideline of "catching up based on self-reliance," which was a wise move at a time when the relation with the Soviet Union went south and China had no other possible sources of assistance. With regard to self-reliance, Nie Rongzhen stressed that "(we should) depend on ourselves and foster a socialist-minded and professionally competent S&T team to achieve the S&T tasks in socialist construction. We must catch up with world-leading level not for the advancement itself, but in order to build the socialist country."[101] This could be generalized as "self-reliance based on domestic conditions." At that time, China had certain S&T accumulation and academic foundation and could be self-reliant proactively, while the worsening relation with the Soviet Union and the contention with the United States also forced us to passively be self-reliant. The hardworking spirit was a valuable treasure of the Chinese nation. The idea of "indigenous innovation" we put forth in the 21st century was actually an upgrade of the philosophy of "self-reliance based on domestic conditions" back then.

"Self-reliance based on domestic conditions" was the philosophy of S&T development established in the *10-year S&T Program*. The Program pointed out that we must concentrate all forces on a war of annihilation. Due to limited resources, to catch up, we had to "have priorities and identify the most critical areas."[102] We should focus on the strategic key issues of national construction and modern S&T development and concentrate all forces on a war of annihilation, so as to approach or catch up with the world level in a relatively short period of time in several important and urgently needed aspects. As to key technologies, the *Program* made it clear that during socialist construction, there were many subjects that we had to study with complete independence. These subjects were usually critical in our independent economic and national defense construction, so we had to mobilize forces to make breakthroughs in them. The *Program* warned against any idea of dependence, but the S&T development after the reform and opening-up was launched seemed to have ignored the concept of self-reliance. As we review this concept today, we have new interpretations. The *10-year S&T Program* also called for "comprehensive arrangement and solid foundation". "We must prevent the trend of theories being divorced from reality, prevent narrow understanding of combining theories with reality, and prevent the neglect of basic and theoretical research. We should take a longer view, not only catching up with the S&T level of the 1960s but also getting ready for the development in the 1970s. The more we emphasize self-reliance, the more we should study and grasp the advanced S&T achievements in foreign

[100]Technical Direction and Technical Policies in the *10-year S&T Program, Selected Articles of Nie Rongzhen on Science and Technology* (Beijing: National Defense Industry Press, 1999), 405–415.

[101]Speech at the national work meeting of agricultural science and technology (1999, 378).

[102]Outline of Science and Technology Development Progam 1963–1972, published on the website of the Ministry of Science and Technology: http://www.most.gov.cn/kjgh/lskjgh.

countries."[103] It's clear that the concept of self-reliance then didn't mean closing our doors to the outside world. Regarding advanced foreign technologies, the *Program* said "while introducing foreign technologies, we should first strengthen our own S&T research in that aspect and try to become able to independently solve problems as soon as possible. Besides, even if we introduce foreign technologies, we still have to allocate corresponding research and experimental personnel to digest them and grasp and develop them based on China's specific conditions."[104] After the reform and opening-up was launched, some industrial sectors adopted the tactic of "trading market for technology", which didn't prove quite successful in the end. This made us better understand the profound insight and vision of the *10-year S&T Program*. Apparently, we were more interested in introducing foreign technologies than digesting them because introduction could take immediate effect while digestion and absorption took a long process. That was because a highly utilitarian planning concept emerged after the reform and opening-up began.

The *10-year S&T Program* also paid close attention to the public's active role in promoting S&T development and to combining professional research with popular scientific experiments, as the latter could solve a lot of technical problems in production on site and in time. There was a special understanding and handling method for a special historical event, like the Great Leap Forward, in a special period. That was actually a reflection of the self-reliance concept as, to depend on ourselves; we had to mobilize more forces to engage in S&T work. Public scientific work, especially technical work, could help implement the S&T plan to some extent. As far as the scientific method was concerned, the "trial-and-error method" required a lot of people to try. In reality, many practical technical reforms back then were indeed derived from public S&T practices.

[103]Ibid.

[104]Ibid.

The first S&T plan implemented the planning philosophy of "technology comes first in S&T planning," "national core interests come first," and "enhanced political functions". These three concepts were carried on in the 1963–1972 S&T program with the addition of "self-reliance based on domestic conditions."

IV. Some achievements of the *10-year S&T Program*

In general, the 2nd long-term S&T plan had a clear direction. It was formulated on the basis of the adjustment, amendment, supplement, and development of the 1956–1967 S&T program in light of the new tasks of China's socialist construction and the new trend of world S&T development. If it could be implemented as planned, we could imagine that our overall S&T strength would take a big step forward from the successful implementation of the *12-year S&T Program*. It would give a strong impetus to our economic construction, national defense and people's living standards. But we cannot make assumptions in history. The "Cultural Revolution" that broke out in June 1966 lent a heavy blow to the implementation of the *10-year S&T Program*, basically putting a stop to it. Many tasks in the plan were not finished, and China's optimistic S&T development came to an abrupt halt.

But the plan was smoothly implemented in the first three years (1963–1966) and yielded some results. In October 1964, China successfully exploded its first atomic bomb, followed by its first hydrogen bomb less than three years later in June 1967. In the mid-1960s, we made breakthroughs in rocket engine and other technologies. In August 1965, China made the plan for our first artificial satellite, and on April 24, 1970, we successfully launched it as the music of Dongfanghong (East Red) invigorated the whole nation. The great success of the two bombs and one satellite stemmed from the continued and consistent efforts of the *12-year S&T Program* and the *10-year S&T Program*, a process that later led to the "manned spaceflight" and "lunar exploration" programs in the 21st century.

The *10-year S&T Program* made other achievements too, the most important ones including completing the national survey of arable soil, soil improvement, rational fertilization, prevention and treatment of plant diseases and insect pests, improvement of plant species and cultivating technologies, control of desertification and alkalization, and other research and experimental projects. Extensive surveys were conducted in the Yellow River Basin, Yangtze River Basin and Yellow Sea and Huaihai Sea plains, and governing and development plans were drafted. In support for the research and testing of the atomic bomb, hydrogen bomb and missile, we developed a large variety of new materials, instruments and apparatus, precision machinery and large equipment with special specifications and rigorous technical conditions, which was a considerable contribution to the independent development of China's national defense research and industry. To meet the needs of economic construction, we designed and trial-produced a group of high-tech, sophisticated equipment, such as a computer, electron microscope, radio telescope, high-speed camera, ammonia clock, and double water inner cooled turbogenerator. We also built plants, mines and railways such as the Panzhihua iron and steel base, second automobile works, Chengdu-Kunming Railway, 10,000-ton ocean liner, large coal mines, large hydropower and thermal power plants, and heavy machinery works, as well as complete equipment for them.[105] These achievements were crucial for enhancing the national strength and resisting external aggression.

In terms of basic theoretical research, advanced findings that received close attention from the international S&T community were achieved in areas such as mathematics, computational mathematics, elementary particles, nuclear physics and structural geology. China's research on synthetic insulin was a typical case. The Shanghai Institute of Biochemistry of CAS began this work in the summer of 1958. Later other units including Peking University and CAS Shanghai Institute of Organic Chemistry joined the research too. After more than seven years of unremitting efforts, the research personnel succeeded on September 17, 1965, and the world's first synthetic protein was created in China.

These achievements, especially the success of the "two bombs and one satellite", significantly raised China's political status on the international stage, consolidated our national defense capability and safeguarded our national interests. The achievements of the *10-year S&T Program* promoted domestic economic construction and made great contributions to our socialist undertakings.

3.3 S&T Plan and Difficult S&T Development During the "Cultural Revolution"

The *10-year S&T Program* gradually came to a stop in 1967 due to the disruption from the "Cultural Revolution". The State S&T Commission developed the *1967 Plan for Interim Experiments for Industrial Science and Technology* and *1968 Plan for S&T Interim Experiments and Trial Production of New Products*, but they were not well implemented. Relevant state departments planned to formulate an S&T plan during the "Cultural Revolution". In August 1971, the Education and Science Group of the State Council and CAS asked for the State Council's permission to "formulate an S&T plan and convene an S&T meeting." In May 1972, CPC Central Committee decided to convene the national S&T meeting. The State Council issued the "No. 56 Notice of 1972," which decided to "hold a national S&T work conference in Beijing on August 10, 1972, and an important content is discussing the formulation of a new S&T plan. The meeting lasted till early 1973, but no official document was issued concerning the S&T plan."[106]

The CAS also attempted to formulate an S&T plan. In 1973, it proposed to compile the *Opinions about Formulating the Long-term Plan for 1973–1980*. On the principle of combining

[105]Jun (2001, 224–226).

[106]Lifan (2007).

theories with the realistic conditions, CAS strove to lay a solid foundation for certain major subjects in the field of theoretical research on basic sciences and to make remarkable achievements in the accumulation of basic materials of natural sciences. Therefore, the CAS proposed plans in five fields: "cutting-edge basic theories of modern scientific research, several important emerging technologies, basic subjects, and interdisciplinary ones, accumulation of basic materials of natural sciences, and supporting the technological revolution in agriculture." In July 1974, the CAS drafted the *Work Arrangements for Formulating CAS' 10-year Plan*, in which it set the following goal: "make scientific preparations for realizing the four modernizations and solving some key research subjects."[107] In 1974, the National Planning Commission also proposed to formulate the 10-year plan for 1976–1985, but neither of them was completed in the end.[108] Systematic S&T research was rare during the "Cultural Revolution", but S&T in that period was exactly the tortuous development and continuation of the 1963–1972 S&T program. We have to have a clear perception of S&T during the "Cultural Revolution" in order to correctly grasp the general history of S&T plans, not to mention that the S&T development in that period had a major bearing on the formulation of S&T plans after the "Cultural Revolution", especially the one in 1978.

Economic construction progressed slowly during the "Cultural Revolution", so did S&T, but a number of substantial S&T achievements were still registered thanks to the efforts of S&T workers and the support of senior CPC leaders represented by Zhou Enlai. Building on the past foundation, Chinese S&T workers continued to work hard and achieved high-level S&T results, including launching and recovering artificial satellites, exploding the hydrogen bomb, prospecting and exploiting oil and gas resources in north China, Panjin, Jianghan, and the central plains, breeding and promoting Indica hybrid rice, and completing the Nanjing Yangtze River Bridge. Progress was made or the foundation laid for other important S&T results in this period too. The *10-year S&T Program* was partially put into practice during the "Cultural Revolution" with many twists and turns, especially the extraordinary achievement of the "two bombs and one satellite" in military S&T.

I. Social development during the "Cultural Revolution"

1. The 3rd and 4th five-year plans for national economic development

The 10-year "Cultural Revolution" went through two 5-year plans. The 3rd 5-year plan for national economic development, which began to be studied and developed in early 1964, lasted from 1966 to 1970. Basic tasks in it were prioritized as follows: agriculture, light industry, and heavy industry. To be more specific, it was set to strongly develop agriculture; appropriately strengthen national defense construction and make breakthroughs in sophisticated technologies; intensify basic industries, continue to improve product quality, increase product varieties and output, and develop transportation, business, culture, education and scientific research accordingly, with a view to developing the national economy with focus and at a certain proportion.[109] However, according to the strategic guideline of "preparing for war, preparing for famine and serving the people" put forth by CPC Central Committee, the National Planning Commission in early September 1965 drafted the *Outline of the Report on Arrangements of the 3rd Five-year Plan*, which expressly vowed to focus on ware preparedness, prioritize national defense construction, and accelerate the Third-Front Movement. According to the *Outline of the Report*, all main economic indicators during the 3rd five-year period met the target, but the blind pursuit for high speed and high stock left problems for future economic development.

The compilation of the 4th five-year plan (1971–1975) started in 1970. In March 1971, CPC Central Committee adjusted the economic

[107] Cang (2006).

[108] Jianxin (1994, 252–254).

[109] Party group of National Planning Commission (1996).

indicators, lowering the steel output target to 32–35 million tons and then further down to 30 million tons. The national economic situation turned for the better in 1972 and 1973, and main economic indicators met and exceeded the targets in 1973. Those two years saw the fastest economic growth since the 1st five-year plan. However, the plan wasn't well implemented later due to the disruption by the "Cultural Revolution". In July 1973, the National Planning Commission developed the *Outline of the 4th Five-year Plan (revised draft)*, which set the target for 1975 to check the results of the implementation of the 4th Five-year Plan, and most economic indicators met the target.

2. Resource survey

Chinese scientists conducted a scientific survey of Mount Zhumulangma in 1959–1960. On November 8, 1965, Vice Premier Nie Rongzhen approved the *Report on Making Scientific Survey of the Mount Zhumulangma Region* submitted by the CPC Committee of CAS. From 1966 to 1968, with the strong support of the People's Liberation Army (PLA) and the Tibetan people, science workers conducted a large-scale multidisciplinary scientific survey of Mount Zhumulangma. The result of the survey—*Report on Scientific Survey of the Mount Zhumulangma* (1966–1968)—was published by the Science Press in 1975. Another survey was conducted in 1975–1976. The survey of Mount Zhumulangma was a comprehensive scientific survey of an alpine region conducted by China with the largest scale and most disciplines at that time.

The CAS began to organize the study of glaciers and permafrost in 1958 and formed a specific research institute for that purpose. The glacier survey covered Mount Qilian, Mount Tianshan, Altay Mountains, west Mount Kunlun, the Himalaya, Nyainqentanglha, Hengduan Mountains and other mountains on the Qinghai-Tibet plateau. In 1978, China officially participated in the work of international glacier cataloging. The permafrost survey covered the Qinghai-Tibet plateau, Hengduan Mountains, west Mount Kunlun, Altay Mountains and the Greater Khingan Mountains, and a primary understanding of the basic situation nationwide was formed. The results of the glacier survey were extensively applied to economic construction and atmospheric research.

The CAS formed the editing committee of the *Flora Reipublicae Popularis Sinicae* in 1959 and the editing committee of the *Fauna Sinica* in 1963, and organized biologists both inside and outside the CAS to compile the two magazines. With the approval of the State Council, CAS presided over the work conference on the compilation of China's fauna and flora sinica in Guangzhou in 1973. The plans for compiling *Flora Reipublicae Popularis Sinicae*, *Fauna Sinica* and *Spore Plants Sinicae* were drafted at the meeting, and they would be implemented as official plans after obtaining the agreement of the competent departments and the task undertakers.[110] The plans for specific disciplines could be viewed as the partial expansion and implementation of the 1963–1972 S&T Plan in the special context of the "Cultural Revolution".

3. Progress of medical technology

Chinese scientists found in 1969 that the fundamental reason for the prevalence of Keshan disease and Kashin-Beck disease was human deficiency in selenium,[111] and selenium supplement had conspicuous effect in preventing and curing the disease. These two research findings captured close attention in the academic circle. Starting from 1959, the CAS Shanghai Institute of Organic Chemistry began the research and development of artificial blood. After ten years of efforts, a new type of blood substitute capable of oxygen therapy was successfully developed in 1969.

[110]History of Chinese Academy of Sciences, 1973, published on CAS website: http://www.cas.ac.cn.

[111]Keshan disease is a local disease first discovered in Keshan county of Heilongjiang province, hence the name. Symptoms include nausea, vomiting of yellow water, chest distress, cold limbs, low blood pressure, occasional dyspnea and edema of lower limbs. The patient is in great pain. The low-selenium environment has something to do with Keshan disease and low-selenium regions coincide with the incidence of the disease, providing the environmental evidences to figuring the cause and the method of prevention and treatment—selenium supplement.

In June 1970, Shanghai Institute of Experimental Biology and Shanghai No. 2 Hospital took the initiative in China to jointly study human AFP, and they developed the AFP antiserum in August that year, which could help diagnose 65% of liver cancer cases. This result was used in the anti-cancer survey in the first quarter of 1971, during which two early-stage liver cancer patients were discovered who had no clinical symptoms yet, attesting to the clinical application of AFP. This research result was a huge contribution to promoting the early survey and diagnosis of liver cancer and to improving the treatment of it in China.

Artemisinin is a new type of anti-malaria medicine found in the treasure house of traditional Chinese medicine that can be used to save pernicious malaria patients. In February 1974, CAS Shanghai Institute of Organic Chemistry used new physical technology and chemical reaction to determine the structure and absolute configuration of artemisinin, and used photochemical reaction to realize its total synthesis as well as the total synthesis of a series of compounds, including arteannuin and derivatives. This achievement won the second prize of national invention award in 1982 and the Lasker Awards in clinical research in 2011 more than 30 years later. Artemether was the first internationally recognized synthetic medicine by China. It was listed by the WHO as the top choice for treating pernicious malaria and was included in the *International Pharmacopoeia* in 1995.

Despite the chaos during the "Cultural Revolution," Chinese leaders began to pay attention to the population. In March 1964, Nie Rongzhen made the *Report on Scientific Research of Family Planning* to Premier Zhou Enlai and said that the S&T work of family planning was a key task of the *10-year S&T Program.* "It will be very effective for family planning."[112] On July 8, 1971, the State Council forwarded the *Report on Well Implementing Family Planning* submitted by the Ministry of Health, Ministry of Commerce and Ministry of Fuel & Chemical Industry, demanding governments of all levels to strengthen the leadership of this work except in scarcely populated areas. The Report said China would reduce the natural population growth rate year by year during the 4th Five-year period and strive to lower it to 10‰ in regular cities and below 15‰ in the countryside in 1975. But in reality, family planning wasn't implemented very well. Today China has an extremely large population, which creates immense environmental and resource pressure in the next few decades or even century to come.

The fast population growth and excessive pressure gave rise to the mounting demand for planned use of resources and for mitigating the population pressure through S&T. Therefore, China's S&T plan was not only an effective way of consolidating the regime, but might become an effective guarantee for meeting the basic living needs under the population pressure.

II. Progress in industrial technology during the "Cultural Revolution"

On December 29, 1968, the Nanjing Yangtze River Bridge was completed and put into use ahead of schedule. It was the largest bridge independently designed and built by China at that time. A lot of technical personnel and workers continued with industrial construction steadfastly in spite of the interferences of the "Cultural Revolution".

On December 25, 1970, Zhou Enlai proposed and Mao Zedong agreed to build the first gigantic dam on the Yangtze River—the Gezhouba hydro-junction project. In November 1972, Zhou Enlai demanded a suspension of the project after debriefing its major technical issues and formed the technical committee of Gezhouba project to redo the survey and design. In October 1974, the main construction of the project re-started with the approval of the State Council. The first phase was completed and put into operation in 1983 and the second phase was completed in 1988.

[112] Report on Scientific Research of Family Planning, *Selected Articles of Nie Rongzhen on Science and Technology* (Beijing: National Defense Industry Press, 1999), 513–520.

After the September 13th Incident, Zhou Enlai seized the moment and adopted various measures to adjust the national economy. From December 16, 1971 to February 12, 1972, the national planning conference was held in Beijing. Before the conference, Zhou Enlai debriefed the National Planning Commission on the conference on December 5, saying, "The management of our enterprises is a mess now. It has to be straightened out." Later the State Council drafted the *Minutes of 1972 National Planning Conference* and put forth several corrective measures. In 1973, the main indicators of the national economic plan all met or exceeded the target and state revenues and expenditures were in balance. This year marked the best economic situation since the "Cultural Revolution" broke out and laid a solid foundation for future rectification.

While adjusting the national economy, Zhou Enlai opened the door of foreign cooperation. On February 5, 1972, he approved the *Report on Importing Complete Sets of Chemical Fiber and Fertilizer Technologies and Equipment* by the National Planning Commission, which was then submitted to and approved by Mao Zedong. Later the National Planning Commission proposed an import plan worth USD4.3 billion, which would expand China's foreign economic exchange and break its isolated state since the "Cultural Revolution". The plan was put into action in early 1973 after being reviewed and approved by Mao Zedong and Zhou Enlai. On August 21, 1972, CPC Central Committee and the State Council approved the installation of the 1700 mm rolling mill imported from the Federal Republic of Germany and Japan at Wuhan Iron and Steel Company. It was one of the largest import projects since the new China was founded and was of advanced level in the 1970s. It was also new China's second round of large-scale technical import, which enhanced the technical content of Chinese industry and fortified the momentum of our future economic development.

In 1974, the new process of producing vitamin C through two-stage fermentation developed by the CAS Institute of Microbiology was put into industrial use. This new process won the second prize of the national invention award in 1980 and was an initiative in the world.[113] In September 1974, China announced a new mega oilfield in the Bohai Bay area of Shandong province—the Shengli oilfield; in February 1975, Liujiaxia hydropower station, the largest of its kind at the time, was completed.

In 1975, to develop the iron and steel industry and significantly increase steel output, the State Council asked the National Planning Commission, CAS, Ministry of Metallurgical Industry and State Administration of Geology to find 2–3 large high-grade iron ore deposits in five years. Therefore, the CAS formed the Office of High-grade Iron Ore in March 1975 and officially began the ore-seeking campaign. During the three years from 1976 to 1978, CAS organized hundreds of research personnel from its own institutes as well as from six universities under the Ministry of Education to carry out research on a wide range of subjects. This included theoretical studies of the distribution pattern, conditions for the formation and deposit indicator of a high-grade iron core, basic studies including detection technology, and new technologies and methods such as remote sensing technology, image and data processing and mathematical geology. It also organized six field teams of geological survey. The State Administration of Geology conducted drilling in 16 provinces and autonomous regions and the geological section of the Ministry of Metallurgical Industry also joined the campaign. Due to the tight schedule and arduous tasks, the studies could not meet the prospecting requirement and the campaign in search for high-grade iron ore deposit failed to meet the goals. In October 1978, the Office of High-grade Iron Ore was revoked.

The high-grade iron ore campaign was an attempt during the "Cultural Revolution" to advance the S&T plan in a military approach and through large-scale united efforts. Although it ended up a failure, it did indicate that there was a weak sign of S&T planning in that period.

[113] This method is still used to produce vitamin C in China. It was transferred to Roche of Switzerland for USD5.5 million in 1985, the largest technical export project in China then.

III. S&T achievements during the "Cultural Revolution"

1. Development of satellite and research on the nuclear bomb

The successful development and launch of the satellite was a significant technological achievement during the "Cultural Revolution". The CAS Institute of Satellite Design (Institute 651) was officially founded in January 1966, with meteorologist Zhao Jiuzhang being president. After the Central Special Committee[114] approved CAS' plan for satellite work, the Institute 651 made arrangements for the research and development of Dongfanghong-1 in early 1966, under the leadership of Zhao Jiuzhang and Qian Ji.

After the demonstration meeting of the satellite work plan, simulated computation ensued. In February 1966, Chen Fangyun took the lead in compiling the report titled "plan for satellite ground observation system and suggestion on the division of work." CAS was responsible for building the ground observation system. On May 3, 1966, CAS decided to form a bureau to be in charge of the ground observation system and to carry out its general design, installation, adjustment, connection and operation. The new bureau was headed by Yang Jiade and Chen Fangyun.

From May 10 to 25, 1966, based on the work done by Institute 651, CAS systematically demonstrated the plan for China's satellite series. The meeting proposed to start with scientific experiment satellite to lay the foundation and then develop applied satellites for communication, meteorology, geodetic purpose, etc. It also suggested scientific research projects that must be carried out in order to develop various applied satellites. This meeting had a guiding and stimulating effect on the development of satellite series in China. It not only demonstrated but also arranged research projects for scientific experiment satellite, recoverable remote sensing satellite and geostationary communication satellite. As satellite research and development concerned national security, it was protected by the military and could therefore proceed in a stable way. Research work in this area during the "Cultural Revolution" had strong attributes of S&T planning. It was a good individual S&T plan for a specific field.

On April 24, 1970, the artificial earth satellite was successfully launched. On the afternoon the next day, Xinhua News Agency was authorized to announce to the world that China had successfully launched its first artificial earth satellite. On March 3, 1971, the second artificial earth satellite, Shijian-1, was launched into orbit. The satellite series titled Shijian were tasked to carry out space exploration and experiment on new space technologies. The Shijian-1 was the first satellite in this series. The recoverable satellite was also successfully developed later. In early 1966, Institute 651 began to consider the plan for recoverable satellite and conducted research on individual technologies. On November 26, 1975, China launched a recoverable satellite that returned as planned.

In the meantime, the development of hydrogen bomb succeeded too. At the end of 1960, Qian Sanqiang of China Institute of Atomic Energy led the theoretical studies of hydrogen bomb. In January 1965, a group of theoretical researchers, including Huang Zuqia and Yu Min, began explorations for the principles, structure and materials of hydrogen bomb. While they conducted theoretical research on light-nuclear reaction, the China Institute of Atomic Energy and the Institute of Modern Physics in Lanzhou also conducted light-nuclear reaction experiments. The Computer 119 independently designed and manufactured by CAS Institute of Computing Technology and Computer J501 by East China Institute of Computing Technology ensured all the computations necessary for designing China's first hydrogen bomb. On June 17, 1967, the first hydrogen bomb test in China was a success, making it the fourth country in the

[114]In November 1962, a special committee was formed under the leadership of CPC Central Committee. With Zhou Enlai being director, the special committee (known as Central Special Committee) included 7 vice premiers and 7 ministers: He Long, Li Fuchun, Li Xiannian, Bo Yibo, Lu Dingyi, Nie Rongzhen, Luo Ruiqing, Zhao Erlu, Zhang Aiping, Wang Heshou, Liu Jie, Sun Zhiyuan, Duan Junyi and Gao Yang. The Central Special Committee headed up the development of atomic energy industry and work related with nuclear technology in China.

world that had grasped the hydrogen bomb manufacturing technology. It took the US seven years and four months to move from the 1st atomic bomb test to the 1st hydrogen bomb test, more than four years for the Soviet Union and United Kingdom, and only two years and eight months for China. The successful hydrogen bomb explosion earned China respect in the world and it was acknowledged as one of the countries with advanced nuclear technology.

Underground nuclear explosion was also a success. The first underground nuclear explosion in China was originally scheduled after 1966, but was postponed due to the "Cultural Revolution". On September 23, 1969, China had the first underground nuclear explosion. In September 1971, the first nuclear submarine it built on its own had a successful trial trip, indicating that China had grasped the nuclear power technology. It was a new breakthrough in China's nuclear technology and nuclear industry after the successful test of atomic bomb and hydrogen bomb, making contributions to empowering the Chinese navy and consolidating our national defense.

The successful development of military satellite and nuclear bombs during the "Cultural Revolution" was almost the most important S&T achievement in China before the reform and opening-up. S&T plan in this field could succeed because these research works concerned the nation's core interests and were implemented with the assurance of the national regime. In spite of the chaos and power struggles during the "Cultural Revolution", the CPC was of one mind on this issue. The success of individual S&T plan in such a special context once again attested to the concept of "national core interests come first" and "enhanced political functions" in China's S&T plan.

2. Progress in computer research

In March 1964, the Second Ministry of Machinery Industry said it needed to use a large computer that could perform 200,000 calculations per second. In early August 1965, the computer 109-C was primarily developed with an average calculating speed of 115,000 times per second. It was a fairly mature large computer independently designed by China in the mid-1960s and made remarkable contributions to the development of the "two bombs and one satellite," being honored as a "meritorious computer" by leaders of the National Defense Science Committee. In 1965, East China Institute of Computing Technology began to develop the Type 655 IC computer, which was successfully trial-manufactured in 1973 with the calculating speed of 1 million times per second. It was the IC computer with the fastest calculating speed and largest memory capacity in China at that time. The Institute also developed the Type 013 IC computer in 1976 that could perform 2 million calculations per second, which, combined with Type 655, replaced 109-C to perform the mega-calculations for the development of atomic bomb, missile, and artificial satellite.

In 1968, Premier Zhou Enlai approved the Program 905, a large program tasked to specifically develop computers for the military. In 1974, he approved the Program 748, meaning that the nation's top leaders paid close attention to computerized Chinese information processing technology, which entered a new stage—the era of Chinese information processing. In the winter of 1975, Yu Qiuli and Gu Mu, leaders of National Planning Commission, visited East Europe and made a report to Premier Zhon Enlai on building the computer network. This suggestion was studied and approved by Zhou, marking the beginning of the construction of the first national computer network in China.

We couldn't have achieved such rapid development and outstanding results in computer after the "Cultural Revolution" without the solid foundation laid during that period. Therefore, it was inappropriate to say that S&T was at a total standstill during the "Cultural Revolution", but it probably would have developed faster without the disruption.

3. Basic research

In May 1966, Chen Jingrun announced on the 17th issue of *Chinese Science Bulletin* that he had proved (1, 2), which meant that any sufficiently large even number could be expressed as the sum of two numbers, one being a prime number, the other being either a prime number or the product of two prime numbers. In 1973, he published a paper that proved (1, 2) in great detail and improved the numerical results he announced in 1966. Chen Jingrun obtained the best result (closest to "1 + 1") about Goldbach conjecture so far, which made a great sensation in the world mathematical community. His result was called "Chen's theorem" and won the honor for China. The results achieved by Chen Jingrun, Wang Yuan and Pan Chengtong in the research on Goldbach conjecture won the first prize of the national award for natural sciences in 1982.

In 1966, Zhu Hongyuan, He Zuoxiu, Dai Yuanben and Hu Ning put forth a new thought about the study of particle physics in China—straton model, which was an important achievement in the theoretical study of hadron structure. The Beijing taskforce of elementary particle theory comprised of people from CAS Institute of Atomic Energy, Institute of Mathematics, Peking University and University of Science and Technology proposed the straton model to study the dynamic properties of particles. The main idea of straton model is as follows: every substance has infinite layers in structure; its particle layer is composed of stratons, but stratons are not the smallest particle; they only constitute one of the infinite layers of the structure of a substance; before the kinetic theory between layers is finally established, people can study the structure of substance by expressing the wave function of stratons' motion in hadrons; it is necessary to develop a method for calculating matrix elements in the bound state and dealing with the wave function of the internal motion in the bound state; the motion of stratons inside the hadron can be regarded as a non-relativistic approximation, the whole motion of hadrons must have relativistic covariance.[115] Straton model was a good theory of hadron structure model before the basic theory of straton dynamics was established. It won the second prize of national award for natural sciences in 1982 for its contribution to particle physics.

American scientists determined in 1965 that the nucleic acid of primary structure was yeast alanine transfer RNA, which consisted of 76 nucleotides. The US was the only country that figured out this structure then. During a discussion on August 24, 1964, Mao Zedong said "we should study the origin of life," so joint research began on the artificial synthesis of yeast alanine transfer RNA, and the joint research team was named "Team 824" at first. The joint research progressed very slowly in the first eight years and only 12 nucleotide fragments were synthesized. After the "Gang of Four" was smashed, a taskforce of the artificial synthesis of nucleotide was formed in 1977, paving the way for realizing the artificial total synthesis of yeast alanine transfer RNA at the end of 1981. Moreover, in August 1973, the Beijing research group of the crystal structure of insulin comprising personnel from the CAS Institute of Physics, CAS Institute of Biophysics and Peking University completed the determination of 1.8A crystal structure of insulin, which won the second prize of national award for natural sciences in 1982.

In December 1972, the resumed *Acta Archaeologica Sinica* published in its 1st issue Zhu Kezen's paper *Primary Study of Climate Changes in China over the Past 5,000 Years* in full length, which, after important revisions, was published again on the 1st of the resumed *Scientia Sinica* in February next year, both on the foreign and Chinese editions. Zhu Kezhen collected a massive amount of historical literature, compared the climate between ancient and modern times, referred to the abundant phenological materials of each period, and used temperature as an indicator. In this way, he outlined the climate changes in China over nearly 5,000 years from 3,000BC to early 1970s, and pointed out that climate fluctuations were global. Zhu Kezhen's quantitative analysis of climate changes, which spanned a long period, covered a wide coverage of areas and was based on extremely complicated historical literature, yielded successful results. He was acknowledged by the international meteorological community as making remarkable contributions to the study of historical climate changes worldwide.

[115]History of Chinese Academy of Sciences, 1966, published on CAS website: http://www.cas.ac.cn.

Guo Musun, a research fellow at CAS Institute of Chemical Metallurgy, put forth the theoretical system of particulate fluidization of chemical metallurgy in his article *Particulate Fluidization in Chemical Metallurgy*. The article was published on the 3rd issue of *Scientia Sinica* in 1973 and announced at the 1st Chemical Engineering Conference in Iran in May that year. This research result won the second prize of the national award for natural sciences in 1982.

Even during the "Cultural Revolution", Premier Zhou Enlai inquired about work related with basic sciences and theoretical research several times. After the Lin Biao-led counter-revolutionary group was exposed and defeated, Zhou Enlai repeatedly emphasized the teaching and scientific research of basic sciences since 1972. When he met with Chen-Ning Yang on July 2, 1972, he asked Zhou Peiyuan, who was present at the meeting, to advocate the improvement of basic theories. On September 5 that year Zhou Enlai told Salam, scientific advisor to Pakistani president who was visiting China, that he was dissatisfied with the neglect of theoretical studies. On September 11, Zhou wrote in the letter to Zhang Wenyu that was also copied to Zhu Guangya that CAS should strengthen the work in basic sciences and theoretical research and combine theoretical research with scientific experiments. He also stressed the research on high-energy physics and on the prefabrication of high-energy accelerator must be one of CAS' key projects and it should not be put off any longer. To act on Zhou Enlai's instructions, the CPC core group of CAS drafted the *Report on Strengthening the Work in Basic Sciences and Theoretical Research*, which, however, was criticized shortly after as a typical example of the "theoretical wind".

In 1972, Chen-Ning Yang advocated the study and research of basic theories. When meeting with Chen-Ning Yang in 1973, Mao Zedong re-stressed the importance of scientific experiment (Fig. 3.2) and discussed philosophical issues in scientific development.

From mid-May to early July 1973, CAS assigned a delegation of high-energy physicists led by Zhang Wenyu to visit the United States and the Conseil Européen pour la Recherche Nucléaire (CERN) in Geneva. In September, CAS submitted to the State Council a report on high-energy physics research and the prefabrication of high-energy accelerator. The National Planning Commission agreed to establish a project for the research and fabrication of high-energy accelerator and called it "Program 753", but this work was stopped due to the movement of "condemning Lin Biao and Confucianism" and of "condemning Deng Xiaoping and resisting the right-leaning trend of reversing previous convictions."

Fig. 3.2 Mao Zedong meet with Chen-Ning Yang in 1973[116]

Zhou Enlai was a CPC leader that had a profound understanding of basic sciences. He studied in France as a young man and had personal experience of the global science development. But his understanding of natural sciences didn't become a consensus in the whole Party, and this situation led to the fickle policies on intellectuals.

IV. **Changes in the S&T system**

1. Reform of the S&T research system

The reform of scientific research system during the "Cultural Revolution" wasn't an S&T system reform when all the conditions were ready, but a kind of special protection of S&T research institutes and scientists given by proletarian revolutionaries of the older generation, such as

[116]Xinhua News Agency: http://www.xinhuanet.com.

Nie Rongzhen and Zhou Enlai, in that special period.

In 1967, Nie Rongzhen made the *Report on the Plan for Adjusting and Reorganizing the Scientific Research System for National Defense*, which suggested further organizing the scientific research forces for national defense and setting up 18 institutes. Mao Zedong approved the Report on October 25 that year. The purpose of the institutional adjustment was, in addition to further concentrating the forces, to protect the research and technical personnel for national defense in the chaotic period of the "Cultural Revolution". However, even the department of national defense research was embroiled in the movement soon, and the institutional adjustment stopped as Nie Rongzhen came under criticism.[117]

According to the multiple instructions given by Premier Zhou Enlai, the Military Regulation Committee of State S&T Commission and CAS Revolutionary Committee worked together in the same office building for a year, laying the foundation for their integration. After the State S&T Commission and CAS were officially integrated in July 1970, a new CAS Revolutionary Committee was formed, with Guo Muoruo being director and Li Siguang and Liu Xiyao being deputy directors. In July 1972, CPC Central Committee appointed Guo Muoruo as head of the CPC core group in CAS, and Liu Xiyao, Wu Heng and Wang Jianzhong were deputy heads of the group. This arrangement protected Guo Muoruo and lots of other scientists.

On August 11, 1973, the Science & Education Group of the State Council and CAS issued a joint notice to local areas and various departments to the effect that CAS would be responsible for organizing the formulation of long-term S&T plan and annual plans in the civilian sector. On August 24, CAS formed the "S&T Office".

After State S&T Commission and CAS were integrated, CAS set up the No. 1 Business Group to be in charge of the work of the former State S&T Commission. After the national work conference on education in July 1971, the State Council decided to set up a Science & Education Group to be in charge of the work of former State S&T Commission and the Ministry of Education, with Li Siguang being group leader and Liu Xiyao and Chi Qun being deputy group leaders. After Li Siguang passed away in 1971, Liu Xiyao succeeded him as the group leader.

In July and August 1972, the Science & Education Group and CAS discussed and decided that the Science & Education Group would be responsible for national S&T management, and personnel engaged in overall planning and scientific experiment in CAS' No. 1 Business Group would be transferred to the Science & Education Group as S&T Group. In August 1973, the Science & Education Group of the State Council continued to be responsible for national S&T management in the civilian sector in the name of the S&T Office, and this situation lasted till the reestablishment of State S&T Commission.

Strictly speaking, such institutional changes during the "Cultural Revolution" couldn't be called the reform of S&T system, but just an expedient in special conditions. But these changes implied the possibility of S&T system reform and served as a forerunner for the formal reform of the S&T system after the reform and opening-up were launched.

2. International S&T exchanges during the "Cultural Revolution"

Foreign S&T exchanges and cooperation almost stopped completely during the "Cultural Revolution". After the American table tennis team visited China in 1971, some American scientists also came in succession. American people and groups in the S&T circle repeatedly expressed their wishes, through correspondences or in other ways, to forge connections with China, and invited Chinese scientists to visit the US. In that period, CAS contended with Taiwan's Academia Sinica on the world scientific stage. Furthermore,

[117] Report on the Plan for Adjusting and Reorganizing the Scientific Research System for National Defense, *Selected Articles of Nie Rongzhen on Science and Technology* (Beijing: National Defense Industry Press, 1999), 616–621. Nie Rongzhen came under criticism in 1967, and his articles in this book jumped from 1967 directly to 1979, indicating the "marginalization" of him during the Cultural Revolution.

Sweden, the United Kingdom and Canada, which had formed diplomatic ties with China, also invited Chinese scientists several times. Thanks to the "Ping Pong diplomacy" in 1972 and with the approval of the central government, a Chinese scientist delegation with Bei Shizhang as the leader and Bai Jiefu as deputy leader visited the United Kingdom, Sweden, Canada and the United States from October 6 to December 16. Members of the delegation included Zhang Wenyu, Qian Weichang, Qian Renyuan, Hu Shiquan and Li Fusheng. Premier Zhou Enlai gave detailed instructions before they set off.

The delegation went to 23 cities in the four countries, and visited 68 universities, research institutes and research organizations affiliated to plants and enterprises, as well as government bodies, academic groups and museums. They were warmly received by foreign scientists. Through the visits and discussions, the Chinese delegation learnt about the status quo and latest trend on a number of subjects, including high-energy physics, nuclear physics, plasma and controlled thermonuclear reaction, computer, molecular biology and environmental science, and intensified academic exchanges. When returning to China, the delegation reported to the CPC core group of CAS, and made suggestions on China's scientific research work and international scientific cooperation and exchange. It was the first scientist delegation sent abroad by CAS since the "Cultural Revolution" began.

It was worth noting that the scientific circle played a part in the establishment of Sino-US diplomatic ties. After the two countries officially forged the diplomatic relation, bilateral relations went through ups and downs, but their scientific circles always maintained contact even when bilateral relations were in the ebb. They became an important bridge of Sino-US relation and the political function of science was brought into full play in China.

3. National S&T work conference

On August 24, 1971, the Science & Education Group of the State Council and CAS asked for the State Council's permission to formulate S&T plan and hold a national S&T work conference. From the end of 1972 to January 10, 1973, the national S&T work conference was held in Beijing. The condemnation of the Lin Biao-Chen Boda anti-Party group in 1972 created the conditions for people to realistically comment on the work before the "Cultural Revolution". Most attendees at the meeting urgently hoped to see the messed-up thoughts sorted out, rules and systems reestablished, the wrongly abolished research institutes resumed, the normal order of research work restored, and S&T personnel returning to labs. These requests were stated in the draft meeting minutes. The meeting lasted more than 20 days, some representatives stayed to wait for the CPC Central Committees' comments on the meeting minutes and for Premier Zhou Enlai to speak at the meeting.

However, as the "Gang of Four" was extremely opposed to criticizing the ultra-left force, and also because Mao Zedong mistakenly believed that the top priority then was opposing the Lin Biao-led "ultra-right" force, Vice Premier Ji Dengkui of the State Council denied the deliverables of the meeting when meeting with the representatives after the New Year of 1973. The revised *Minutes of National Science and Technology Meeting* was approved by the State Council and distributed in January 1973. The Minutes failed to faithfully reflect the actual situation of the meeting and the opinions of most representatives. Soon later, the campaign of "condemning Lin Biao and Confucianism" was waged across the country. In December 1973, the extended meeting of the CPC core group of CAS labeled and denounced the national S&T work conference in 1972–1973 as a typical example of "resurgence". This attempt on the S&T front at correcting the leftish inclination met a serious setback.[118]

4. Rectification of the CAS

After Deng Xiaoping came back to power, Hu Yaobang, Li Chang and Wang Guangwei were assigned to head up the CAS on July 18, 1974

[118]History of Chinese Academy of Sciences, 1972, published on CAS website: http://www.cas.ac.cn.

and began to rectify the academy. Soon after he was assigned there, Hu Yaobang submitted the *Several Issues Concerning Science and Technology Work* (hereinafter referred to as "Report Outline") to CPC Central Committee and the State Council. The Report Outline was revised several times before finally submitted to Deng Xiaoping on August 17.

Main contents of the Report Outline included the following: it affirmed the S&T performances since the new China was founded; it re-assigned the work of former State S&T Commission; it put forth several key concepts, including "science and technology are productive forces" and "carry out scientific research with open doors"; it clarified the policy on intellectuals on the S&T front, and asserted that the majority of the 4 million S&T personnel were good or quite good; it made five points regarding the primary conception of a 10-year S&T program; and it mentioned the rectification of the Head Office of CAS and units directly under it.[119] This Report Outline systematically summarized past experience and aimed to systematically rectify mistakes on the S&T front. It paved the way for formulating and implementing new S&T plans after the "Cultural Revolution". This showed that Deng Xiaoping attached great importance to S&T planning. After the "Cultural Revolution", he resumed duties as a top leader for the third time and pushed for the formulation of the 1978 S&T plan.

On September 26, 1975, the State Council meeting chaired by Deng Xiaoping discussed and primarily passed the "Report Outline". Hu Qiaomu made many revisions and renamed it from *Several Issues Concerning Science and Technology Work* to *Outline of the Work Report by Chinese Academy of Sciences*. Its contents mainly consisted of three parts: direction and tasks of CAS' scientific research work; resolutely and comprehensively implementing Chairman Mao's revolutionary S&T line and formulating the ten principles of S&T line based on Mao Zedong's relevant remarks; and rectification of CAS.

To restore order in scientific research work as soon as possible, the then Vice Premier Hua Guofeng, according to Deng Xiaoping's instructions, presided over a symposium of CAS scientists on August 25 in the theme of "all schools of thoughts prosper." Afterward, CAS held four symposiums by itself entrusted by Hua Guofeng, and each symposium was reported to the higher level in the form of *CAS Bulletin*.

Regarding the leadership rectification, Deng Xiaoping upheld the three important instructions made by Chairman Mao as the guideline: (1) Study theories and oppose and prevent revisionism; (2) Pursue stability and unity; and (3) Strongly develop national economy. The whole set of guiding principles and policies regarding the rectification formed by Deng Xiaoping in that period was collectively reflected in the *On the General Guideline for All Works of the Whole Party and the Whole Nation* (*On the General Guideline*), *Outline of the Work Report by Chinese Academy of Sciences* and *Several Issues Concerning the Acceleration of Industrial Development*. These three documents systematically condemned the "left-leaning" reactionary views disseminated by the Gang of Four. In September 1975, the movement of "condemning Deng Xiaoping and resisting the right-leaning trend of reversing previous convictions" emerged, during which the Gang of Four mobilized strong forces to criticize the three documents and called them "three poisonous weeds" aimed to counterattack the "Cultural Revolution" in revenge and resurrect capitalism across the board.[120]

The rectification of CAS by Deng Xiaoping, Hu Yaobang, and other leaders failed. But the rectification of the S&T and educational fronts, including the CAS, made Deng Xiaoping realize that the quickest way of stepping out of the post-"Cultural Revolution" chaos was straightening up those two fronts. The rectification also paved the way for quickly formulating the S&T plan after the "Cultural Revolution".

[119]History of Chinese Academy of Sciences, 1975, published on CAS website: http://www.cas.ac.cn.

[120]Yuzhi (2004).

Table 3.1 Main industrial indicators before and after the 10-year "Cultural Revolution"[121]

National economy in 1966	National economy in 1976	Average annual growth
Industrial & agricultural output value: RMB253.4 billion	Industrial & agricultural output value: RMB453.6 billion	RMB20.02 billion
Grain: 214 million tons	Grain: 286.31 million tons	7.231 million tons
Cotton: 2.337 million tons	Cotton: 2.055 million tons	−28,200 tons
Steel: 15.32 million tons	Steel: 20.46 million tons	514,000 tons
Raw coal: 252 million tons	Raw coal: 483 million tons	23.1 million tons
Crude oil: 14.55 million tons	Crude oil: 87.16 million tons	7.261 million tons
Power generation: 82.5 TWh	Power generation: 203.1 TWh	12.06 TWh
Railway freight: 549 million tons	Railway freight: 840.66 million tons	29.166 million tons
Total investment in capital construction: RMB20,942 million	Total investment in capital construction: RMB37,644 million	RMB1670.2 million
Total volume of retail sales: RMB73.28 billion	Total volume of retail sales: RMB133.94 billion	RMB6,066 million
Total import & export: RMB12.71 billion	Total import & export: RMB26.41 billion	RMB1.37 billion
Total national fiscal revenues: RMB55.87 billion	Total national fiscal revenues: RMB77.66 billion	RMB2,179 million
Total expenditures: RMB54.16 billion	Total expenditures: RMB80.62 billion	RMB2,646 million

V. Impacts of "Cultural Revolution" on S&T Planning

The ten-year "Cultural Revolution" exerted tremendous negative effects on national construction and S&T planning, but there were still signs of economic construction and S&T planning during that period.

Some progress was made in economic construction during those ten years (table one). Seen from Table 3.1, the national economy developed slowly during the ten-year "Cultural Revolution", but most productive sectors saw improvements and some even developed quite rapidly, such as capital construction, power generation and steel production. This was to some extent attributed to S&T and in a way proved China's S&T progress during the "Cultural Revolution", displaying the situation of "slow development in general, improvement in specific areas." The saying that there was no S&T planning at all during that period is untenable as individual S&T plan did exist. On one hand, it was the continuation of the S&T plan in 1956 and especially the one in 1963; on the other hand, it provided the historical basis and practical experience for making new S&T plans after the "Cultural Revolution".

As there was only weak and individual S&T planning during the "Cultural Revolution", the national will and domination wasn't distinctly reflected, and protecting and increasing the people's interests naturally became the S&T planning concept in that period. In sum, in the special context of the "Cultural Revolution", thanks to the painstaking efforts of the older-generation revolutionaries and the S&T personnel, S&T in some areas, especially military and national defense, made substantial progress in spite of the generally slow progress. This was ascribed to the foundation laid by the previous two S&T plans, and it also made the requirement of "catching up" and "making up for losses" on new S&T plans after the "Cultural Revolution".

[121]Data in the table are based on the 1966–1976 bulletin published at the CPC news website: http://cpc.people.com.cn/GB/64162/64164/4416081.html.

In that special period, the special S&T plans concerning the "two bombs and one satellite" were implemented with good effect because, as stated earlier, they concerned national core interests. S&T planning philosophy was a system with different levels and fields, and the concept of plan formulation varied from that of implementation, but there were some top-level concepts at the core. While S&T plan was seriously damaged by external factors, one of the top-level concepts—"national core interests come first"—was maintained, without which there would be no S&T plan at all.

3.4 Review and Enlightenment

This chapter discusses the historical evolution of S&T plans before the reform and opening-up and analyzes the planning concepts, including the 1956–1967 S&T program, the 1963–1972 S&T program and the difficult S&T development during the "Cultural Revolution".

The 1956–1967 S&T program was the first of its kind in New China and initiated many traditions. The *12-year S&T Program* was a huge success and completed five years ahead of time. It laid the foundation for China's S&T development and created the big scientific development system with Chinese characteristics. It is still considered one of the most successful S&T plans by the academic circle and Chinese government to date, and many of its achievements are still taking effect today, especially military achievements represented by the "two bombs and one satellite". The manned spaceflight today and the moon landing program in the future are both developed on that basis.

The traditions formed by the first S&T plan had profound and far-reaching impacts. The mode and way of execution of the first S&T plan was supported by its planning philosophy. The word philosophy was the generalization of the essential ideas and thoughts concerning and regulating the formulation of S&T plan. Its formation was limited by a series of conditions of the time, including political, economic, military and S&T strength, and once formed, it came back to affect the environment than through the S&T plan. In history, scholars of different subjects had different understandings and definitions of concept, that is philosophy. Plato understood concept as an independent general notion that was "true reality".[122] The modern understanding of concept obviously differs from that in ancient Greece and tends to be a succinct generalization of the subject. "Concept is a philosophical notion on the spiritual and ideological level concerning the epistatic and overall structure. It is a scientific reflection of the objective (laws, presence) by the subjective (perception, idea), and an abstract generalization of the thoughts, ideas, aspirations, ideals and philosophical viewpoints formed by people through rational thinking and practices over a long period. It is a navigation mark that guides the people to carry out theoretical explorations and practices, and a theoretical and systematic system of perceptions and ideals with relative stability, continuity and directivity. In sum, concept refers to people's rational perception and ideal of a matter or phenomenon and the system of ideas formed on that basis."[123] It is an epistatic and highly general notion.

Concept is a system and the concept of plan formulation can be understood on several levels. The most central is top-level concept, which is similar to the guideline of S&T planning; then comes the selection and proportion of specific projects; and the last but not least is S&T subprograms worked out based on the plan. This book studies S&T plan as a whole and focuses on the most central philosophy, namely the top-level concept.

The *12-year S&T Program* upheld three top-level concepts. The first one was "technology comes first in S&T planning," which meant that during the formulation and implementation of medium to long-term S&T plan, technical contents and projects were naturally prioritized. Some scholars criticized our S&T plans as technology plans. The CPC's understanding of science and technology went through a process. Our S&T practices in the warring period before New China was founded made us believed that technology had the most direct impact on

122 Zenggu (1983, 134).

123 Yanming (2003, 58).

winning a war. This was elaborated in the first chapter. The large group of S&T cadres cultivated during the revolution were the backbone of S&T plan formulation after New China was founded, and they endowed China's S&T plans with the preference and inclination toward technology. While the plan was implemented, technological tasks were the easiest to accomplish and took effect the soonest, and the process could be managed in a military mode, a management method we were familiar with. With the continuous development of technology, especially the immense success of high technologies such as the "two bombs and one satellite", more priority was given it. As a result, the concept of "technology comes first in S&T planning" became a tradition that was carried on by future S&T plans.

The second top-level concept was "national core interests come first," which meant S&T plan was tasked and aimed to serve the nation's core interests. National interests involved too many aspects, some of which were even in conflict. For instance, fast development and environmental protection were both national interests, but they conflicted with each other. But S&T planning was a national action rather than the spontaneous action of the scientific community, and it represented national core interests. The "two bombs and one satellite" could succeed because they complied with the nation's core interests. In a sense, S&T plan was used as a means to realizing the interests of the ruling class and everything had the same goal—national core interests. When New China was just founded, its core interests was laying a solid foundation and consolidating the regime, so the first S&T plan was worked out on this concept.

The third top-level concept was "enhanced political functions", which meant S&T plan was bestowed with political functions other than the task of S&T development, and its formulation and implementation implied and had to reach certain political purposes. Although it was an S&T plan aimed at S&T development, it wasn't a research outline for the scientific community or research conceptions by famous scientists, but an instrument of state governance used to fulfill the national intents. When New China was first founded, everything centered on the political goal that "the people are masters of the country," S&T not being an exception. As the scientific community in China was rather weak and not independent enough, its "enhanced political functions" were more conspicuous. As a result, sometimes requirements and projects that didn't consist with the laws of S&T development appeared in the S&T plan, such as the frequent statement that China should reach a specific level or catch up with and surpass a specific country within a specific period of time. This was more a political demand rather than the demand for S&T development. Another pressing task of the newly founded People's Republic of China was to prove to the world its legitimacy and rationality, which was done in multiple ways, such as armed conflicts, political negotiations and economic construction. Since S&T had to serve the political center, it was used to prove the regime's rationality. As long as S&T developed fast and was able to enhance the national strength quickly and significantly, it could prove the legitimacy and rationality of the new regime. That was why our S&T plan could receive such close attention from the top leaders when New China was first founded. At that time, the political function of S&T and S&T plan was highlighted.

These three top-level concepts gave our S&T plans strong Chinese characteristics and ensured that our first S&T plan was a huge success, which in turn reinforced these concepts and they were inherited in the future. The three concepts complemented each other. "Enhanced political functions" made sure that "national core interests come first," and to achieve that goal, "technology comes first in S&T planning."

What with the changed environment and with the halt of Soviet assistance, the second S&T plan added a new concept—"self-reliance based on domestic conditions"—to the original three. It meant that we had to make the S&T plan based on our own abilities and resources and rely on ourselves to develop S&T. Therefore, the S&T plan for 1963–1972 made arrangements for how to improve the research ability and technical level in China, and the country might have been better at indigenous innovation but for the Cultural Revolution.

No clear S&T plan was made during the "Cultural Revolution" like the previous two ones, but there were weak S&T planning and individual S&T plans, which led to the achievements in satellite and nuclear bombs. The concept of "national core interests come first" was fully demonstrated during the "Cultural Revolution". For instance, specific plans that concerned national core interests, including computer, missile, satellite and nuclear bomb, were carried out with good effect.

China's S&T development was promoted and elevated by S&T plans, and S&T plans were formulated and implemented based on planning concepts. This is the inherent logic and meaning of studying the historical evolution of S&T programs and the planning philosophy.

References

Baichun, Zhang et al. 2004. Soviet Technology Transfer to China 1949–1966. Jinan: Shandong Education Press.

Cang, Liu. 2006. The Flexural Development History of Basic Research on Natural Sciences in the New China, General Review of the Communist Party of China, (11).

Chengsi, Wen. 1993. Thinking of Science in Contemporary China 1949–1991. Shanghai: Shanghai Branch of Joint Publishing.

Chengsi, Wei. 1993. Trend of Thoughts on Science in Contemporary China 1949–1991. Shanghai: Shanghai Branch of SDX Joint Publishing Company.

Chunyi, Liang. 1983. Ъ.Р.Лазаренко, Inventor of EDM, Electromachining & Mould, (5).

Collection of Laws and Regulations of the People's Republic of China (Jan.–June, 1956). Beijing: Law Press, 1956.

Enlai, Zhou. 1957. Government Work Report of the State Council. People's Daily, June 27, 1957.

Heng, Wu. 2006. *Zhou Enlai's Care and Guidance for the S&T Cause in China, Selected Important Literature Since the Foundation of New China*. Beijing: Central Party Literature Press.

Huanfen, Lin. 1995. *Basic Formation of China's Planned Economy and its Historical Features*. Literature of Chinese Communist Party, (2).

Huayuan, Xiong. 1991. *A Meeting of Profound Significance—The Meeting of Matters Concerning Intellectuals Held by CPC Central Committee in 1956, The Road Walked by the People's Republic of China (1953–1956)*. Beijing: Central Party Literature Press.

Jianxin, Chen. 1994. *Contemporary History of Science and Technology Development in China*. Wuhan: Hubei Education Press.

Jun, Wang. 2001. *Mao zedong and Industrialization of China*. Fuzhou: Fujian Education Press.

Lifan, Yang. 2007. *Formulation and Adjustment of the Outline of National Science and Technology Development Program 1978–1985, Research on China's S&T Plans and Policies*. Jinan: Shandong Education Press.

Party group of National Planning Commission. 1996. The Third Five-year Plan (1966–1970), Literature of Chinese Communist Party, (3).

Party Literature Research Center of CPC Central Committee, ed.. 1992. Selected Important Literature Since the Foundation of New China (Vol. 2). Beijing: Central Party Literature Press.

Party Literature Research Center of CPC Central Committee, ed. 1993. Selected Important Literature Since the Foundation of New China (Vol. 5). Beijing: Central Party Literature Press.

Party Literature Research Center of CPC Central Committee, ed. 1994. Selected Important Literature Since the Foundation of New China (Vol. 9). Beijing: Central Party Literature Press.

Party Literature Research Center of CPC Central Committee, et al. 2004. Portrait of Deng Xiaoping. Chengdu: Sichuan People's Publishing House.

Qingquan, Xu. 2007. The hard journey of CPC Determining the Class Nature of Intellectuals after the New China was Founded, Xiangchao, (5).

Qizhi, Zhang, and Yang Xiancai. 2002. *The People's Republic of China*. Taipei: Wunan Publishing Co., Ltd.

Selected Works of Mao Zedong. 1966. (Vol. 1). Beijing: People's Publishing House.

Selected Works of Mao Zedong. 1977. (Vol. 5). Beijing: People's Publishing House.

Selected Works of Mao Zedong. 1999. (Vol. 8). Beijing: People's Publishing House.

Selected Works of Zhou Enlai. 1984. (Vol. 2). Beijing: People's Publishing House.

Speech at the National Work Meeting of Agricultural Science and Technology, Selected Articles of Nie Rongzhen on Science and Technology. Beijing: National Defense Industry Press, 1999.

Weijia, Hu. 2006. *Selected Materials About China's S&T Policies (I)*. Jinan: Shandong Education Press.

Weijia, Hu. 2007. *From Planned S&T Work to the First S&T Plan, Research on China's S&T Plans and Policies—Research Series of Contemporary and Modern S&T History in China*. Jinan: Shandong Education Press.

Wenli, Yang, Zhang Meng. 2007. Formulation and Implementation of the First S&T Development Plan

of New China and Historical Experience from It, CPC History Studies, (6).
Yanming, Han. 2003. *Synopsis of Idealism—A College Course*. Beijing: People's Education Press.
Yanxia, Zhao. 2003. Overview of Academic Studies in China of the Reasons for the Deterioration of Sino-Soviet Ties. *Journal of Zhuzhou Teachers College* (6).
Yibo, Bo. 1991. *AReview of Several Major Decisions and Events*. Beijing: Central Party School Press.
Yuzhi, Gong. 2004. A Historical Complicated Case and a Starting Point of Research. Chinese Cadres Tribune.
Yuzhi, Gong. 2006. The Lu Dingyi I Know, Study Times.
Zedong, Mao. 1977. Selected Works of Mao Zedong (Vol. 5). Beijing: People's Publishing House.
Zedong, Mao. 1999. The Purpose of Socialist Revolution is Liberating the Productive Forces, Selected Works of Mao Zedong (Vol. 7). Beijing: Feople's Publishing House.
Zedong's, Mao. 1990. Articles after the People's Republic of China was Founded (Vol. 4). Beijing: Central Party Literature Press.
Zenggu, Quan. 1983. *A history of Western Philosophy (I)*. Shanghai: Shanghai People's Publishing House.
Zitong, Gong et al. KobДa. 2004. Envoy of Soil Science Exchanges between China and Former Soviet Union - in Celebration of the 100th Anniversary of the Birth of B.A. KobДa. Chinese Journal of Soil Science, (1).
Zuoxiu, He, Zhang Boling. 2006. A review of the 12-year S&T Program, S&T in China, (2).

4 History and Philosophy of Science and Technology Planning in the Transitional Period

Abstract

This chapter describes the evolution of China's science and technology planning and philosophy from the Cultural Revolution to the end of the 20th century. The philosophy of planning in this period includes "technology comes first in S&T planning", "utilitarianism tendency" and "strengthening economic function". The Medium and Long-Term Science and Technology Programs have greatly promoted China's economic development in this period, but it has neglected or even had a negative impact on basic research. With the development of market-oriented reform, the reform of science and technology system was developing continuously.

Keywords

Reform and Opening-up · Tackling key problems in S&T · Utilitarianism tendency · Strengthening economic function · Science and technology system reform · Ignoring basic research

4.1 Idealistic Science and Technology Program Shortly After the "Cultural Revolution"

I Macro background of the *Science and Technology Program in 1978*

1. Better international environment

From the early 1970s, China's diplomacy made substantial progress although domestic economic construction was still off track due to the "Cultural Revolution". In 1971, China resumed its legal position in the United Nations, ending its 20-plus-year-long hostility towards the United States. It also formed official diplomatic ties with Japan and improved its relations with other Western countries in this period, broke the West's long-term economic and technological blockade, and continued to develop friendly cooperation with third-world countries. By the time the third plenary session of the 11th CPC Central Committee was held at the end of 1978, Chinese leaders had visited a long string of countries in Southeast Asia, Europe and Latin America, the *Sino-Japanese Treaty of Peace and Friendship* was signed, and negotiations on the normalization of Sino-US relationship were completed, all indicating a better international political environment for China.

Z. Chen, *China's Medium and Long-Term Science and Technology Program*,
https://doi.org/10.1007/978-3-030-75146-3_4

The international and domestic situations provided the conditions for China to change its development strategy. The third plenary session of the 11th CPC Central Committee decided to shift the Party's focus and the people's attention to the socialist modernization drive, with the aim of quickly improving productivity, obviously bettering people's lives and strengthening national defense. The national science conference held in 1978 greatly liberated the mind and mobilized the Chinese people to strive for S&T modernization, and the theory that "science and technology are a productive force" provided the theoretical basis for future S&T guidelines and policies. However, the *Outline of Science and Technology Development Program 1978–1985* (known as the *Outline of the 8-year Program*) announced at the science conference reflected a strategy that was basically a continuation of the S&T development strategy disrupted by the Cultural Revolution and more like a "restoration". At that time, the whole nation was eager to retrieve the time lost in the Cultural Revolution and had a strong desire to quickly realize the four modernizations. Therefore, a rash and impetuous sentiment was inevitably conveyed in the *Outline of the 8-year Program*, which was too eager to "catch up with and surpass" Western powers.

2. New guideline for economic adjustment

In 1975, the CPC Central Committee formulated the *Outline of Ten-year National Economic Development Program 1976–1985* and made the "5th Five-year" Plan, in the last three years (1978–1980) of which an independent and quite complete industrial system and national economic system were to be established. In March 1978, the outline of the ten-year program was revised, demanding that steel output reach 60 million tons and oil output 250 million tons by 1985; the nation planned to build or continue to build 120 large projects, including 10 large iron and steel bases, 9 non-ferrous metals bases, 8 coal bases and 10 oil & gas fields. Capital construction investment in 1978–1985 would amount to the combined amount of the past 28 years if the plan were followed, averaging RMB70 billion every year, but that was impractical and went against the laws of economic development. Therefore, the CPC Central Committee revised the outline again soon after the third plenary session. The reason for the rash plan was that the Cultural Revolution was just over and the long-suppressed enthusiasm for socialist construction burst forth all of a sudden, and the nation was eager to recover the losses caused by the Cultural Revolution and catch up with developed countries. Likewise, the 1978–1985 S&T program was also a little rash. According to the implementation of the "5th Five-year Plan", the total social output value and national revenues increased dramatically in 1977 and 1978, and the output of main industrial and agricultural products came back to or exceeded the highest level in history, but the too radical demand on economic development and too high targets led to new mistakes in our economic construction. The third plenary session of the 11th CPC Central Committee held in December 1978 shifted the Party's focus to the socialist modernization drive, emphasized the necessity of following the objective economic laws, and primarily rolled out the tasks and measures of economic adjustment and reform. In April 1979, the CPC Central Committee officially put forth the guideline of "adjustment, reform, rectification, and improvement" (the new guideline) at its work meeting and began to adjust the national economy that year.

The 6th Five-year Plan for National Economic and Social Development (1980–1985) was formulated according to the strategic deployments made by the 12th CPC National Congress for economic construction by the end of the 20th century and was a quite rounded five-year plan after the 1st one. The 6th five-year plan made the following requirements: continue to implement the guideline of "adjustment, reform, rectification, and improvement," score a decisive victory in fundamentally improving the finance and economy, and lay a better foundation for economic and social development in the 7th five-year period. It also set requirements for S&T development: mobilize S&T forces nationwide to overcome key difficulties and promote the application of S&T results, strongly develop education, science, and cultural causes, promote socialist material and cultural-ethical progress, and intensify defensive forces by strengthening national defense and related industries.[1] The 6th Five-year Plan made remarkable achievements. Total industrial and agricultural output value increased by 11% and gross national product by 10% every year on average. The output of important products rose by a large margin. From 1980 to 1985, steel output increased by 26.1%, coal 37.1%, power generation 35.8%, crude oil 17.9%, grain 21.4% per annum on average and cotton 92.8%. Capital construction and technical reform made remarkable progress. Fixed asset investment in units owned by the whole people reached RMB530 billion, RMB380 billion fixed assets were added, 496 medium-to-large-scale projects were put into operation, and 200,000 technical reform projects were completed.[2] The constructive effect of S&T on economic construction was more clearly reflected in the 5th and 6th five-year plans, so the 1978–1985 S&T program and the S&T challenge tackling programs adapted from it later all reflected this.

3. Fast-developing new technologies in the world

The Cultural Revolution widened the S&T gap, which had been narrowed before the campaign, between China and world advanced level. Many key S&T problems in economic construction remained unsolved for a long time, and emerging S&T wasn't applied in economy; theoretical research on natural sciences stopped, S&T personnel were mostly old without enough young successors; and scientific research equipment and experimental approaches were very backward.

In comparison, dramatic S&T progress was achieved worldwide after WWII and the human society experienced its 3rd S&T revolution. Progress in quantum mechanics, molecular biology and systems science led to amazing inventions and breakthroughs in a spate of R&D fields, especially atomic energy, computer, aerospace engineering and bio-engineering. The computer showed a new trend in this period—downsizing. Thanks to the painstaking efforts of Chinese computer scientists during the Cultural Revolution, China didn't lag very far behind in this field. We had a great foundation in aerospace engineering too, but the Cultural Revolution took a toll here. On February 20, 1962, American astronaut John Glenn had a successful flight around the earth; in July 1969, the US launched the Apollo 11 and sent the first human beings to walk on the moon. The Soviet Union, on the other hand, made efforts in another direction. It launched the first space station in 1971. In 1962, the US launched Mariner 1 to Venus; in 1965, it launched the first commercial communication satellite between itself and Europe. After 1972, the US and the Soviet Union launched multiple earth resource satellites.

During the ten-plus lost years, almost every S&T field experienced profound changes, saw epoch-making breakthroughs, and gave birth to a series of emerging technologies. Some people held that new S&T results in the 1960s–1970s surpassed those in the past more than 2,000 years combined. American scientist James Martin estimated that knowledge doubled every 50 years in the 19th century, every 10 years in the 1940s–1950s and every 5 years in the 1970s. If this is true, China had at least 50% less knowledge creation than other countries because of the Cultural Revolution.

The new technological revolution significantly improved human lives. Per capita GDP in developed countries was around USD1, 000 before WWII, but it shot up to about USD10, 000 in the 1980s, growing by 5–10 times in less than half a century, whereas China was still a traditional agricultural country at that time.

The new technological revolution also changed the demographic structure and people's living status. Urban population went up fast and salary for employees in various countries increased considerably. The average weekly wage in private non-agricultural sectors in the US rose by 3.4 times in 30 years from USD53.13 in 1950 to USD235.1 in 1980, while the monthly wage for Soviet employees rose from 64.2 to

[1]The 6th Five-year Plan for National Economic and Social Development (1983).

[2]Ibid.

168.9 Rubles in the same period. What was more, residents had reliable social security. The US had a fairly well-developed social insurance system at the end of the 1960s, but China was considered on the verge of economic collapse.[3] The new technological revolution widened the economic gap between China and the world that was originally narrowed. In some emerging fields, "first comer is king". and the late comers could hardly overtake those in front if not blazing new fields.

In sum, the ten-year chaos and some wrong policies in the first few years afterward widened the once narrowed S&T gap between China and the world. It's understandable that the 1978 S&T plan formulated against such a background carried many heavy historical tasks. China had to catch up both economically and in S&T, and the latter obviously had to undertake the historical task. As a result, the post-Cultural Revolution S&T program adopted some top-level concepts never seen before.

II Formation of the *Outline of Science and Technology Development Program 1978–1985*

1. Preparation for the national science conference in 1978

After the "Gang of Four" was smashed, the CPC Central Committee made achieving four modernizations by the end of the 20th century the national goal. Having held the national conferences respectively themed with "Learning from Dazhai in Agriculture" and "Learning from Daqing in Industry," it prepared to hold a national science conference as well.

In May 1977, Fang Yi, Li Chang and Wu Heng, leaders in the CAS Party group, reported to the Political Bureau of CPC Central Committee about the work at CAS, and Hua Guofeng suggested holding a national science conference. In June, Fang Yi convened a meeting of leaders of CAS, the National Defense Science Committee and the Ministry of Education to discuss the preparations for the national science conference. Soon the CPC Central Committee approved a "16-member leading group for the preparations led by Fang Yi and including Li Chang, Wu Heng, Zhang Aiping, Yao Yilin and Liu Xirao."[4] Preparations for the national science conference officially started.

The actions at the central level triggered a science surge at the local level. In many provinces, cities and State Council departments, it was the Party secretary that was personally in charge of this work, mobilizing the people to embrace the national science conference. For example, the Ministry of Education issued a notice in early July 1977, requiring the educational front to make good preparations for the conference. Another important political change in this period was that the third plenary session of the 10th CPC Central Committee held in July 1977 decided to reinstate Deng Xiaoping in all his positions both inside and outside the Party, and the reinstated leader agreed to resume the college entrance exam,[5] which was a turning point that changed the social ethos and emancipated people's mind. As education and S&T were closely related, efforts would soon begin to bring the S&T field back into order too.

On the morning of August 8, 1977, Deng Xiaoping gave an important speech at the end of a symposium, in which he objectively evaluated the scientific and educational performance in the 17 years before the Cultural Revolution. He said that "in the 17th years after New China was founded, the education front, like the scientific research front, mainly followed the red line. The majority of intellectuals in China willingly and conscientiously serve the socialist cause. They are workers." The topic of S&T plan was discussed, and Deng made the following remarks about medium/long-term S&T plans.

"The State S&T Commission formulated a 12-year S&T program for 1956–1967, which was basically completed in 1962, and then developed a 10-year program. I feel that our science and education isn't in a good state now. We need an organization to make plans, distribute resources,

[3]Lianzhong et al. (2000, 73–81).

[4]Research group of the History of Chinese Academy of Sciences (2008).

[5]Mingyu (2008).

make arrangements and guide coordination all in a centralized fashion. I personally agree to re-create the State S&T Commission. When? How? Should we put military scientific research under its unified lead too? The CPC Central Committee and State Council will study these issues and I won't give a conclusion now. Whatever form of organization, unified planning is necessary. The plan will not only determine the research projects, but has to consider which research institutes have to be combined and which to be separated, which disciplines in colleges to be combined and separated, added, reduced or canceled. Military research has to be included in the plan too."[6]

Deng's thoughts were the guiding thoughts for S&T planning at that time and played a positive role in making the 1978–1985 Program. On the one hand, the idea that there should be a plan to guide S&T technology was consistent with Zhou Enlai's idea; on the other hand, detailed studies had to be conducted as to how to make the plan.

In September 1977, Deng Xiaoping, when talking with main leaders of the Ministry of Education, pointed out that the "two evaluations"[7] of education work were untrue. "How can we put such a label on the tens of millions of intellectuals? Isn't it true that most of our talents now were cultivated in those 17 years? We must respect knowledge and talents, create conditions, and stimulate the enthusiasm of scientific and educational workers."[8]

The preparation office for the national science conference issued the *Key Points in the Publicity of National Science Conference* on August 29, 1977 to build up public opinions for the conference. A representative event was the publication of the sensational reportage *Goldbach Conjecture*. As part of the publicity efforts for the science conference, the *People's Literature* magazine invited famous writer Xu Chi to interview Chen Jingrun at the CAS and then write this reportage, which was published on the 1st issue of the magazine in 1978 and reprinted on *People's Daily* on February 17, quickly establishing a glorious image for S&T workers in the science circle and among the general readers. The fact that Chen Jingrun and Xu Chi got famous overnight had typical characteristics of the time. On the one hand, it was to create a science-respecting atmosphere to welcome the national science conference; on the other hand, it reflected the recovery of the long-suppressed science community. However, it also in a way indicated that S&T, as a tool for bringing the country back to order, once again became an instrument in the special context of China, which implied the utilitarian side of the planning philosophy.

In September 1977, a preliminary meeting was held for the national science conference, which discussed the *Notice of Holding the National Science Conference* drafted on behalf of the CPC Central Committee. On September 18, a meeting of the political bureau of CPC Central Committee reviewed and passed the notice, which stated, "Science and technology modernization holds the key to the four modernizations. Whether we can uplift our S&T is a significant issue that concerns the destiny and future of our nation."[9] S&T development didn't just involve the scientific community but was a necessary condition for national development.

While preparing for the science conference, all provinces and departments began to correct things on the scientific front, especially working out new medium/long-term S&T programs. In June 1977, CAS held a symposium to discuss the long-term plan and took the initiative to develop its own strategic plan and the 3-year and 8-year plans for each discipline.[10] In mid- and late-August, CAS

[6]Selected Works of Deng Xiaoping (1994), 52.

[7]"The bourgeoisie was dominant on the educational front in the 17 years before the Cultural Revolution, which was considered the 'dictatorship of the black line'; most intellectual had a bourgeois world outlook and they were bourgeois intellectuals".

[8]Memorabilia of the Communist Party of China—1977. http://cpc.people.com.cn/GB/64162/64164/4416108.html.

[9]Research group of the History of Chinese Academy of Sciences (2008).

[10]The State S&T Commission was re-established in 1977 and CAS no longer performed its functions.

held a 10-day symposium on the S&T plans for each department. On September 18, 1977, the CPC Central Committee decided to re-establish the State S&T Commission, comprising 15 members with Fang Yi being the director.

From September 27 to October 31, 1977, CAS presided over the national meeting on planning for disciplines of basic sciences. More than 1,000 experts, scholars and managerial personnel from CAS institutes, other research institutes and universities nationwide held discussions and worked out the development plans for math, physics, chemistry, astronomy, geoscience and biology, based on which the *Outline of National Plan for Disciplines of Natural Sciences* was formulated.

From December 12, 1977 to January 16, 1978, more than 1,000 leaders from State Council departments, all provinces, municipalities, and autonomous regions, and the military commission gathered in Beijing for the national meeting on S&T planning, at which the *Outline of National Science and Technology Program 1978–1985 (draft)* was adopted, which was then carefully reviewed and passed at the national science conference later. In October that year, the CPC Central Committee officially forwarded the *Outline of Science and Technology Development Program 1978–1985* (known as the *Outline of the 8-year Program*).

2. National science conference and Deng Xiaoping's speech

On March 18–31, 1978, the first national science conference after the Cultural Revolution was held. It was a grand event. A total of 5,586 representatives from 30 provinces, including Taiwan, and various departments gathered in Beijing, the oldest representative being a 90-year-old professor He Jie, a distinguished luminary in China's geological circle.

The conference received close attention across the country (Figs. 4.1. and 4.2). During the 20 days from March 12 to 31, it received 15,000 congratulatory letters and telegrams from around the country and from overseas compatriots, as well as more than 4,000 gifts, ranging from technological results, academic papers, calligraphy and paintings to pennants.[11] The unprecedented scale and grandeur of the science conference signified a surging enthusiasm for science that had been suppressed by the Cultural Revolution. If the resumption of the college entrance exam was a turning point that changed the social ethos, the conference marked the beginning of an overall transformation of the whole nation.

Fig. 4.1 Deng Xiaoping shakes hands with Chen Jingrun at the national science conference[12]

Fig. 4.2 Young Pioneers present flowers to scientists attending the national science conference[13]

[11]Yao Kunlun, "A Look back on the national science conference in 1978," 50 years of S&T development in the People's Republic of China. http://www.tech110.net/html/?uid-1429-action-viewspace-itemid-19334.

[12]Source: *People's Daily*, March 20, 2008; http://news.xinhuanet.com.

[13]Ibid.

When addressing the conference, Deng Xiaoping emphasized that "of the four modernizations, the modernization of S&T is the key. The CPC Central Committee decided to hold this national science conference with the aim of calling on the whole Party and whole nation to pay close attention to S&T, work out plans, commend advanced science workers, and come up with measures to develop S&T more quickly."[14] He continued to expatiate on three issues. The first was the perception that science was a productive force. Deng held the Marxist view that S&T was productive forces. The development of modern S&T tightened the connection between them and production. If we upheld this view, we then had to answer a question—how to look at the mental work of scientific research? Were those engaged in S&T work workers too? Deng Xiaoping pointed out that "generally speaking, most of them (intellectuals) are part of the proletarians. Their difference from physical workers lies in the division of work."[15] His remark that intellectuals were part of the working class was a huge push for the liberation of mind at the time and indicated that the CPC Central Committee had resumed the right perception of science, technology, and S&T workers that they had before the Cultural Revolution. It was not just a political statement or tactic of struggle, but the result of Deng Xiaoping's long-term thinking, which might mark the maturity of his thought on science and technology. Deng's speech made the scientists attending the conference extremely excited and full of tears. "When I heard at the conference that 'intellectuals are part of the working class and are mental workers', I knew the shackles were smashed and the nation was liberated in the real sense!" exclaimed Chen Jiaer with excited tears.[16]

The second issue was about the S&T team and the definition of being "red and expert" (socialist-minded and professionally competent). According to Deng Xiaoping, "the fundamental question about world outlook is who we serve. If a person loves our socialist country and conscientiously and willingly serves socialism, workers, farmers and soldiers, he has primarily established the proletarian world outlook and should, from a political point of view, be considered 'red' rather than 'white'."[17] At that time, the ultra-left thought was still prevalent, but Deng's understanding of "red and expert" conformed to the actual situations and could be accepted by all parties. It was a political statement and pointed out the political direction for scientists.

The third issue was how to realize divided responsibility under the leadership of the Party Committee in S&T departments. Deng Xiaoping made it clear that CAS should be staffed with three key leaders—a secretary of Party committee who was passionate about science and education, mostly a layman but preferably a maven; a director who was able to lead and organize scientific work and must be a professional; and a leader responsible for logistics. He even volunteered to be the logistics head.

On the afternoon of March 24, Hua Guofeng gave a speech titled "raising the scientific and cultural level of the whole Chinese nation," in which he said that raising the scientific and cultural level of the whole nation was a cause concerning hundreds of millions of people, and called on all the people to strive for S&T modernization. However, as Wu Mingyu, the writer of central documents, recalled, Hua Guofeng's speech didn't reflect the idea of "strengthening the country through science and education." Instead, he continued to stress the necessity of helping and guiding most intellectuals to foster a proletarian world outlook, and only mentioned "a red and expert team of intellectuals."[18] This meant that top leaders in the central government still had major divergences on science, technology, and intellectuals. The pro-scientist political forces and the scientists themselves hoped to set more store by science to prove their argument. As a result, the 1978–1985 S&T program carried many political functions and pressure, which led

[14]Xiaoping (1978).

[15]*People's Daily*, March 19, 1978.

[16]*Science and Technology Daily*, March 17, 2008.

[17]*People's Daily*, March 19, 1978.

[18]Mingyu (2008).

to the much higher targets and requirements. This, combined with the unleashed enthusiasm of the scientific community that had been suppressed for so many years, made the *Outline of the 8-year Program* too rash and overreaching.

In March 27–30, there were break-out activities in the morning and speeches by representatives in the afternoon every day. Apart from leaders who had to give speeches according to conventions, a group of scientists also addressed the conference, including physicist Zhou Peiyuan, mathematician Chen Jingrun and bio-agronomist Jin Shanbao.

According to the agenda, Fang Yi, the then vice premier, explained the *Outline of Science and Technology Development Program 1978–1985* (*Outline of the 8-year Program*) in his report. Calling the plan an advanced one aimed to achieve the four modernizations and high-speed development, he said the *Outline of the 8-year Program* made all-round arrangements for S&T tasks in such fields as natural resources, agriculture, industry, national defense, transportation, and marine science as well as in the two major categories of basic sciences and technological sciences, and also determined key projects. It also highlighted eight comprehensive fields with overall influence (agriculture, energy, material, computer, laser, space, high-energy physics, and genetic engineering), major emerging technologies and leading disciplines, in a bid to push forward the whole science, technology, and national economy at a high speed. If this plan was implemented successfully, China was able to approach or reach the world advanced level of the 1970s in some critical technological fields.

This S&T plan was still formulated in a "top-down" approach revolving around the central theme of national development. Given the massive damages caused by the ten-year Cultural Revolution, it was unrealistic to reach the world advanced level of the 1970s in a short period of time, and setting such a goal revealed that we didn't fully apprehend the fast-developing S&T then. China had accomplished the 1956 and 1963 S&T programs ahead of schedule and achieved remarkable S&T progress in general, and this made those who formulated the 1978 S&T program believe that the goals they set would be achieved soon too. This might be true when the scientific system and college development were in the foundation-laying stage, but substantial progress became extremely difficult when S&T reached a certain level.

On March 31, 1978, the National Science Conference held the closing and awarding ceremony at the Great Hall of the People, at which 862 advanced collectives, 1,192 advanced S&T workers and 7,675 excellent research results were commended. Before the closing ceremony, a speech written by Guo Muoruo, president of CAS, was read. In this speech titled *The Spring of Science*, he announced in poetic words, "This is the spring of revolution, of the people, and of science! Let's open our arms and warmly embrace this spring!"[19] Guo Muoruo was in bad health then, but his words were passionate and inspiring and they expressed the simple and best wishes of the scientists.

III **The idealistic *Outline of the 8-year Program***

1. Analysis of the contents of the Outline of the 8-year Program

The *Outline of the 8-year Program* set the guideline of "all-round arrangement with clear priorities" and consisted of several parts—foreword, goals, key S&T research projects, research team and organizations, specific measures, execution of the plan, and inspection of it. It continued the slogan of class struggle—"we will hold up high the great flag of Mao Zedong Thought and continue the three revolutionary movements of class struggle, production struggle and scientific experiment."[20] But it also vowed to "build China into a great, strong socialist country with all-round modernization in agriculture, industry, national defense and S&T in this century. Science and technology are a productive

[19]Muoruo (1978).

[20]Outline of National Science and Technology Development Plan 1978–1985. Website of the Ministry of Science and Technology: http://www.most.gov.cn.

force and the key to the four modernizations is S&T modernization."[21] This showed that around the time when the *Outline of the 8-year Program* was rolled out, top Party leaders didn't reach a consensus on some fundamental issues about S&T, and the *Outline* was a transitional document. However, it was the first time that "science and technology are a productive force" was stated in an S&T plan, which marked a change of the role of S&T plans from consolidating the regime to developing the economy.

The *Outline of the 8-year Program* set the grand goal of realizing modernization in the 20th century: "In the 20th century, we must use machinery in all sectors and places where possible, electrify both urban and rural areas, realize automated production in main sectors of industrial production, significantly raise labor productivity and quickly improve social productive forces. We will change our social and economic outlook across the board, substantially enhance our national defense, and make sure our proletarian dictatorship will be invincible. In science and technology, we must cultivate a massive red and expert S&T team of the working class, including a group of world-class scientists and technical experts, and have the most advanced experimental equipment. On this basis, we will strive to approach world advanced level in most S&T fields, catch up with that level in quite a few, and stand in a leading position in certain fields."[22]

Grand as they were, this goal was beyond China's scientific and economic capability of that time. The drafters apprehended science and technology from a non-scientific point of view, viewed and planned science development from a political perspective, but didn't take the laws governing S&T development into full consideration. This, while reflecting the good wishes of the decision makers and the scientific community then, also showed their linear understanding of the laws of science development.

The *Outline of the 8-year Program* set the following specific targets: (1) Approach or reach world advanced level of the 1970s in some important S&T fields; (2) Have 800,000 professional scientific research personnel; (3) Build a batch of modernized bases for scientific experiments; (4) Establish a national S&T research system.

These targets didn't seem very "out of reach", but they were at the executive level. The *Outline* made comprehensive arrangements for S&T research tasks in 20 fields, including natural resources, agriculture, industry, transportation, marine science, environmental protection, medicine, finance and trade, and culture and education, and in two major categories—basic sciences and technological sciences, and also determined 108 key S&T research projects out of them. If this plan was accomplished, China could approach or reach the world advanced level of 1970s in some important S&T fields and the gap could be narrowed to about 10 years, but the results were not as good as wished.

The *Outline* listed 27 key projects in basic sciences, including math, physics, chemistry, astronomy, geoscience and biology. It's worth noting that the 108th key project was "dialectics of nature and study of S&T history."

The *Outline* took agriculture, energy, materials, computer, laser, space science, high-energy physics and genetic engineering as comprehensive S&T fields with overall influence, major emerging technologies, and leading disciplines, and assigned the following main tasks in those eight fields.

(1) Agriculture: guarantee high and stable yield according to the guideline of "all-round development with the focus on grain"; resolve the south-to-north water diversion project and related S&T issues; build comprehensive experimental bases for agricultural modernization; step up basic theoretical studies for agricultural science.
(2) Energy: adopt new exploration and exploitation technologies, detect more oil & gas reserves, and improve drilling and extraction quality; realize mechanization in key coal mines; conduct research on coal gasification and liquefaction.
(3) Materials: develop intensified mining technology with the focus on steel and grasp a series of

[21] Ibid.
[22] Ibid.

modern metallurgic technologies; develop special and compound materials needed for national defense industries and emerging technologies; make breakthroughs in material science and experimenting and testing technology.

(4) Computer: conduct basic studies in computer science and related disciplines; intensify studies in peripheral equipment, software, and applied math; solve S&T issues for industrial production of large-scale IC and make technical breakthroughs in ultra-large IC; successfully develop large computers capable of 10 million calculations per second and huge computers capable of 100 million calculations per second; foster the manufacturing capability of computer series and strongly promote applied computers and microcomputers.

(5) Laser: quickly improve regular laser; conduct basic laser studies and make considerable progress in exploring new laser, developing new laser wavelengths, and using laser to study material structure; build an advanced laser industry.

(6) Space: carry out research on space science, remote sensing and satellite application; develop carrier rocket series to launch a variety of scientific and applied satellites for astronomical, communication, meteorological, navigation, broadcast and resource detection purposes; actively conduct studies for launching space lab and space probe; build modern space research center and satellite application system.

(7) High-energy physics: build a proton accelerator of about 30 giga EV within five years and one of about 400 giga EV within 10 years; make good preparations and cultivate the team for high-energy physics experiments; make theoretical breakthroughs in this field.

(8) Genetic engineering: build and reinforce relevant labs, conduct basic research on genetic engineering and combine it with the research on molecular biology, molecular genetics, and cell biology; approach or reach the world advanced level in some important aspects of biosciences.[23]

These goals indeed look idealistic today and some of them still haven't been achieved. Enthusiasm and increased investment cannot guarantee rapid S&T development. As we analyzed before, a top-level concept of China's S&T plans was "strengthening political functions". The chaos and ideological problems caused by the Cultural Revolution had to be rectified with some aggressive measures. The S&T plan assigned many lofty tasks to the scientific community, both scientific and implicit political tasks.

The *Outline of the 8-year Program* also set other goals beyond that period. For instance, in eight years, "the number of S&T personnel will increase from 360,000 in 1977 to 800,000 in 1985. All scientific research institutes, factories and mining enterprises, design institutes and colleges must take steps and make plans to improve the S&T competence and foreign language skills of young S&T personnel. In two or three years, most of them should be able to read professional books in more than one foreign language…"[24] It takes a long time to study foreign languages. College students nowadays are not necessarily able to read foreign professional books, not to mention that back then many people didn't know the first thing about English and some even had difficulty in studying Chinese. It was obviously a goal out of place.

Besides, some matters that shouldn't have been included in an S&T plan appeared in it anyway, such as strongly supporting scientific farming and technical renovation among the public, holding science contests among young students, and organizing excellent scientists, engineers, and skilled technicians to participate in the compilation and review of teaching materials for colleges, middle and primary schools. This in a way indicated that this S&T program wasn't scientific and comprehensive enough.

2. Continuation of the pre-Cultural Revolution planning philosophy

The 1978–1985 S&T program was developed in a hurry in a sub-standard style and was of a much

[23]Ibid.

[24]Ibid.

lower quality than the S&T programs launched in 1956 and 1963. It was obviously a transitional program.

Since the *Outline of the 8-year Program* was beyond China's national strength and scientific condition, it went through massive adjustment during implementation; and since the key areas and research projects it listed were not specific enough, the guideline of "priority development" wasn't well observed during execution either. Later as the nation gradually shifted the work focus, the S&T community further specified the strategic guideline that S&T development must be oriented toward economic construction.

In 1982, the National Planning Committee and State S&T Commission reduced the number of key projects in the *Outline* to 38, which were carried out in the form of the *6th Five-year National S&T Challenge Tackling Program* and achieved good results.

Although the *Outline of the 8-year Program* had many deficiencies, the 8-year program itself and the adjusted *6th Five-year National S&T Challenge Tackling Program* later still made important achievements. During the 8 years from 1978 to 1985, China's S&T progressed quickly and the number of S&T achievements increased dramatically. In the period of 1981–1985, 98, 245 major S&T results of provincial and ministerial level were achieved nationwide.

In agriculture, the new species of indica hybrid rice and "Lu Mian No. 1" new cotton species of high and steady yield were successfully cultivated and promoted on a large scale; in industry, we grasped the petroleum digital seismic survey technology, the technology of building semi-submersible offshore oil drilling platform, and the gentle-dip mining technology of medium and thick deposits. Breakthroughs were made in lean hematite beneficiation, blast-furnace smelting of high-titanium magnetite, vanadium-titanium overall utilization and recovery, separation and extraction of rare earth elements, and the exploitation, beneficiation, processing and recovery of paragenic ore of precious metals like copper and nickel.

In emerging technologies and basic theoretical studies, we successfully developed the large computer of 100 million calculations per second, solve the problems in carrier communication of medium coaxial cables, grasped the radio measurement and control system, experimental communication satellite and microwave measurement and control system for satellite carriers, successfully launched the "Long March 3" rocket, created the two-step fermentation technology for vitamin C production with bioengineering, developed the preparation technology for HBV core antigen and antigen, developed the potassium phosphate oxide, which was a new non-linear laser material, and the bismuth germanate crystal, succeeded in the hot forging of non-metallic IOL materials, and artificially synthesized mica and RNA.[25] In a technological basis, we established 140 national measurement standards, including 6 basic standards.

These achievements were not as brilliant as was expected when the program was formulated, and this program continued the philosophy of the two S&T programs formulated before the Cultural Revolution, namely "technology first", "national interests paramount" and "intensifying political function". This S&T program displayed strong features of a technological plan with a lot of technological projects. As reform and opening-up began to receive more attention in 1978, the planning philosophy of "independence based on domestic conditions" wasn't distinctly reflected. Besides, since the whole Party and nation shifted the work focus to economic construction, this S&T program and the adjusted S&T program showed the tendency of serving economic construction, but a clear economy-centered planning philosophy didn't emerge yet. Economic factors were to weigh heavily in future S&T programs.

This S&T program was too idealistic, and the *Outline of the 8-year Program* had irrational elements typical of that special historical period. Some American scholar believed "the (8-year program) set ambitious yet unrealistic targets. All the suppressed energy and conceptions found a vent in the plan and were included in the plan or projects for China's S&T modernization. It was a

[25]Science and Technology Development Plans and Programs of the People's Republic of China (2008).

grand program that not only clearly reflected the nation's priorities but also revealed the special interests of certain scientists or groups of scientists."[26] One of the differences of the *Outline* from the 1956 and 1963 S&T programs was that scientists had larger room to express their ambitions while politicians intervened less in the contents, but it was exactly the over-high ambitions that made the *Outline* deviate from reality farther than the previous two programs. This set this thinking: unlike the pure and free research plans of scientists, the S&T plan was a national action that conveyed the nation's will and needed the constructive opinions of politicians. The question is how to combine politicians' intention of state governance with scientists' creativity in the best way so that the S&T plan can play its role to the fullest.

IV Shift of focus to economic construction and new S&T planning philosophy

After the Cultural Revolution, China gradually shifted its work focus to economic construction, and the function of the S&T plan also changed from consolidating the regime to promoting economic development. This was a process of slow changes, during which S&T development and economic development became combined more closely. From December 25, 1980 to January 1981, the State S&T Commission held the national S&T work meeting, at which the attendees discussed the unrealistic ambition of the S&T community and they came to realize that S&T must be consistent with economic construction. "Science and technology should develop in coordination with economy and society and take promoting economic development as its primary task. We should strengthen the research on production technology, choose the right technologies and form a reasonable technology structure…The center of this guideline is that S&T should first and foremost promote national economic development."[27] This wasn't a statement from the top leadership, but it presaged the change in the function and concept of S&T plans.

The 12th CPC National Congress was held in Beijing from September 1 to 11, 1982, and Deng Xiaoping at the opening ceremony put forth the important thought of "going our own way and building socialism with Chinese characteristics." In his report titled *Creating a New Situation of Socialism Modernization in All Respects*, Hu Yaobang pointed out that "the key of the four modernizations is the modernization of S&T. Going forward, we should push forward large-scale technological renovation in a planned way, promote all sorts of technical results with good economic benefits, and actively adopt new technologies, equipment, processes and materials. We have to emphasize the research in applied sciences and basic sciences, and organize all parties to work toward major progress in critical research projects."[28]

This marked the top leadership's change in the attitude toward S&T, directly led to the change of the 1978 S&T program into a challenging-tackling program, and would influence the formulation of future S&T programs. The then premier Zhao Ziyang made a report titled *A Strategic Issue about Economic Vitalization* at the national S&T reward conference held on October 24, 1982, in which he said "economic construction has to depend on S&T and scientific and technological work must help economic construction."[29] This became the general requirement for S&T development in the 1980s after the reform and opening-up began.

As S&T development reached this point after the founding of new China, an institutional reform was inevitable. At the end of 1984, the CPC Central Committee drafted the *Decision on*

[26]Joint Economic Committee of the United States Congress, *China Marching Toward Four Modernization (Vol.1-II)*, translated by Guo Zhongyan, et al. (Beijing: China Foreign Economic Relations and Trade Publishing House, 1986), 263–280.

[27]"Further clarifying the guideline for science and technology development," *People's Daily*, April 7, 1981.

[28]Report made by Hu Yaobang at the 12th CPC National Congress, September 8, 1982. Website of CPC News: http://cpc.people.com.cn/GB/64162/64168/64565/65448/4526430.html.

[29]State Science and Technology Commission (1986, 280–290).

Reforming the Science and Technology System, which was released in March 1985, since when S&T system reform became an important content in medium/long-term S&T programs. As mentioned before, the huge success of the first S&T program (1956–1967) created some traditions in China's S&T programs. It is known to all that once traditions are formed in China, they are hard to change. The planned S&T system in China that was created by the first S&T plan was no longer suitable for the development in the 1980s and had to be reformed, and one round of reform wasn't enough.

After the *Outline of the 8-year Program* was changed into a challenge-tackling plan, a new S&T program was put on the agenda. At the end of 1982, the State Council approved the proposal to formulate a 15-year S&T program, the *National Medium-and Long-Term Science and Technology Development Program 1986–2000*, and set up an S&T leading group led by Premier Zhao Ziyang as the group leader. The old S&T system that only S&T plans were made to guide S&T development gradually disintegrated, and the S&T system featuring the combination of "S&T plans" and "S&T programs" began to take shape.

4.2 Guiding Principle for National Medium and Long-Term Science and Technology Development Program 1986–2000

I Reform creates the new background for S&T plan

1. Change of political direction and reform of S&T system

When the reform and opening-up came to the mid-1980s, many new problems emerged in our socialist construction, and the CPC was trying to find answers. Deng Xiaoping, as the master designer of the reform and opening-up campaign, made deep thinking about many fundamental issues in China. For instance, from the winter of 1983, the CPC spend three years overhauling the Party's work style and organization batch by batch and stagey by stage. In 1984, Deng Xiaoping proposed to solve the Hong Kong issue and Taiwan question on the "one country, two systems" principle. When he met with Tanzanian Vice President Mwinyi on April 15, 1985, he said poverty wasn't socialism, socialism had to eradicate poverty, and that it didn't meet the requirements of socialism if we didn't develop productive forces or improve people's lives. In 1985, the Chinese government made the strategic decision to cut the military by one million troops, indicating that the government's view about the world situation had changed from the Mao Zedong era. In July 1983, *Selected Works of Deng Xiaoping* (1975–1982) was published nationwide, meaning Marxism had evolved to the stage of Deng Xiaoping Theory in China. Deng's many thoughts on S&T development had a direct bearing on future S&T programs.

In 1986, Deng Xiaoping talked about the political system reform four times. According to him, when we rolled out reform, it included political system reform, without which we couldn't protect the results of economic system reform, couldn't keep it forward, the development of productive forces would be obstructed, and so would the realization of four modernizations. There were three general goals for the political system reform: (1) consolidate the socialist system; (2) develop socialist productive forces; and (3) advocate socialist democracy and mobilize the initiative of all people. The key to mobilizing people's initiative was developing productive forces and improving their lives.[30] The political system reform was aimed to develop productive forces, while S&T was the most direct driving force for that. S&T had assumed a role in socialist construction that was beyond its role in the convention fields—it became an important force for supporting political reform. As a result, S&T was endowed with more political significance during reform and opening-up, so was S&T plan in the economic transition period.

[30]Memorabilia of the Communist Party of China 1986. Website of CPC News: http://cpc.people.com.cn/GB.

中共中央关于科学技术体制改革的决定

（一九八五年三月十三日）

Fig. 4.3 Decision on Reforming the Science and Technology System by the CPC Central Committee[31]

The political system reform vitalized institutional reforms in other fields, especially the S&T field, and scientists and S&T workers enjoyed an ever higher social position. On March 13, 1983, the CPC Central Committee held a large congress in Beijing celebrating the 100th anniversary of the death of Marx, at which Hu Yaobang stressed how to correctly treat knowledge and intellectuals, calling it a significant and pressing issue of combining the universal truth of Marxism with the concrete practice of China's socialism modernization drive. On February 26, 1985, the General Office of CPC Central Committee forwarded the *Report on Recruiting Excellent Intellectuals as Party Members on a Large Scale* submitted by the Central Department of Organization. The report asked Party Committees of all levels to pay close attention to the situation that the Party members had a low cultural and scientific level in general, and to recruit on a large scale excellent intellectuals who were willing to devote themselves to socialist modernization and communism as Party members. This ensured the political position of excellent intellectuals from an organizational-procedural perspective, reflected the CPC and the government's care for scientists, and brought science and politics closer to each other.

On March 2–7, 1985, the national S&T work meeting was held in Beijing, which discussed the reform of the S&T system. When addressing the meeting, Deng Xiaoping said the reform of both the economic system and the S&T system was aimed to emancipate productive forces. The new economic system must be beneficial for technological progress and economic development at the same time. Soon afterward, the CPC Central Committee issued the *Decision on Reforming the Science and Technology System* (Fig. 4.3) on March 13, 1985, indicating that the S&T system itself became the target and focus of reform. The goal was that S&T must adapt to the development of the market economy and serve national economic construction.

The issuance of the *Decision on Reforming the Science and Technology System* was of great importance in the science history because it was the beginning of China's S&T system reform for many years to come. What was most critical was that the *Decision* demanded that "the appropriation system for research institutions be reformed. Classified management will be implemented for funds for different types of S&T activities according to their respective features." For scientific research institutes mainly engaged in

[31]*People's Daily*, March 13, 1985.

technology development, the technology contract system was promoted, and fiscal appropriation from governance finance would be reduced or even stopped in about three or five years. As to basic research and some applied research, the national natural science foundation and other S&T foundations would be set up to gradually implement the science fund system, which would be mainly sponsored by state budgetary appropriation. The *Decision* also included other contents: promote the commercialization of technological results and explore the technology market to adapt to the development of socialist commodity economy; adjust the organizational structure of S&T system, encourage research, educational and design institutes to work with productive units, and enhance the enterprises' ability of technology absorption and development; reform the agricultural S&T system to help adjust the rural economic structure, and push rural economy to be more professional, commercialized and modernized; reasonably deploy scientific researches to ensure lasting momentum for economic as well as S&T development; give more autonomy to research institutes and improve the government's macromanagement of S&T work; open up so that our scientific cause could go to the world; reform the S&T personnel management system and create a good environment where talents could come forth constantly and fully exert themselves.[32] The S&T system reform was part of the reform of the general national system. It sometimes lagged behind the general institutional reform, but most times it was expected to serve as an example for and one of the drivers of it. S&T system reform was also a key content in the 1986–2000 medium/long-term S&T program and future programs.

2. New situations during the reform

The reform and opening-up had made remarkable achievements by 1986. The 6th Five-year Plan (1980–1985) was highly fruitful, national economy increased steadily across the board, total industrial and agricultural output value increased by 11% per year on average and GNP by 10%, and capital construction and technological renovation gained headway. State revenue and expenditure were in balance, and S&T, education and cultural sectors began to thrive. But economic development did see some difficulties and problems during the 6th five-year period, such as the excessive fixed asset investment, too-fast growth of consumption funds, and excessive currency issuance, which had adverse effects on stable economic growth. In 1983, the State Council set about drafting the 7th Five-year Plan, and the *Suggestion of the CPC Central Committee on Formulating the 7th Five-year Plan for National Economic and Social Development* was finalized in the first half of 1985. In September 18–23 that year, the CPC National Congress was held in Beijing, at which Deng Xiaoping said the nearly 7 years since the 3rd plenary session of the 11th CPC Central Committee was one of the best and critical times since the founding of new China, during which we mainly did two things—putting wrongs to rights and conducting comprehensive reform. Reform was the self-improvement of the socialism system, whereby revolutionary reforms did take place within a certain scope and to a certain extent. The congress passed the 7th Five-year Plan, which made a special point that "we will put developing scientific and educational undertakings in a strategic position, push for S&T progress and accelerate intellectual development."[33] The main tasks of economic and social development during the 7th Five-year period included: continue to create a sound economic and social environment for economic system reform, work hard to guarantee the basic balance between total social demand and supply; step up technological renovation and intellectual development, and make necessary material, technological and talent preparations and reserves for the continued economic and social development in the 1990s. The achievements of the 6th Five-year Plan and the targets of the 7th constituted

[32]Chinese Academy of Science and technology for Development, *Research Report on the Reform and Development of China's Science and Technology System*, 2002.

[33]Ziyang (1986).

the general economic background for the 1986 S&T program.

On January 2, 1983, the CPC Central Committee issued the *Several Issues Concerning the Current Rural Economic Policies* (No. 1 central document of 1983), which affirmed the household contracting system as a great creation of Chinese farmers under CPC's leadership and discussed how to blaze out a path of socialist agricultural development with Chinese characteristics. From then on, the CPC Central Committee issued No. 1 document on promoting agricultural development for three years in a row. That agriculture, as the foundation for social construction, received close attention was reflected in the 1986 S&T program.

In industry, China began to carry out the reform of "tax payment in replace of profit submission" in State-Owned Enterprises (SOEs) in 1983, and faster reform steps were taken to straighten out and adjust the SOEs with the aim of improving economic benefits. Urban private economy and collectively owned economy became an important component of socialist public economy, while other non-public economic forms and components also appeared in reform with their position recognized, which made the formulation of various S&T programs other than S&T plans inevitable. In 1985, the State Council approved the Spark Program (developing a group of S&T projects with a short cycle, small investment and quick returns to promote local economy) formulated by the State S&T Commission. According to the Program, State S&T Commission would help local areas to develop a batch of "inconspicuous" S&T projects that were suitable and demonstrative for Small and Medium-sized Enterprises (SMEs), especially township enterprises and that featured the close connection of S&T and economy. The purpose was to raise the S&T level of SMEs, township enterprises and rural construction and inject new vitality to further develop local economy.

Economic system reform went deeper continuously. In 1984, the Chinese government designated 14 coastal cities, including Dalian, for opening up. On October 20 that year, the 3rd plenary of the 12th CPC Central Committee was held in Beijing, before which a 6-day preparatory meeting was held. The meeting passed the *Decision of the CPC Central Committee on Economic System Reform*, which expressly stated that the current development status in China urgently demanded that we further implement the guideline of vitalizing the economy internally and opening up externally and move faster to reform the general economic system focused on cities.[34] The basic goal of the reform was to establish a socialist economic system with Chinese characteristics and full of vigor and vitality and promote the development of social productive forces.

In May 1985, the CPC Central Committee and the State Council held the national education work meeting in Beijing, at which the attendees discussed the *Decision of CPC Central Committee on Reforming the Educational System* as well as the steps and measures to do that. On January 24, 1986, the CPC Central Committee and the State Council forwarded the *Report on Reforming the Professional Title Appraisal System and Adopting the Appointment System for Professional and Technical Positions* submitted by the central leading group of professional title system reform. It was a significant reform of China's professional and technical personnel management system, which would break the "big pot" system (which featured the equalitarian distribution of resources and rewards) for S&T workers. In 1986, the State Council issued the *Notice on Promoting the Reasonable Flow of Science and Technology Personnel*, asking leading local organs to create conditions for the reasonable flow of talents. All these meant that market factors began to be introduced to the intellectual and scientific circles.

A major S&T event worth mentioning at this time was that on March 3, 1986, scientist Wang Daheng, Wang Ganchang, Chen Fangyun and Yang Jiachi submitted a suggestion to the CPC Central Committee on developing high technologies. Deng Xiaoping gave the following instructions on this suggestion: "this should be decided as soon as possible without delay." The suggestion led to the creation of the "863

[34]Memorabilia of the Communist Party of China 1986. Website of CPC News: http://cpc.people.com.cn/GB.

Program." On November 18 that year, the CPC Central Committee and State Council forwarded the *Outline of High Technology Research and Development Program (the "863" Program)*, which listed 15 projects in 7 fields, namely biotechnology, aviation, information technology, advanced defense technology, automation, energy technology and new materials, as the key areas of future high-tech development. These had direct and indirect impacts on future S&T plans.

II Utilitarian S&T plan in pursuit of economic benefits

1. Process of formulation

In September 1982, CPC held its 12th National Congress, which determined the general task of the Party in the new historical period: by 2000, China's total agricultural and industrial output value will be quadrupled from RMB710 billion in 1980 to RMB2.8 trillion. During the congress, Hu Yaobang in his political report reiterated that S&T modernization held the key to realizing the four modernizations, and specified for the first time that S&T was one of the strategic priorities in national economic construction. "The key to the four modernizations is the modernization of science and technology. In the future, we should push forward large-scale technological renovation in a planned way, promote all sorts of technical results with good economic benefits, and actively adopt new technologies, equipment, processes and materials. We have to emphasize the research in applied sciences and basic sciences, and organize all parties to work toward substantial progress in critical research projects."[35] The position and function of S&T in socialist construction were further clarified with a growing utilitarian role.

In October that year, the national science and technology reward meeting was held, at which Zhao Ziyang, on behalf of the CPC Central Committee and State Council, gave an important speech titled *A Strategic Issue about Economic Vitalization*. He expounded on the guiding thought of S&T development that "economic construction has to depend on S&T and scientific and technological work must help economic construction." This guiding thought directly affected the formulation of the 1986 S&T program. Meanwhile, China began to launch extensive reforms in a wide range of aspects, covering the economic system, planned management system, price system, fiscal system, investment system and S&T system, which created lots of new situations for the new S&T plan.

In October 1980, the State S&T Commission held the national S&T work meeting, at which the attendees discussed the guideline for China's S&T development and formed the *Outline of the Report on China's S&T Development Guideline*. The report outline put forth the guideline for our S&T development for some time to come. First, S&T should develop in coordination with economy and society and take promoting economic development as the primary task; second, strengthen the research on production technology, choose the right technologies and form a reasonable technology structure; third, intensify technological development and promotion at mining enterprises and factories; fourth, make sure basic research progress steadily; and fifth, make studying, absorbing and digesting foreign S&T achievements an important way of developing our own S&T.[36] The *Outline* also raised conceptions and made arrangements for the 1986 S&T plan. "In addition to actively participating in the formulation of the 5-year economic and social development plan, the State S&T Commission and S&T organizations of various local areas and departments should also work with the planning committee and other relevant departments to formulate the 5-year plan and 10-year proposal for S&T development according to the requirements made by economic and social development as well as the S&T trend and

[35]Hu Yaobang, "Creating a New Situation of Socialism Modernization in All Respects," report at the 12th CPC National Congress, September 1, 1982. Database of CPC National Congresses: http://cpc.people.com.cn/GB.

[36]Notice of the CPC Central Committee and State Council on forwarding the *Outline of the Report on China's S&T Development Guideline* submitted by the Party Group of State S&T Commission, February 23, 1981, Zhong Fa [1981] No.14.

possibilities. A 20-year long-term plan can be considered on that basis."[37] In April 1981, The CPC Central Committee and State Council approved and forwarded the *Outline of the Report on China's S&T Development Guideline*, and demanded that the State S&T Commission work with relevant departments to draft the S&T development plan for the next 10 to 20 years.

At the end of 1982, the State Council approved the *Report on Formulating the 15-year Science and Technology Development Program (1986–2000)* submitted by National Planning Committee and State S&T Commission, and set up the S&T leading group to lead the formulation of the long-term S&T plan, the study of major technology policies, etc. On January 18, 1983, the S&T leading group of the State Council[38] held a mobilization meeting in Beijing for the formulation of long-term S&T plan and invited more than 1,000 scientists and engineering and technical experts to it. Later the "office of long-term S&T plan" was formed under the joint leadership of State S&T Commission, National Planning Committee and Economic Committee, which had 19 specialized planning groups and more than 200 experts were gathered in Beijing to work on the plan. Altogether over 3,000 people participated in this job nationwide.[39] The "office of long-term S&T plan" also invited famous figures and engineering and technical experts from West Germany, Japan, the European Community and the US for symposiums and discussions in order to understand the international S&T trend and some countries' lessons learnt during their development. These discussions provided good references for China's S&T planning. The formulation of the long-term S&T plan was conducted in two stages: from February to September 1982, background information on 26 industries was collected and the 1986–2000 framework for 27 industrial sectors was established; from March to October 1984, the "7th Five-year S&T plan and outline of the 15-year S&T plan" were formulated along with technology policies for 12 fields. Actually, several plans or outlines were developed this time, including the *7th Five-year National S&T Development Plan*, *Framework Outline of the 1986–2000 National S&T Development Program*, and the *Outline of the 1986–1990 National S&T Development Plan* for various key traditional industries and 6 emerging technological fields. We'll mainly discuss the medium/long-term S&T program for 1986–2000.

2. Analysis of the plan's characteristics and contents

The program wasn't so focused on S&T itself as on some peripheral issues, such as system reform, international exchanges, and policy measures, while the S&T part accounted for about 1/4. This reflected the infiltration of market factors and the trend of direct management being replaced by indirect and macromanagement.

Compared with previous programs, the 1986–2000 S&T program paid far more attention to combining it with the technological renovation plan and economic development plan, with targets of economic development being the restrictions. "This guideline is formulated in view of the decision made by the 13th CPC National Congress and the spirit of plenary sessions of the CPC Central Committee since that congress. It is aimed to explain the strategy, guideline, policies and priorities of our natural science and technology development in the medium and long term, and guide the coordinated development of science and technology, economy and society through 2000 or even 2020."[40] Economic factors

[37]Ibid.

[38]The S&T leading group of the State Council was formed in Beijing in 1981, with Zhao Ziyang being the group leader and director of State S&T Commission Fang Yi and deputy director of National Planning Committee Song Ping as deputy group leaders. The S&T leading group had the following tasks: organize and manage the national S&T team in a centralized way and allocate them as needed; lead the formulation of long-term S&T plan; study the decisions on major technology policies; decide the introduction and digestion of major technologies; and coordinate S&T work among different departments. *People's Daily*, January 28, 1983.

[39]Science and Technology Daily (2006).

[40]*Guideline for National Medium/long-term Science and Technology Development (1986–2000)*. Website of the

were a main reference for S&T plan in the transitional period.

The 1986–2000 S&T program explained and arranged specific scientific programs, and China's S&T system began to adopt the "science plan—science program" model. This S&T plan actively advanced high-tech research and industrial development programs such as the "Challenge-tackling Program", the "863 Program" and the "Torch Program", striving to make breakthroughs in certain fields in order to accelerate the commercialization, industrialization and internationalization of high and new technologies. After the plan was rolled out, a string of S&T programs were launched one after another to ensure the plan's implementation, including the high-tech research development program (863 Program), the Torch Program aimed at the industrialization of high technologies, the Spark Program for the countryside, and the National Natural Science Foundation. It's worth noting that the "trunk airliner program" was proposed. "We will develop the design, manufacturing and mass-production technology for China-made trunk airliners and strengthen the research on air safety technology." It took large aircraft as an example to explain how complicated it was to make S&T plan and carry out scientific projects during the transitional period. This meant that developing large aircraft was a national scientific project long ago.

The *1986–2000 S&T Development Program* consisted of six aspects: situation and choice; strategy and guideline; key areas; deepening the reform to establish a new system beneficial for economic development and S&T progress; adhering to opening-up and actively advance international S&T cooperation and exchanges; and policies and measures. In Part 1 "Situation & Choice", the idea that "science and technology are productive force and the primary productive force" was mentioned in an S&T plan for the first time. "Science and technology are primary productive force and a great revolutionary force that pushes economic and social development. Under the superior socialist system, S&T progress and better S&T management will play a decisive role in our modernization drive."[41] It also stated for the first time that economic construction should depend on S&T. "We should work harder to steer economic construction to the track of depending on S&T progress and the improvement of workers' capability, so as to ensure the successful accomplishment of our strategic goals in the second stage and lay a solid foundation for accomplishing those in the third stage." The plan summarized China's S&T achievements, experience and lessons learnt in the past, analyzed realistically the resource restrictions on achieving the four modernizations, and had a deeper and clearer understanding of the importance of S&T toward that goal.

Part 2, "Strategy & Guideline", elaborated on the guiding thought for China's S&T development in the next 15 years or further. The basic guideline that "economic construction has to depend on science and technology and scientific and technological work must help economic construction" was expressly put forth, and "science and technology should develop through reform and opening-up and S&T system reform should coordinate with economic and political system reforms," once again strengthening the political function of S&T. Historical practices later proved that S&T system reform was sometimes restricted by political system reform and sometimes served the economic system reform. Compared with the rash S&T program made after 1978 when the Cultural Revolution just ended, this medium/long-term program was more practical. It proposed that China would basically reach the technological level of developed countries in the 1970s or early 1980s by 2000 and reach their technological level in the early 21st century by 2020, thus getting closer to the world advanced level in general. The plan had clear goals for the key areas—promoting industrial development through S&T. It called for using S&T to promote agricultural development and upgrading. "We must implement the guideline of 'developing agriculture with S&T',

Ministry of Science and Technology: http://www.most.gov.cn/kjgh/lskjgh.

[41]Ibid.

and transform traditional agriculture to modern intensive agriculture based on modern S&T."[42]

Regarding industry, S&T was also given a special mission to boost industrial upgrading and benefits. The main task of developing industrial S&T was to conduct technological renovation of main industrial sectors in China with modern S&T. we needed to adjust the optimize the industrial structure, upgrade equipment, processes and products, and make large-scale production more automated and intelligent. The plan hoped to utilize S&T to improve the efficiency of key and bottleneck industries that affected national economic development, such as energy, transportation, communication, material and mechatronics. S&T was considered an effective tool to raise economic benefits.

The plan also made arrangements for the development of high technologies. It pointed out that high-tech research had to be closely related with the technological renovation of traditional industries from the very beginning, and called for strong development of high-tech industrial development zones and promoting the commercialization, industrialization and internationalization of those technologies. The plan vowed to make breakthroughs in a range of fields, including microelectronics, information, biology, new materials, aviation and aerospace, automation, new energy, laser and marine science, so that China could take a position in the world high-tech industry. This led to the "863" high-tech program. Boosting the development of high technologies would give a strong push to economic construction on one hand and continue and reinforce the planning concept that "technology comes first in S&T plans."

Compared with the previous three S&T programs, this one went much lighter on basic research either in terms of the proportion of contents or descriptions, and there was more about applied basic research than pure basic research. In other words, scientific research was focused on the Pasteur's Quadrant. "While making overall arrangements for cutting-edge disciplines, we should carry out multidisciplinary comprehensive research in strategic key areas concerning economic development, such as agriculture, energy, transportation, information and materials, as well as in such important areas as population, medical care, resources and ecology."[43] The request for a close science-economy connection, however, resulted in the neglect of basic sciences that required input over a long period but had little economic output, which might be one of the reasons on the state level why our basic sciences received little attention for a long time. In the 20th century, the highest national S&T award was mostly conferred to scientific achievements made before the 1986 S&T program, which perhaps partially indicated that that program made a mistake with basic sciences.

Unlike previous civilian S&T programs that didn't have a word about national defense S&T, the 1986 program made arrangements in the national defense sector as well in hopes that S&T could play a critical role in improving national defense technology on the condition of sufficient economic ability. The focus was on weapon and equipment systems that were of great importance for enhancing our military's combat capability and that could significantly boost our national defense S&T or even the general S&T level in the country. To that end, we had to step up research on key technologies that were crucial for the development of these weapon and equipment systems. Given the special nature of national defense, S&T program was divided into the civilian type and national defense type from the first one ever. Although they had overlapping and repetition, both types were successfully implemented before the reform thanks to the planning philosophy of "national core interests paramount." National core interests referred to interests concerning national security and sovereignty, such as national unity, nuclear weapons and satellite. After the reform and opening-up, however, everything was centered on economic construction, so the separate formulation and implementation of civilian and national defense S&T programs could no longer adapt to the new situation as the two sectors were vying for

[42]Ibid.

[43]Ibid.

resources. Therefore, there was the sign of integration between civilian and military sectors after the reform and opening-up.

The biggest difference of the 1986 S&T program from the previous three S&T programs was that it allocated a large part to the S&T system reform, which had something to do with the background then of the surge of system reforms. The plan made the following statement about the overall goal of China's S&T system reform: establish and improve a new system that complies with the objective laws governing S&T development, that adapts to the socialist planned commodity economy, and that features the combination and mutual promotion of S&T and economy; propel S&T work to actively serve the socialist planned commodity economy, steer economic construction to the track of depending on S&T progress and better quality of workers, and give full play to S&T as the primary productive force. The nation would provide steady support for basic, public welfare research and research on major technical difficulties, while research institutes specializing in market development and applications had to fight their own ways in the market, or the main economic battlefield. The nation gave key support to a group of R&D institutes that undertook basic research, applied basic research and public welfare S&T work, as well as substantial, comprehensive, long-term research tasks, or undertook the R&D of general technologies for the whole industry. The remaining majority of scientific research units owned by the whole people, especially the large number of medium and small-sized research institutes focused on technological development, would be gradually restructured according to their specific conditions under the guidance of national policies and based on market adjustment. Later reality told us that most research institutes that were thrown to the market survived after experiencing painful transformation. They made growing profits, played a positive role in socioeconomic construction, and their research personnel's personal income increased dramatically from before the reform. But basic and public welfare research was seriously affected, and many excellent research personnel didn't receive stable support, which was why the overall quality of our basic sciences was improved so slowly in the 1980s–1990s.

The fifth part of the program expatiated on the insistence on opening-up and the promotion of international S&T cooperation and exchanges. Based on past experience that S&T communication could be the prelude to and important pillar of political exchanges, China believed international S&T cooperation could give a strong boost to our S&T development and economic construction. "In international relations, S&T communication is usually the prelude to international exchanges and one of the most stable areas of cooperation. Therefore, actively promoting international S&T cooperation and communication should be an important long-term guideline for developing S&T undertakings in China."[44] The program expressly vowed to help enterprises digest new technologies through international communication. "Doing a good job at the introduction, digestion and innovation of technologies is a long-term, strategic measure to speed up our modernization drive…We should work out the rules, sequences and standards for technology introduction, intensify macro control and management, prevent blind and repeated introduction, and pay close attention to IPR protection…The nation should legislate to guarantee the funding for the digestion and innovation of introduced technologies, and make sure research institutes get involved in the whole process of introduction, digestion and innovation."[45] Nevertheless, although enterprises were active to introduce technologies, they were not enthusiastic enough about digestion and innovation, which resulted in high repetition and poor innovation capability. After we came to the 21st century, the "2006–2020 medium/long-term S&T program" had a special part on how to foster enterprises' innovation capability.

The last part of the program elaborated on policies and measures. It said that by the end of the 20th century, funding for basic research and applied basic research should account for 10% or so of the total R&D funding, a goal that was not

[44]Ibid.

[45]Ibid.

attained though. China's S&T plans and programs tended to set high goals but ended in poor execution.

3. Conspicuous boost to economic development

Since China formulated and carried out in 1990 the 5th medium/long-term S&T plan, The *Ten-year Science and Technology Development Program and Outline of the 8th Five-year Program of the People's Republic of China (1991–1995–2000)*, the effect of the 1986 S&T program after 1990 could only be estimated in reference to later S&T programs. Based on the relation between future medium/long-term S&T program and five-year plans, the former, if they spanned 10–15 years or more, were broke down into each five-year plan that would be adjusted dynamically according to actual conditions. In fact, the 1986 S&T program was mainly implemented through the 7th five-year S&T plan. To realize the philosophy of the *Science and Technology Development Program 1986–2000* and arrange S&T work for the 7th five-year period, the Planning Office organized the National Planning Committee and State S&T Commission to work out the *National Science and Technology Development Program in the 7th Five-year Period*, which consisted of the following five parts:

- Key S&T projects: 168 projects in total, including 40 priority ones;
- Key industrial experimental projects: 47 in total;
- Key technology introduction and digestion projects: 98 in total;
- Key construction projects for S&T undertakings: 45 in total;
- Key basic research projects: 43 in total.[46]

At last, the *National Science and Technology Development Program in the 7th Five-year Period* was officially formulated, including the *Science and Technology Challenge Tackling Program in the 7th Five-year Period*, which consisted of 200 technology development, promotion, and application projects and 76 key S&T challenge-tackling projects. There were 34 projects regarding seed and grain transformation and the digestion of foreign mature technologies; 16 projects regarding the processing of light industry, textile and agricultural products and deep-processing of metals and chemical materials; 11 projects of microelectronics and information technology in emerging fields; and 15 social development projects covering resources, ecology, environment and medical care. All these projects were basically completed at the end of 1990.

The *National Science and Technology Development Program 1986–2000* and the "7th five-year" S&T plan made breakthroughs in many aspects, the most important being in the reform of the S&T system. In 1985, the CPC Central Commission issued the *Decision on Reforming the Science and Technology System*, marking the beginning of a new period when our S&T front had comprehensive reform and opening-up. The reform of the S&T appropriation system injected vitality into R&D institutions. By means of technology contract, transfer of technological results, investment with technology as capital stock, co-establishment of enterprise and various S&T funds and loans, scientific research institutes no longer relied on government appropriation alone but had more channels of S&T funding. The implementation of *Technology Contract Law* and *Patent Law* and the cultivation of the technology market after 1985 secured the position of technology results as commodities, and the Chinese technology market was gradually created and strengthened. S&T began to "grow into" economy. Main S&T forces nationwide turned to serve the main task of economic construction and fostered specialists and S&T managerial talents of all sorts. Breakthroughs were also made in international S&T cooperation and technology introduction. However, practices showed that it was a commonplace phenomenon that enterprises attached more importance to introduction than digestion, and technology repetition was serious, while there were not enough specialists and skilled technicians either.

The number of research personnel increased dramatically. In 1990, there were more than

[46]Department of Development Planning of the Ministry of Science and Technology (2008).

600,000 people across the country engaged in R&D activities, including 400,000 scientists and engineers. The number of scientists and engineers engaged in R&D per million people was 357, taking the 4th place worldwide after the Soviet Union, the US, and Japan, which was a huge advancement. However, this number was far behind that in developed countries and was only equivalent to the middle-upper level of developing countries. The national S&T funding stood at RMB42.16 billion in 1990.

As to S&T achievements, China registered more than 110,000 major scientific achievements of provincial, ministerial and national level during the 7th five-year period, including about 26,000 in 1990 alone. From 1987 to 1990, 3,645 National S&T Progress Awards, 816 National Invention Awards, and 237 National Natural Science Awards were granted, and 62,000 Chinese and foreign patents were certificated. On October 16, 1988, the Beijing electron-positron collider, the first high-energy accelerator in China, had a successful first collision, marking yet another breakthrough in high technology as its performance reached the world advanced level of the 1980s.

Regarding S&T thesis, SCI, ISR, ISTP and EI, the four most famous search systems in the world, took in 47,000 Chinese S&T theses in 1987–1990, and China's ranking in terms of the number of theses was up a little.

S&T also achieved a lot in promoting economic development. China exported USD7.46 billion worth of high-tech products in the 7th five-year period, with the export volume rising from USD1.29 billion in 1988 to USD2.69 billion in 1990. In general, this S&T program was very effective in boosting economic construction, far more so than previous S&T programs. However, the over-emphasis on S&T as a means or tool to stimulate economy had its side effects, and the utilitarian planning concept was quite conspicuous.

III The "S&T program—S&T plan" system

Another major change resulting from the implementation of 1986 S&T program was the formation of the "S&T program—S&T plan" system. As China's S&T development was more oriented toward economic construction, a group of S&T development programs consistent with the S&T plan came into being. This part will give a brief overview of S&T programs before, during and after the 7th five-year period.[47]

(1) National S&T challenge tackling program

The *Science and Technology Challenge Tackling Program in the 7th Five-year Period* consisted of 200 technology development, promotion and application projects, 76 key S&T challenge-tackling projects and 349 topics, with about 108,000 joining the program every year on average. The program achieved 10,462 results, and all projects were basically completed at the end of 1990.

(2) National program of key industrial experimental projects

The arrangements in this program were aligned with S&T challenging tackling, research, and technology development programs on the upstream and were rolling on an annual basis. In 1987, the nation began to build industrial experimental bases for some key industrial and technological fields in major institutes and key universities that boasted strong technological power and ample achievements.

(3) Spark Program

The Spark Program was an S&T program approved by the CPC Central Committee and State Council in 1986 to vitalize the rural economy and an important part of China's S&T program system. After the Spark Program was implemented, it was warmly welcomed in local areas, especially the countryside. The nation invested RMB17.08 billion in the program over five years, including nearly RMB120 million from the central finance, nearly RMB1.21 billion from local revenues, about RMB6.38 billion of bank loans, and RMB9.37 billion self-raised capital. Through the implementation of this program, S&T gave a strong impetus to rural

[47]Science and Technology Development Plans and Programs of the People's Republic of China (2008).

economic development. It well reflected the guideline that "economic construction has to depend on science and technology and scientific and technological work must help economic construction" stated in the 1986 S&T program.

(4) National program of promoting key S&T achievements

In 1989, the State S&T Commission carried out the national program of promoting key S&T achievements across the country on a trial basis, involving 37 achievements and the total investment of RMB179 million. This program was officially launched in 1990, involving 92 large projects and the total investment of RMB641 million. The industrial sector involved 7 fields with a total of 196 achievements, focused primarily on the improvement of efficiency and technology and secondly on energy-saving technologies. The agricultural sector involved 14 fields with a total of 289 achievements, focused primarily on superior crop varieties and breeding technology and secondly on increasing the yield of superior timber varieties and cultivation technology.

(5) High-tech research (863) program

The 863 program was an S&T program that kicked off in November 1986 to develop high technologies and was officially implemented in 1987. During the 7th five-year period, the program achieved more than 400 achievements in the five fields of biotechnology, information technology, automation technology, energy technology, and new materials, all of which played a leading and central role in high-tech development in China. Deng Xiaoping wrote the inscription of "Develop High Technology, Realize Industrialization" for China's high-tech undertaking, which expressed his wish for S&T, especially high technology, to play a bigger role in boosting the economy.

(6) Torch Program

The Torch Program was a guiding high-tech industrial development program approved by the State Council in 1988 to be implemented nationwide. On August 5–8, 1988, State S&T Commission held the first work meeting on Torch Program, marking its official implementation. The Torch Program was the outcome of further developing the high-tech results achieved by the 7th five-year S&T challenge tackling program and the 863 program, with the aim of producing high-tech products that would be competitive in the domestic and foreign markets. From 1988 to 1990, Torch Program increased the output value by nearly RMB12.07 billion, increased profits and taxes by RMB3.18 billion, and earned nearly USD700 million foreign exchanges.

(7) National Natural Science Foundation program

The National Natural Science Foundation began to fund basic scientific research in 1986 and the amount was rising to RMB158 million in 1990. The Foundation sponsored about 60,000 research personnel and more than 20,000 graduate students to conduct basic research and has become one of the main funding channels for basic research in China. During the 7th five-year period, the National Natural Science Foundation approved 263 key projects and provided RMB19, 207, 500 for them.

(8) State Key Laboratory program

In 1984, the State Key Laboratory program kicked off at colleges and research institutes affiliated to the State Education Commission, Chinese Academy of Sciences, Ministry of Agriculture and Ministry of Health. In 1990, 70 State Key Laboratories were founded across the country, which created fairly good research environment and conditions and fostered a stable, high-level basic research team by continuously recruiting talented scientists to carry out cutting-edge scientific research.

Later the nation rolled out many other specific S&T programs. Since the first national S&T program, the *National Science and Technology Challenge Tackling Program in the 6th Five-year Period* came out, more than 30 S&T programs were launched during the 30-plus years of reform and opening-up, establishing a multi-functioned, multi-tiered national S&T program system. The "S&T plan—S&T program" system was perfected (see Table 4.1).[48]

Table 4.1 shows that the nation rolled out a string of specific S&T programs to push the implementation of the medium/long-term S&T programs. The programs carried out the projects and principles stated in the plan and consequently established a fairly complete S&T program system in China. Since the S&T plan was in the center of the S&T program system, its philosophy was expanded to the entire system.

Most of these S&T programs were aimed to directly serve economic construction and reflected the guideline that "economic construction has to depend on science and technology and scientific and technological work must help economic construction." But what was the planning philosophy behind this guideline? It was a utilitarian concept. According to the views of sociology of science, science, technology and society are interactive with each other, and science, especially technology, can promote the development of all social aspects, including military, politics, economy and culture, pushing the society forward both in good ways and bad ways. But in the formulation and implementation of China's S&T plans, we took it for granted that S&T would take a good effect on social development, especially economic construction, and often over-used its "utilitarian functions" for more economic values. The most obvious representation was the sole pursuit for high-speed GDP growth while giving little heed (planning, analysis) to environmental protection, social equity and S&T's negative effect. Such a utilitarian concept guided the formulation of S&T plans and penetrated the implementation as well.

[48]The table is prepared in reference to relevant documents of the Ministry of Science and Technology and Li Liya, Li Ying, "National Science and Technology Program System and Evolution of Its Management," *Forum on Science and Technology in China 8* (2008).

4.3 Guideline for National Medium and Long-Term Science and Technology Development Program 1990–2000–2020

I An S&T guideline spanning 30 years

Five years after the 1986 medium/long-term S&T program was made, or around 1990, the Chinese government made a long-term S&T program spanning 30 years—the *Guideline for National Medium/long-term Science and Technology Development Program 1990–2000–2020*, which was basically overlapping with the *1986–2000 Medium/long-term S&T Program* and the *Ten-year S&T Development Plan and Outline of the 8th Five-year Program of the People's Republic of China (1991–1995–2000)* later. Since this program spanned too long a time and was not implemented through specific five-year programs, it was basically useless. Some scholars didn't even think it was an S&T program in history. However, this book holds that although the program wasn't really put into practice, its basic thought was reflected in future S&T programs, and it was listed as an S&T program on the website of the Ministry of Science and Technology. For a historical review of S&T programs, it should not be omitted, and we'll give it a brief introduction here.

From October 25 to November 1, 1987, the 13th CPC National Congress was held in Beijing, with 1,936 deputies and 61 invited guests representing the more than 46 million Party members nationwide. Zhao Ziyang gave a report titled *Marching ahead along the Socialist Path with Chinese Characteristics*. The congress put forth and systematically elaborated on the theory of the primary stage of socialism, determined the CPC's basic line in this stage, and decided the "three-step" development strategy and various reform tasks. In the report, Zhao Ziyang suggested "the State Council work out the guideline for medium/long-term science and technology development, reasonably organize S&T forces in the nation and set about formulating the

Table 4.1 National S&T programs launched since the 6th five-year period

Period	National S&T programs
6th five-year period: 1981–1985	National S&T challenge tackling program
	6th five-year national S&T challenge tackling program
	Major technology & equipment development program
	National major scientific project
	National technology development program
	National key laboratory program
	National key industrial experimental program
	National key, new technology promotion program
7th five-year period: 1986–1990	Spark Program
	National Natural Science Foundation
	863 program
	S&T-based poverty alleviation
	Military-to-civilian S&T development program
	National key, new product program
	Torch Program
	National key S&T achievement promotion program
	National soft science research program
8th five-year period: 1991–1995	National engineering (technology) research center program
	National basic research priorities program (Climbing Program)
	National demonstration project for industrialization of major S&T achievements
	Productivity promotion center
9th five-year period: 1996–2000	Social development S&T program
	National technology innovation project
	National key basic research development program (973 program)
	S&T SMEs technology innovation fund
	Knowledge innovation program
	Basic S&T work fund for central research institutes
	Technology R&D fund for research institutes
	Action Plan "prospering trade with S&T "
	Public welfare research fund for research institutes
	S&T development fund for relocation of the Three Gorges
	National university science park
	S&T program for China Western Development
10th five-year period: 2001–2006	International S&T cooperation priorities program
	National basic S&T platform construction program
	Fund for translation of agricultural S&T achievements
	National agricultural science park
	Olympic S&T (2008) action plan
	National S&T support program

(continued)

Table 4.1 (continued)

Period	National S&T programs
11th five-year period: 2006–2010	Technology innovation guidance program
	Action to promote sustainable regional development
	National engineering technology research program
	(The 11th five-year national S&T program system was adjusted significantly and mainly consisted of major special projects and basic programs.)

guideline through concerted efforts."[49] The *Guideline for National Medium/long-term Science and Technology Development Program (1990–2020)* was developed according to the decision of the 13th CPC National Congress and the spirit of relevant speeches. It was intended to explain the strategy, guideline, policies and priorities of China's medium/long-term S&T development, and was a macro guide and framework of our S&T development in 1990–2000 and further to 2020. To implement the guideline, the former State S&T Commission, under the leadership of the State Council, formulated the *Outline of Medium/long-term S&T Development Program 1990–2000–2020*, which was more detailed than the "guideline", but neither of them was really put into practice. In a sense, this medium/long-term S&T program carried a strong political import and was more concerned about the action of "making the plan" itself than about how to implement its contents.

All sectors paid close attention to making the "guideline" and "outline" and formed leading groups (or coordinating group) comprised of main leaders of the sector and drafting groups comprised of experts. Incomplete statistics showed that various departments held 113 workshops successively and experts participated in the compilation and demonstration for 4,787 person-times, including 98 person-times for ministerial leaders, 1,331 person-times for departmental and section-level cadres and 3,342 person-times for scientists and experts. The "guideline" drafting team at the State Council also reviewed and consulted on the 27 fields and industries of the "outline" one by one. In 1991, the State S&T Commission, after collecting and organizing all the contents of different fields (industries), officially printed a book of about 350,000 characters.[50] On March 8, 1992, the State Council promulgated the *Guideline for National Medium/long-term Science and Technology Development Program.*

The formulation and many contents of this *Guideline* reflected the spirit and contents of Zhao Ziyang's report at the 13th CPC National Congress, in which he said, "we should put S&T development and education in the highest position and steer economic construction to the track of depending on S&T progress and the improvement of the workers' capability."[51] The *Guideline* held that S&T progress and higher management level would fundamentally decide our progress of modernization drive. To accelerate S&T progress, we should respect actual conditions and choose the right directions and priorities. In particular, the primary task of S&T work was boosting national economy, so we had to speed up the reform, create a mechanism where S&T was closely combined with economy, give enterprises, especially large leading ones, more motivation and pressure to apply S&T results, promote the development of technology market and shorten the cycle of commercializing technological results. All in all, the prevalent principle of the *Guideline* was that S&T should serve economic construction.

[49]Marching ahead along the Socialist Path with Chinese Characteristics, report by Zhao Ziyang at the 13th CPC National Congress, October 25, 1987.

[50]Science and Technology Development Plans and Programs of the People's Republic of China (2008).

[51]Marching ahead along the Socialist Path with Chinese Characteristics, report by Zhao Ziyang at the 13th CPC National Congress, October 25, 1987.

II Textual analysis of the long-term S&T development guideline

Since the *Guideline for National Medium/long-term Science and Technology Development Program (1990–2020)* was really put into force, we could only analyze the text. The *Guideline* consisted of six parts: situation and choice; strategy and guideline; key areas; deepening the reform to establish a new system beneficial for economic development and S&T progress; adhering to opening-up and actively advance international S&T cooperation and exchanges; and policies and measures. Part 1, "Situation & Choice", mentioned that science and technology were primary productive force. "Making general arrangements for China's S&T development in the next 10 to 30 years according to the three-step strategic deployments of our economic construction and in reference to international S&T trend is a significant historical mission."[52] Affirming China's remarkable S&T achievements after the founding of the People's Republic of China, the Guideline also pointed out the main problems and lessons in our S&T development. There were several representations: neither the guiding thought nor actual work of economic and social development consistently put promoting S&T progress in an important strategic position; there was neither the mechanism nor the inherent motive for economic development to rely on S&T progress, and S&T development had no enthusiasm to serve economic construction either; sometimes we were blindly anxious for quick success regardless of national conditions; we made serious mistakes about intellectual policies, which hurt the initiative and creativity of S&T personnel and undermined the S&T team's healthy growth; a social atmosphere of respecting knowledge and intellectuals wasn't formed yet. Pointing out these problems was good for deepening our understanding of the laws governing S&T development and making the S&T plan more scientific.

After analyzing the hard restrictions on our economic and social development, including population, resource and environment, the *1990–2020 Guideline* emphasized that the key to realizing S&T modernization was achieving economic modernization and considerably improving the labor productivity. We should give the fullest play to S&T as primary productive force, respect knowledge and talents, and more conscientiously steer economic construction to the track of depending on S&T progress and higher worker quality. This clearly showed that science and education were the two fundamental pillars of prospering the nation. This momentous decision signified a historic change in our S&T, economic and social development strategy and was of great importance for achieving our goals for the end of the 20th century and the mid-21st century.

Part 2 expounded on the strategy and guideline for S&T planning and development in the next 30 years. The basic strategy for China's S&T development was to enhance the whole nation's S&T awareness, improve workers' quality, mobilize and lure most S&T forces to devote themselves to economic construction, pay close attention to technological innovation, and carry out technological upgrading in all economic sectors more quickly. For a very long period to come, S&T development would be aimed at the modernization of industrial technologies and equipment for large-scale production. Meanwhile, high technologies and related industries would be developed in a planned way with priorities, and basic research would be enhanced to increase scientific reserve.

The *1990–2020 Guideline* set the following strategic goals: by 2000, main industrial sectors of China will roughly reach the technical level of economically developed countries in the 1970s or early 1980s; by 2020, they will reach their technical level in the early 21st century and get closer to world advanced level in general. The Guideline also made macro plans for S&T in different fields.

[52]Notice of the State Council on issuing the *Guideline for National Medium/long-term Science and Technology Development Program*, Bulletin of the State Council, 1992, (7).

Regarding agricultural S&T development, the "Spark Program", "Harvest Program" and "Prairie Fire Program" aimed at developing the rural economy continued to be implemented; regarding industrial S&T development, economic benefit should be improved and S&T progress in enterprises, especially large and medium-sized ones, boosted.[53] Modern S&T and management technologies should be applied, especially electronic information technology should be utilized for the technological renovation of industrial sectors. Industrial structure and product structure must be optimized and efforts must be made to transform the Chinese economy from the pattern of high consumption and low benefit to one of low consumption and high benefit.

Regarding S&T for social development, R&D will be stepped up in a wide range of areas, including population, medical health, social service, public infrastructure, environmental and ecological protection, and disaster monitoring and prevention. But this was rather empty and general compared with previous S&T programs.

Regarding the development of high technologies, the principle of "moderate goals, clear priorities" will continue to be followed. The nation should specifically support research on those high technologies where major breakthroughs are possible and that have extensive application prospects, and strongly bolster the development of high-tech industries. The nation actively promoted the "national S&T challenge tackling program," "863 program", "Torch Program" and other high-tech research and industrial development programs, with a view to making breakthroughs in certain areas and speeding up the commercialization and internationalization of high technologies.

Regarding basic research, the *1990–2020 Guideline* held that we should observe the law inherent in science development, reinforce explorations and tracking in a number of cutting-edge areas, and strive to reach world advanced level. Applied basic research should be aimed to solve major S&T problems in economic and social development, but it cannot be completely separated from pure basic research—this indicated an obvious state orientation in basic research.

Regarding national defense S&T development, the guideline of "shortening the front, highlighting priorities, strengthening scientific research and realizing tier-by-tier upgrade" will be implemented. Efforts will be continued to push the modernization of weaponry, equipment and technologies, so as to make substantial progress in high-tech regular weapons, maintain the validity of our self-defense nuclear deterrence, shorten our distance to world advanced level and reach that level in certain aspects, and make sure national defense S&T continues to have a place in world advanced S&T field.

The *1990—2020 Guideline* stressed over and over the basic guideline that "economic construction has to depend on science and technology and scientific and technological work must help economic construction," in order to promote the combination of S&T and economy. At the 5th session of the 7th National People's Congress (NPC) held on March 20, 1992, the then Chinese premier Li Peng gave the government work report, in which he emphasized again that "we should actively develop S&T and education and improve the overall quality of the national economy. We will earnestly carry out the guiding thought raised by Comrade Deng Xiaoping that science and technology are primary productive force, and boost economic development through S&T progress and better worker quality. We must adhere to the guideline that economic construction has to depend on science and technology and scientific and technological work must help economic construction, and promote their combination." It was clear that "economic construction has to depend on science and technology and scientific and technological work must help economic construction" had become a consensus among China's top leaders in the early 1990s. However, the development of science and technology per se didn't receive enough attention, so the "scaling heights" guideline was proposed in later S&T programs.

[53]Guideline for National Medium/long-term Science and Technology Development Program. A historical review of S&T plans of the Ministry of Science and Technology: http://www.most.gov.cn/ztzl.

To fully implement the idea that "economic construction has to depend on science and technology and scientific and technological work must help economic construction," the Guideline also put forth some specific sub-guidelines, such as earnestly implementing the idea of "science and technology are primary productive force"; creating favorable environment and conditions for S&T development; adhering to reform and opening-up; developing S&T through reform and opening-up; combining independent development with technology introduction and taking the introduction, digestion and innovation of foreign advanced technologies as an important way of accelerating our S&T development; upholding the guideline of "all flowers bloom together, all schools of thoughts prosper," fully exerting socialist democracy, guarantee academic freedom and encourage exploration and innovation; combining "improvement" with "popularization"; carrying out public technical renovation activities while making in-depth deployments for S&T development; popularizing science and knowledge to constantly raise the scientific quality of workers and fight consistently against ignorance and superstition.

Part 3 was the priorities. The *1990–2020 Guideline* designated a range of key areas, including agricultural S&T, industrial S&T, S&T for social development, high technology and high-tech industry, basic research and applied basic research, and national defense S&T. After it was adopted, the then State S&T Commission further specified its contents according to instructions from State Council leaders, and formulated the *Outline of Medium/long-term S&T Development Program 1990–2000–2020* on that basis, which listed the following 27 key areas (industries):

- Agriculture;
- Forestry;
- Water conservancy;
- Energy;
- Transportation;
- Mechatronics;
- Information & communication;
- Metallic materials;
- Chemical industry;
- Building materials industry;
- New materials;
- Consumer goods;
- Aviation & aerospace;
- National defense;
- Biotechnology;
- Basic research and applied basic research;
- S&T talents;
- Medical health & family planning;
- Construction;
- Geology & minerals;
- Renewable resources;
- Ocean;
- Eco-protection & natural disaster prevention;
- Production safety;
- Public security;
- Surveying & mapping;
- Measuring & standardization.[54]

III **Policy measures**

To make sure that S&T progress was put in a paramount position in the national development strategy and to strengthen the top leading organization's leadership of S&T work, the *Guideline* put forth some policy measures, mainly covering the following aspects:

(1) Make more S&T investment and improve S&T fund management. S&T investment should consist of three pillars—fiscal funding, capital raised by enterprises themselves and loans from financial institutions, and the societal S&T investment should be increased by a large margin. By the end of 20th century, funding for basic research and applied basic research should reach about 10% of total R&D fund. More financial support should be given to intermediate test, development of production process, and commercialization and industrialization of technological results. Several key areas should be picked in high technologies and related industries, where investment should

[54]Department of Development Planning of the Ministry of Science and Technology (2008).

be increased dramatically and VC should be implemented gradually.

(2) Continue to reform the S&T fund management system. Put S&T fund under centralized management and bring in competition. Special funding, foundation, lump-sum appropriation, bidding, or contracting systems will be adopted for different types of research projects. The R&D project appraisal system and independent review organizations will be formed to supervise and inspect the allocation and use of S&T funds.
(3) Enhance the legal work for S&T, fixate guidelines and policies regarding S&T progress in the form of law, so that S&T work will be covered in the legal system and be under the protection and supervision of the law. The *Law of Science and Technology Progress* and other S&T-related laws will be formed as soon as possible. Efforts will be made to make S&T decisions more scientific, democratic, and rules-based.
(4) Develop talents and give full play to the role of S&T personnel. We have to foster new-generation S&T personnel more quickly through various channels, boldly use young people with both integrity and capability, and encourage self-study. Political and ideological work of S&T personnel will be strengthened and great efforts will be made to improve their working conditions and treatment.
(5) Create a stable, democratic and academically free environment for S&T development. Different opinions on academic topics should be solved in the academic circle through scientific practices and discussions rather than determined in administrative ways. A social S&T supportive system should be established and perfected, covering scientific research equipment, S&T intelligence, books and materials, and publication.
(6) Intensify societal support for S&T work. Party and government leaders of all levels should better understand S&T work, and attach great importance and give strong support to it. All departments should also give more support to S&T work and make joint contributions to S&T development.[55]

Generally speaking, this long-term S&T program spanning 30 years, as a master document of China's medium/long-term S&T development, made macroscopic and general descriptions of China's S&T prospects through 2020, and offered guidance and references for making the ten-year S&T development plan and the 8th five-year S&T program later. Compared with previous plans, the 30-year-long one had more political significance, and there weren't many auxiliary regional or departmental S&T plans for it since implementation didn't seem to be most important. Only the State Bureau of Surveying and Mapping formulated in 1990 an *Outline of Medium/long-term S&T Development for the Surveying & Mapping Industry 1990–2000–2020*, which set the following goal: "In the next 20–30 years, Chinese surveying and mapping industry will look to the world and the future. We will not only replace the old management system with a new one, but will also transform the technological system. We will adopt new technologies to ensure more precise geodetic surveying benchmark, to comprehensively update the basic land surveying information and products in China, and to comprehensively conduct the basic survey and special surveying of our sea areas, so as to fill in the blank of cadastral surveying and mapping in China and explore new surveying and mapping service areas at home and abroad."[56] The future development of surveying and mapping has to embrace the challenge of the new technological revolution and industrial revolution in the world. We need to deepen the institutional reform in view of China's real situations and work hard to develop surveying and mapping technologies in order to successfully accomplish all the arduous tasks.

[55]Guideline for National Medium/long-term Science and Technology Development (1992).

[56]Strategic study of surveying and mapping development in China: http://218.244.250.72/xiangguanziliao/D04.htm
.

The S&T development guideline that spanned 30 years continued the utilitarian concept of pursuing economic benefits. On one hand, it was a continuation of the utilitarian philosophy of the previous S&T program; on the other hand, it designed a long-term blueprint for how to better serve economic construction. The next 8th five-year S&T plan continued this utilitarian planning philosophy featuring the pursuit of economic development and growing speed.

4.4 Ten-Year Science and Technology Development Program and Outline of the 8th Five-Year Program (1991–2000)

I **Goal of quadrupling the economy**

1. Science and technology are primary productive force

Bourgeois liberalization was a serious problem in some Chinese cities in 1987. In March 1989, conflicts began to appear in the society. First a small group of separatists created a serious riot in Lhasa, and then after Hu Yaobang passed away in April, large-scale student upheaval and unrest appeared in Beijing and other big cities. But the Chinese government didn't get on the old path of social movement, like what happened during the "Cultural Revolution", to fight against such bourgeois liberalization. Deng Xiaoping stressed that the path, guideline and policies decided at the 3rd plenary session of the 11th CPC National Congress, including our three-step development strategy, remained correct even though the riot had happened. "The biggest mistake we made in that decade was education, by which I mean ideological and political education, not only in schools and of young students but also of the people in general."[57] In other words, the guideline regarding S&T and S&T plans was not wrong, and that political turmoil could no longer exert lethal impacts on S&T like the Cultural Revolution did.

The CPC had an ever deeper understanding of the effect of S&T on national economy. When Deng Xiaoping met with Czechoslovakia President Husak on September 5, 1988, he put forth the important conclusion that "science and technology are primary productive force," and stressed again that Marx was right in calling science and technology productive force, but that might not be enough—they were primary productive force. At the 4th national congress of China Association for Science and Technology in May 1991, Jiang Zemin highly commented on Deng Xiaoping's scientific conclusion that "science and technology are primary productive force," and pointed out that adhering to this conclusion and truly steering economic construction to the track of depending on S&T progress and better worker quality was an extensive and profound reform. The judgment that science and technology were primary productive force had both scientific and political significance, the latter even more prominent in that it expressed the wish to elevate the political position of S&T and S&T workers in order to quickly improve the S&T level and consequently promote economic development more strongly. On December 19, 1989, the national science and technology reward meeting was held at the Great Hall of the People in Beijing. In his speech titled *Promoting Scientific and Technological Progress is the Historical Mission of the Whole Party and All People*, Jiang Zemin called for always giving S&T strategic priority and boosting economic and social benefits through S&T progress. This provided a new historical opportunity for S&T development.

The S&T system reform also went deeper. It was not only necessary for S&T development but also for political system reform. On January 20, 1987, State Council issued the *Several Provisions Regarding Further Reform of Science and Technology System*, which vowed to further liberalize scientific research institutes, promote lateral integration of scientific research and production on multiple layers and in multiple forms, and push for the close connection of S&T and economy. On May 3, 1988, State Council

[57] Memorabilia of the Communist Party of China—1989. http://cpc.people.com.cn/GB.

issued the *Decision on Several Issues Concerning the Reform of Science and Technology System*, which encouraged research institutes to introduce the competition mechanism, implement the contracting responsibility system in various forms, and separate their ownership from the right of operation and management. S&T system reform was an important content of the 8th five-year S&T plan, which dedicated a whole chapter to it for the first time.

Regarding S&T diplomacy, Premier of the State Council Li Peng visited the Soviet Union in April 1990, the first time a Chinese premier visited the country since 1964 after the bilateral relationship was normalized. The two countries signed six agreements, including the outline of their long-term cooperative development of economy, science, and technology. The normalization of Sino-Soviet relation led to frequent S&T communication and exchanges, but it was no longer a "one-way game" that once happened when the new China was just founded.

2. Proposing the goal of quadrupling the economy

When Deng Xiaoping met with Vice President of Spain Guerra on April 30, 1987, he proposed the three-step strategic goal. According to him, the third step was more important—to quadruple our economy through 30–50 years in the next century, and our per capita GDP will roughly hit USD4, 000. Having achieved that, China will have reached the level of moderately developed countries. From October 25 to November 1, 1987, the 13th CPC National Congress was held in Beijing, when Zhao Ziyang, then general secretary of CPC Central Committee, gave a report titled *Marching ahead along the Socialist Path with Chinese Characteristics*. The congress put forth and systematically elaborated on the theory of the primary stage of socialism, determined the CPC's basic line in this stage, and decided the "three-step" development strategy and various reform tasks. In the report, Zhao Ziyang called for efforts "to put developing science, technology and education in a primary position, and steer economic construction to the track of depending on S&T progress and better quality of workers...Especially in the context of the rapid development of new technological revolution worldwide, we won't be able to shorten our economic and technological distance from developed countries if we don't seize the opportunity and quickly catch up."[58] This report showed that S&T progress and the improvement of management level would fundamentally decide the progress of our modernization drive and was a significant matter concerning national rejuvenation. The primary task of S&T work was vitalizing national economy. We should create mechanisms to closely combine S&T with economy, give enterprises, especially large, leading ones, more motivation and pressure to apply S&T results, push the development of technology market and the commercialization of technical results, and shorten the cycle of applying research results to production. The guideline that "economic construction has to depend on science and technology and scientific and technological work must help economic construction" was further confirmed for S&T development.

In March 1988, acting premier Li Peng proposed in the government work report to increase GNP to around RMB1, 550 billion through the 8th five-year economic development plan, with the annual average growth of about 7.5%. In December 25–30, 1990, the 7th plenary session of the 13th CPC Central Committee was held in Beijing, which pointed out that the period from 1991 to 2000 was extremely critical in the historical process of our socialist modernization drive. During this period, we, on the basis of energetically improving economic benefits and optimizing the economic structure, had to quadruple our GNP at the end of 20th century from 1980 at a constant price, and people would move from having enough to eat and wear to moderate prosperity. The plenary session reviewed and passed the *Suggestions of CPC Central Committee on Formulating the Ten-year*

[58]Marching ahead along the Socialist Path with Chinese Characteristics, report by Zhao Ziyang at the 13th CPC National Congress, October 25, 1987.

National Economic and Social Development Plan and the 8th Five-year Program, listed the basic tasks, guideline and policies for economic and social development in the next decade, and generalized the basic theory and practices of building socialism with Chinese characteristics into twelve principles.[59] From the end of March to early April 1991, the 4th session of the 7th NPC was held in Beijing, at which Li Peng, on behalf of the State Council, gave the *Report on the Ten-year National Economic and Social Development Plan and Outline of the 8th Five-year Program.* He once again set the general goal for 1991–2000: achieving the second-step strategic goal of China's socialist modernization drive and elevating the general quality of national economy to a new level; and quadrupling our GNP at the end of 20th century from 1980 at a constant price on the basis of energetically improving economic benefits and optimizing the economic structure. The *"Ten-year Science and Technology Development Plan and Outline of the 8th Five-year Program of the People's Republic of China (1991–1995–2000)"* was formulated based on the report. In fact, China achieved the annual average economic growth rate of about 11% during the 8th five-year period, its GNP reached RMB5,760 billion in 1995, 4.3 times as much as in 1980 in real terms, and China achieved the "quadrupling" task in advance. Science and technology played an important role in this process and serving the "quadrupling" economic task was the main constraint on the 8th five-year S&T program.

II **Process of formulation**

The 4th session of the 7th NPC held in March 1991 adopted the *Ten-year National Economic and Social Development Plan and Outline of the 8th Five-year Program of the People's Republic of China*, which made overall arrangements for achieving the second-step strategic goal of our socialist modernization drive in the 1990s and specified the goals and guidelines in each sector of economic and social development. On that basis, Premier Li Peng asked State S&T Commission to work out the ten-year S&T plan and the S&T development program during the 8th five-year period as an auxiliary document to the *Ten-year National Economic and Social Development Plan and Outline of the 8th Five-year Program*, to guide our S&T development in the 1990s and contribute to economic vitalization and social development.

In March 1991, according to instructions from Premier Li Peng, the Party Group of State S&T Commission decided to form a leading group headed by executive deputy director Li Xu'e responsible for formulating the ten-year S&T development plan and 8th five-year S&T program, along with a drafting group. The drafting group referred to the ten-year S&T development plans and 8th five-year programs of 33 industries and departments, solicited opinions from more than 40 State Council departments and directly affiliated organizations, and revised and perfected the document several times. In June 1991, a general assembly of State S&T Commission chaired by Song Jian reviewed the *Ten-year Science and Technology Development Plan and Outline of the 8th Five-year Program of the People's Republic of China (1991–1995–2000)* (for review), which was then officially submitted to the State Council. In August that year, Premier Li Peng made instructions on the submitted document, and State S&T Commission, after soliciting opinions of relevant parties, revised the plan and program with some additions and submitted it again to the State Council for review and approval before promulgating it nationwide.

III **Promoting economic development but neglecting basic sciences**

The *Ten-year Science and Technology Development Plan and Outline of the 8th Five-year Program of the People's Republic of China (1991–2000)* formulated in 1991 consisted of five parts—goals and guideline, key tasks, S&T system reform, opening up, supportive conditions and policy measures.

Part 1, "Goals and Guideline for S&T Development in the 1990s" put forth the specific

[59]Memorabilia of the Communist Party of China—1990. http://cpc.people.com.cn/GB.

goal of supporting economic development. "In the next ten years, our S&T work will be based on the goals and deployments of economic and social development, and focus on the main battlefield of economic construction according to the laws governing S&T development. We will work hard to scale S&T heights, yield results and cultivate talents, and fully liberate and exert S&T's role as primary productive force in the modernization drive."[60] The guideline of this plan once again specified "economic construction has to depend on science and technology and scientific and technological work must help economic construction," and vowed to industrialize high technologies in hopes of upgrading the economic structure directly through them. "We must carry out the guideline that 'economic construction has to depend on science and technology and scientific and technological work must help economic construction', uphold science and technology as primary productive force and promote their close connection with economy. High-tech industries will be established at a faster pace."[61] It's worth noting that in 1995 at the end of the 8th five-year S&T program, The CPC Central Committee and State Council issued the *Decision on Accelerating Science and Technology Progress*, which adjusted the guideline for S&T work into "adhere to the idea that science and technology are primary productive force; economic construction has to depend on science and technology and scientific and technological work must help economic development; and strive to scale new heights in science and technology."[62] Therefore, after the mid-1990s, the guideline of S&T plan was changed into "economic construction has to depend on science and technology and scientific and technological work must help economic development; and strive to scale new heights in science and technology." However, basic sciences didn't scale as many heights as the decision makers had wished, and the focus still fell on "economic construction has to depend on science and technology and scientific and technological work must help economic development."

Part 2 was "Key S&T Tasks in the 1990s." The plan put fourth main S&T tasks in four aspects for the 1990s. "Our main S&T tasks in the 1990s are as follows. First, aimed at the main battlefield of economic construction, use S&T, especially electronic information and automation technologies, to renovate traditional industries to make their production technology and equipment modernized and their operation and management scientific. Second, develop high technologies with priorities and push for their industrialization. Third, make substantial achievements in several major areas concerning the man-nature relation, especially population control and environmental protection. Fourth, make notable progress in basic research."[63] The four tasks indicated that in this historical period, the main task of S&T was boosting economic development, while efforts were also made to promote S&T development itself. There were provisions regarding basic sciences, but they were too simple compared with other contents. A result of such utilitarian concept was that the overall strength of our basic sciences declined in the 1990s.

This plan allocated a large part to elaborate on how to develop agriculture, energy, transportation, machinery, light textile, and high-tech industrialization. Regarding transportation, it said "we will focus on making breakthroughs in the design and manufacturing technology of trunk airliners and helicopters in the civil aviation sector, and will be basically capable of designing and manufacturing 150–200-seated trunk airliners and light helicopters by the end of the 20th century."[64] This had something to do

[60]Ten-year Science and Technology Development Plan and Outline of the 8th Five-year Program of the People's Republic of China (1991–2000). Website of the Ministry of Science and Technology: http://www.most.gov.cn/ztzl.

[61]Ibid.

[62]Xin (2003, 260).

[63]Ten-year Science and Technology Development Plan and Outline of the 8th Five-year Program of the People's Republic of China (1991–2000). Website of the Ministry of Science and Technology: http://www.most.gov.cn/ztzl.

[64]Ibid.

with the contention after the large aircraft "Y-10" was called off, and implied that S&T plan in the transitional period was the result of complicated games among various interests. Regarding high-tech industrialization, the plan also listed a number of key areas, including electronic information, computer and software, communication, bioengineering, automation, new-generation energy, new materials, superconductors and laser.

As mentioned earlier, this plan touched little upon basic sciences, which took up less than 3% of the contents, perhaps the least of all S&T plans in terms of content and proportion. The plan set the following goal for basic research in the next decade:

"We will carry out multidisciplinary comprehensive research on strategic areas concerning the national economy, such as agriculture, energy, material and information, as well as on major issues such as population, medicine, resource, eco-environment and natural disaster. On that basis, we will provide the theoretical basis and technical foundation for solving problems and achieve a group of major achievements reaching world advanced level. We will foster a group of scientists with high academic attainments and certain influence on world scientific development, and gradually create a reasonably structured basic research team represented by excellent young and middle-aged academic leaders. We will well develop national key labs and key research institutes engaged in basic research, and gradually establish several cross-disciplinary, comprehensive scientific research centers."

During the 8th five-year period, the main contents and direction of basic research were as follows:"We continue to support scientists in choosing topics freely, the number of which shall be kept around 20,000. Based on their achievements in free research, 200–300 key topics will be selected from basic disciplines and applied basic disciplines that have major influence and great application prospects for economic and social development. For cutting-edge areas where we are likely to achieve substantial breakthroughs within the century and where we have advantages or are in a leading position in the world, or for major basic research areas that have important application prospects and can exert our geographical and resource advantages, the nation will make arrangements by making separate major basic research programs."[65]

Stated above is all there was about basic research in this S&T program. Strictly speaking, it was indeed too simple compared with other S&T programs, resulting in the general decline of China's basic sciences as they required long-term input but didn't necessarily yield results in comparison to other scientific projects. In the general context of pursuing high-speed economic development and GDP centrism, basic sciences were put in a secondary position either consciously or unconsciously, and the utilitarian philosophy of over emphasis on economic benefits resulted in the falling strength of basic sciences in general.

The program mentioned soft science research for the first time, whose main tasks were conducting research on the medium/long-term S&T development strategy, major S&T policies and measures in order to provide a scientific basis for the nation to make macro decisions.

Part 3 of the plan elaborated on the "basic goals and main tasks of deepening the S&T system reform in the new period," a part that didn't appear in previous S&T plans, indicating that S&T system reform had become an important aspect of China's S&T work in this period. This had something to do with the general background that China's political system reform went deeper continuously. The direction of S&T system reform was as follows: "we will strive to establish a new system at the end of the 20th century that's beneficial for economic development and S&T progress, push S&T work to better serve the planned commodity economy, more quickly steer economic construction to the track of depending on S&T progress and better worker quality, and further emancipate science and technology, the primary productive force."[66] The S&T system reform had six main tasks:

[65]Ibid.

[66]Ibid.

reforming planned management, observing value laws and following the market rules; reforming research institutes, deepening the appropriation system reform, introducing the foundation system and contracting system, and selecting and carrying out a batch of key projects in the form of state mandate; promoting S&T progress in enterprises and establishing the mechanism for them to depend on S&T; deepening S&T system reform in the countryside and forming the situation in which big S&T supports big agriculture; deepening the reform of S&T personnel management system; strengthening the macro-control system for S&T progress, and creating new S&T operating mechanism, organizational structure and management model. Meanwhile, legal construction for S&T was also intensified, and efforts were made with the aim of primarily establishing a fairly complete S&T legal system in 2000. This reform was pretty intense and comprehensive, and the reform of the appropriation system had far-reaching impacts, not only affecting the S&T system and changing the decades-long "big pot" system but also setting an example for reforms in other fields. All in all, the S&T system reform integrated S&T into every link of economic construction across the board, and S&T evolved from "serving" to "growing into" economic construction.

The plan also laid down specific provisions regarding S&T input. "In the next 10 years, S&T input should increase faster than GNP, and the proportion of our S&T expenditure to GNP should increase from around 0.7% in 1990 to more than 1.35% in 2000, approaching the level in moderately developed countries...More investment will be made in basic research as well. By the end of the 20th century, basic research funding should account for more than 10% of the total R&D funding."[67] Nevertheless, these targets were not met in the end, which meant there was no legal guarantee for S&T input and that we didn't understand the laws governing S&T development deeply enough, as there was no way to fully predict the relation between S&T development and the investment in it. If a specific ratio of funding increase was given in the plan without the corresponding assurance of the national budget, S&T input was very likely to be less than planned since all departments would be scrambling for resources. This situation appeared repeatedly in later S&T plans. Although S&T plan was a national action, it was tasked with the utilitarian goal of pursuing short-term benefits in the transitional period, a reflection of the utilitarian planning philosophy.

IV **Utilitarian S&T program**

The 8th five-year S&T program made a lot of practical achievements. In agriculture, a large group of major S&T results were achieved, represented by the new crop varieties and improvement of medium/low-yield farmland; in industry, we introduced advanced technologies to narrow our distance from world advanced level in the technology and equipment of certain key industries. We successfully developed the SPC exchange, high-performance parallel computer, retrievable artificial earth satellite and intercontinental carrier rocket, and Computer Integrated Manufacturing System (CIMS). More than ten major scientific projects succeeded, such as the electron-positron collider, and basic research results of international influence were achieved, including high-temperature superconductivity and quasi-crystal with five-fold symmetry.

The 8th five-year S&T program was quite fruitful, especially in promoting economic construction. Since S&T plan during this historical period was implemented through specific S&T programs, we could see the results through those programs. The "plan-program" system with Chinese characteristics was further enhanced. Take the national S&T challenge tackling program for example. During the 8th five-year period, the program arranged 181 projects with the total investment of RMB9 billion, and achieved more than 60,000 S&T results, of which 35% reached world advanced level and 36% reached the leading level in China, along with nearly 800 national patents and 5,000 new products and new processes. The Torch Program, for instance, arranged 6,464 projects during the 8th five-year

[67]Ibid.

period with the accumulative loans of RMB8.74 billion, which increased the total industrial output value by RMB189.9 billion, registered RMB35 billion profits and taxes and created USD3.2 billion foreign exchange. In the same period, the 52 national high-tech industrial development zones approved by the State Council registered the total accumulative income of RMB335.3 billion from businesses related with technology, industry and trade, total industrial output value of RMB296.1 billion, profits and taxes of RMB40.2 billion and foreign exchange of USD5.3 billion through export. As to S&T personnel, there were 2,582,000 people engaged in S&T work nationwide in 1995, an increase of 23.0% from 1990 at the end of the 7th five-year period; there were 1,396,000 S&T scientists and engineers in 1995, an increase of 15.0% from 1991. On the contrary, the number of R&D personnel in research institutes was reduced dramatically, standing at 206,000 people in 1995, which was 31% of all R&D personnel in the country, down from 45.3% in 1990. This meant that the appropriation system reform forced many R&D personnel who used to work in research institutes to leave their former workplace in order to "make a living".

During the 8th five-year period, the national S&T fund increased slowly but that in enterprises increased faster. In 1995, the national S&T expenditure stood at RMB77.86 billion, down 2.8% from 1994 in comparable price. The national R&D funding was on the increase, and enterprises took a larger proportion in it, but the ratio of R&D funding in GDP didn't change much. The goal of raising this ratio wasn't achieved for reasons stated above.

In 1995, 83,000 patent applications were filed, up 6.8% from the previous year; and 45,100 applications were granted, up 4.1% from the previous year. Among that, the number of invention patents granted in the 8th five-year period was reduced by 11.6%, and the number of utility model patents granted in 1994 and 1995 was reduced by 29.6% and 7.2% respectively from the previous year. This in a way indicated that there were fewer inventions that could only be achieved over a certain period of time while research projects that featured short period, low investment and fast results received more attention. Such a utilitarian approach had very negative impacts on long-term S&T development. The 8th five-year S&T program boosted the rapid increase of technology trade in China, and technology import increased by several times in 1995. The national import and export volume of high-tech products exceeded USD10 billion in 1991 and reached USD31.92 billion in 1995.

The 8th five-year S&T program was successful from the perspective of S&T promoting economic development, but it sacrificed scientific accumulation, so much so that S&T didn't have lasting momentum to continue promoting economic construction for a long time to time. In terms of S&T theses—the output of S&T, the number of S&T theses published in China increased amid stability but their quality was to be further improved. In 1995, 108,000 S&T theses were published on 1,231 domestic periodicals nationwide, an increase of 21.7% from 1990. According to statistics based on three search tools—SCI, EI, and ISTP, China produced 26,000 S&T theses in 1995, accounting for 1.95% of all such theses and ranking in the 11th position in the world. But of all Chinese theses collected by SCI in 1990–1995, each of the quoted thesis was quoted 1.78 times on average in 1995, lower than the world average of 2.22 times, which pointed out the lower quality of Chinese S&T theses than the world average.

Besides, the utilitarian philosophy that pursued short-term benefits over long-term interests caused the reduction of major S&T results in the country during the 8th five-year period. In 1995, there were 31,000 major S&T results, a quite low level in general, with only 3,000 national projects and 13,000 provincial and ministerial projects. In particular, the number of results winning national S&T rewards and other high-level rewards decreased accordingly. Regarding S&T results that won the national invention award (Fig. 4.4), there were 937 award-winning projects in the 6th five-year period, 846 ones in the 7th five-year period and only 685 ones in the 8th five-year

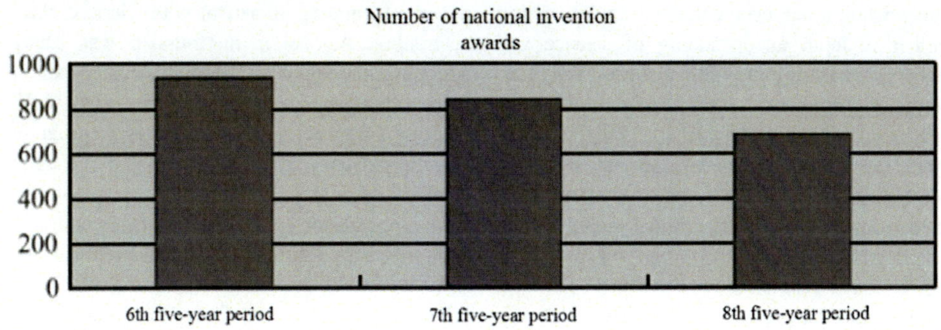

Fig. 4.4 Number of national invention awards from 6th to 8th five-year period

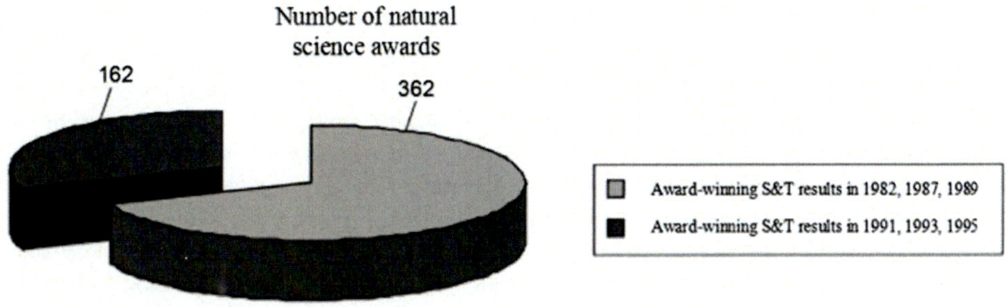

Fig. 4.5 Number of national natural science awards

period. The ratio of S&T results winning first, second and third prizes were also falling over time.

Regarding S&T results that won the national natural science award (Fig. 4.5), there were 362 S&T results winning this award in 1982, 1987, and 1989 in total, but the total number fell to 162 in 1991, 1993, and 1995.

These data showed that China's major S&T results were on the decrease in general, and major theoretical achievements in basic research and breakthroughs in cutting-edge scientific fields were especially rare. This was a clear reflection of the utilitarian philosophy of S&T plans, whereby science and technology were a bit too much taken as a tool to promote economy.

4.5 The "Ninth Five-Year Plan" for National Economic and Social Development and the Outline of Long-Range Goals to 2010

I Deng Xiaoping Theory and Agenda 21

The Eighth Five-Year Plan period has seen breakthroughs made in China's national economy and socialist reform. In terms of political ideas, marked by Deng Xiaoping's "Southern Talks" in 1992 and the 14th National Congress of the Communist Party of China, further in-depth understanding of reform, opening up and modernization has been achieved, thus boosting further in-depth understanding of science and technology. During the "Eighth Five-Year Plan" period, GDP grew by an average rate of 12% a year, reaching more than RMB 5760 billion in 1995, 4.3 times that of 1980 on an inflation-adjusted basis. This was a very important figure at that time, because "the original target of quadrupling GDP in 2000 from 1980 has been achieved five years ahead of schedule." This was an important milestone in the history of China's economic development, as the second strategic goal of modernization was achieved five years ahead of schedule. Significant achievements have been made in key construction projects, with more than 840 large and medium-sized infrastructure projects completed and put into operation. The total rail-laying mileage of the main railway lines exceeded 10,000 km. During the "Eighth Five-Year Plan" period, the construction of water conservancy hubs such as the Three Gorges of the Yangtze River started one after another. A number of water conservancy projects have been completed, including Irrigation in the Huang-Huai-Hai Plain and Diversion of the Datong River to the Qinwangchuan Basin. The rapid development of basic industries and infrastructure has alleviated the bottleneck of economic growth, providing new impetus for future economic development and scientific and technological progress.

From September 12 to 18, 1997, the 15th National Congress of the Communist Party of China was held in Beijing. Jiang Zemin rendered a report entitled "Raise the Great Banner of Deng Xiaoping Theory and Push the Cause of Building Socialism with Chinese Characteristics into the 21st Century." The concept of "Deng Xiaoping Theory" was first used in the report. "Marxism-Leninism has made two historic leaps in combination with China's reality, producing two major theoretical results. The theoretical result of the first leap was the correct theoretical principle and experience summary about Chinese revolution and construction proved by practice. As its main founder was Mao Zedong, the Chinese Communist Party calls it Mao Zedong Thought. The theoretical result of the second leap was the theory of building socialism with Chinese characteristics. As its main founder was Deng Xiaoping, the Chinese Communist Party calls it Deng Xiaoping Theory."[68] The Fifteenth CPC National Congress put forward the basic program for the primary stage of socialism and laid out a strategic plan for cross-century development. One of the major achievements of this congress was the adoption of the Amendment to the Constitution of the Communist Party of China, which clearly stipulates Deng Xiaoping Theory as the guideline of the Party. "The greatest contribution of the congress is to establish Deng Xiaoping Theory as the guideline of the whole party."[69] This showed that China had moved from Mao Zedong's "era of struggle to seize state power" to an "era of building the state". From the theoretical basis, it was further proved that science and technology had shifted from consolidating the new regime to supporting a new historical stage of economic development. The concept of "strengthening political functions" in science and technology planning would gradually transform into "strengthening economic functions".

Discussions on science and technology are found in both Mao Zedong Thought and Deng Xiaoping Theory. One of the important developments is to develop from the notion that "science and technology are productive forces" to the notion that "science and technology are the first productive forces." Science and technology are the first productive forces, and scientific and technological progress is a decisive factor for economic development. We should prioritize the acceleration of scientific and technological progress as a key position in economic and social development so that economic construction really switches to the track of relying on scientific and technological progress and improving the quality of workers. Habermas also put forward the view that "science and technology are

[68]Raise the Great Banner of Deng Xiaoping Theory and Push the Cause of Building Socialism with Chinese Characteristics into the 21st Century, The Fifteenth CPC National Congress, September 12, 1997.

[69]*People's Daily*, September 19, 1997, P. 2.

the first productive forces" almost at the same time, but he looked at the problem more from a scholarly perspective, while Deng Xiaoping drew a conclusion from practice on the height of a statesman.[70]

Economic construction and national development in the transitional period rely more or even largely on science and technology. By the time a national innovation system is built in the 21st century and we entirely rely on the development of science and technology, it would prove that the CPC's understanding and mastery of science and technology have deepened and matured. The proposal to build a national innovation system in the 21st century is a sign that the CCP is maturing in science and technology. Therefore, science and technology and planning for science and technology further serve the national objectives. "We should proceed from the needs of the country's long-term development, formulate a long-term scientific development plan, take a unified view of the overall situation, highlight key points, choose wisely what to do or not to do, strengthen basic research and high-tech research, and accelerate the realization of high-tech industrialization."[71] However, in actual implementation, emphasis on technology is far more important than basic science, because it is difficult for basic scientific research to directly promote economic construction in the short term. This is a utilitarian view on science.

On the economic front, breakthroughs have been made in the reform of the economic system. Financially, a system of tax distribution has been established. Policy finance and commercial finance are initially separated and exchange rates are unified. An overall pattern of opening up to the outside world has basically taken shape. During the "Eighth Five-Year Plan" period, the total import and export volume totaled more than 1 trillion US dollars. At the end of 1995, China's foreign exchange reserves exceeded USD 73 billion. During the "Eighth Five-Year Plan" period, the output of some major products in China increased steadily. Coal, cement, grain, cotton and meat ranked first in the world, steel and chemical fiber ranked second,[72] and electricity generation ranks third. All social undertakings have also developed in an all-round way. Science, technology and education have continued to advance in the reform, with 160,000 national research achievements made in five years. The transformation of scientific and technological achievements into real productive forces has accelerated. Over the 17 years since the reform started in 1978, an economic pattern had been formed in which public ownership was the main body and various economic components were developing together by 1995. This new economic environment put forward different and higher requirements for science and technology planning than before.

On March 25, 1994, the 16th Executive Meeting of the State Council deliberated and adopted the Outline of the National Industrial Policy in the 1990s, and put forward important tasks to be solved by the national industrial policy in the 1990s. Among them, it was proposed to "strengthen the international competitiveness of China's industries; speed up the development of high-tech industries and support the development of emerging industries and new products; and continue to vigorously develop the tertiary industry."[73] This industrial policy exerted a certain impact on the Ninth Five-Year Plan for Science and Technology, as science and technology were further implemented at the industrial level to promote economic development.

In addition, from the international background, China, while developing, is also gradually considering its impact on the world. With the progress of modernization in the world, a series of major global problems concerning human survival have emerged, such as global climate change, the destruction of the atmospheric ozone layer, desertification and the reduction of biodiversity, which have been attracting the attention

[70]Liang (2001).

[71]Zemin (1997).

[72]In 1996, China's steel output exceeded 100 million tons, ranking first in the world.

[73]The Outline of the National Industrial Policy for the 1990 s, adopted by the 16th Executive Meeting of the State Council on March 25, 1994.

of the whole world and the United Nations. The Chinese government took the lead in initiating the Ministerial Conference on Environment and Development of Developing Countries in Beijing in June 1991, at which the Beijing Declaration was adopted. It showed the Chinese government's principled stand on environment and development. At the 1992 United Nations Conference on Environment and Development, a number of agreements including Agenda 21 were adopted. Premier Li Peng made a commitment on behalf of the Chinese government to implement Agenda 21 and other documents.

In August 1992, a leading group headed by the State Planning Commission and the State Science and Technology Commission was set up to organize working groups with 52 departments and more than 300 experts. The China Agenda 21 Management Center[74] was also set up to undertake routine management. In March 1994, the State Council adopted China's Agenda 21, that is, the White Paper on China's Population, Environment and Development in the 21st Century.[75] China's Agenda 21 is an overall strategy and policy plan to promote the coordination and sustainable development of economy, resources and environment based on China's specific reality and construction needs. It is a guiding document for the formulation of China's medium-and long-term programs for national economic and social development. The strategy of sustainable development has become a basic development strategy in China, which has become a moral constraint for the five-year economic plan of the nation and the people and the scientific and technological plan in the future. In fact, with the deepening of economic development and reform, soft and hard constraints are increasing for science and technology planning and the five-year science and technology plan.

II The Strategy of Rejuvenating the Country through Science and Education and the National Conference on Technological Innovation

1. The Decision to Speed up Scientific and Technological Progress

On May 6, 1995, the CPC Central Committee and the State Council made the "Decision of the CPC Central Committee and the State Council on Accelerating the Progress of Science and Technology",[76] which pointed out that the idea that "science and technology are the first productive forces" should be fully implemented while putting forward the strategy of rejuvenating the country through science and education. "The CPC Central Committee and the State Council have decided to unswervingly implement the strategy of rejuvenating the country through science and education, which means to fully implement the idea that science and technology are the first productive forces, to adhere to education as the basis, to prioritize science, technology and education in economic and social development … to shift economic construction to the track of relying on scientific and technological progress and improving the quality of workers, and to accelerate the realization of national prosperity and prosperity."[77] Implementing the strategy of rejuvenating the country through science and education is a strategic decision to fully implement the idea that science and technology are the first productive forces. After the Third Plenary Session of the Eleventh Central Committee, the focus of the Party's work shifted to economic construction as the center,

[74] The China Agenda 21 Management Center has become an institution directly affiliated to the Ministry of Science and Technology. See http://WWW.acca21.org.cn.

[75] China's Agenda 21 consists of 24 chapters and 78 program areas, with more than 200,000 words. There are four parts in total: Part I: the overall strategy of sustainable development. The agenda puts forward the development goals of major industries, social development goals and sustainable development measures for realizing the above goals by the year 2000. Part II: Sustainable Development of Society. Part III: Sustainable Economic Development. Part IV: Rational Utilization of Resources and Environmental Protection. For implementation, the short-term target (1994–2000), medium-term target (2000–2010) and long-term target (after 2010) are proposed are proposed.

[76] Decision of the CPC Central Committee and the State Council on Accelerating Scientific and Technological Progress, Zhongfa [1995] No. 8 Document, May 06, 1995.

[77] Ibid.

and the implementation of the strategy of rejuvenating the country through science and education was deepening this shift and development to a higher stage, which would surely bring about a new leap in productivity.

Through practices in reform, the CPC Central Committee has gradually gained a deeper understanding of the important role of science, technology and education in the country. Therefore, it has put forward the strategy of rejuvenating the country through science and education. Science and education cannot save the nation, but they can rejuvenate it. This document makes more specific arrangements on how science can support economic development in the hope that science and technology will become a main driving force for China's economic construction and national development.

"The basic policy of China's scientific and technological work is to adhere to the idea that science and technology are the first productive forces. Economic construction must rely on science and technology. Scientific and technological work must be addressed to economic construction and aimed at scientific and technological peaks.

In actual work, we should grasp the following principles:

- In economic and social development we should take scientific and technological progress as a main driving force, and in scientific and technological work we should prioritize solving major problems in economic and social development.
- We should take reform as a driving force for scientific and technological development and deepen the reform of the scientific and technological system in the process of development.
- We should integrate independent research and development with the introduction of advanced foreign technologies.

We should combine long-term and short-term goals.

- We should limit goals, highlight key points, concentrate efforts, overcome key problems and dare to innovate.
- We should respect knowledge, respect talents, and create a social environment in which people make the best use of their talents and talents come forth in large numbers.
- We should combine research and development with mass scientific and technological activities.[78]

It is worth noting that although science and technology were highly valued and shouldered an important historical mission, there was obviously not enough emphasis on the development of science itself, especially the development of basic science. Of the nearly 12,000-word document, only more than 800 words were on basic science, accounting for about 6.7%. This shows that the understanding of basic science was not deep enough, while the utilitarian understanding of science made even this small part of the content focus on "catching up with some and not all." "In the current period, basic research should prioritize national goals, take providing power for national economic and social development as its central task, focus on solving basic theoretical and technical problems of future economic and social development, and create new technologies and methods." This statement shows that utilitarian concepts on science and technology planning had a negative impact on the development of basic science. Due to limited resources, it was necessary to have priorities and choices for development. However, excessive pursuit of economic benefits resulted in prioritizing the development of the economic functions of science and technology to the extent of excessive utilitarianism in some places.

2. National Conference on Science and Technology

In order to implement the Decision of the CPC Central Committee and the State Council on Accelerating the Progress of Science and Technology and, more importantly, to promote the strategy of rejuvenating the country through science and education, the CPC Central

[78]Ibid.

Committee and the State Council held a National Conference on Science and Technology in Beijing from 26 to 30 May 1995. Jiang Zemin, then President of China, made a speech at the conference. The spirit of Jiang Zemin's speech was to promote the implementation of the Decision of the Central Committee of the Communist Party of China and the State Council on Accelerating the Progress of Science and Technology. He pointed out, "At present, the full implementation of the idea that science and technology are the first productive forces is a very arduous and urgent strategic task in China's socialist modernization drive, both in terms of international environment and domestic conditions. To rejuvenate the country through science and education is the right choice to successfully achieve the three-step strategic goal."[79] Jiang Zemin believes that the guiding principle of science and technology, that is, "economic construction must rely on science and technology, work of science and technology must address economic development, and efforts must be made to scale the peaks of science and technology," should center on "the close combination of science and technology and economy." What is the most important task of science and technology? At the conference this was defined as that work on science and technology must be more consciously oriented towards economic construction, with promoting economic development as the central task and primary goal. In work on science and technology economic construction should always be taken as the main battlefield and the key issues that need to be solved urgently be taken as main tasks. Nowhere else has a stronger thirst for science and technology than in China. As a result, the instrumental function of science and technology to support the development of national economy, that is, economic function, has been highlighted.

It is worth noting that Jiang Zemin's speech presented a further understanding of innovation. "Innovation is the soul of a nation's progress and an inexhaustible motive force for the country's prosperity. If we do not improve our indigenous innovation capability but rely solely on the introduction of technologies from overseas, we would never be able to get rid of technological backwardness."[80] Despite the fact that it had not been proposed to build a national innovation system then, the CPC Central Committee had a deep understanding of the national innovation system, which laid a foundation for science and technology planning and the construction of the national innovation system at the beginning of the 21st century. This conference required Party committees and governments at all levels to conscientiously implement the "Decision of the Central Committee of the Communist Party of China and the State Council on Accelerating Scientific and Technological Progress"[81] and "Several Opinions of the Central Committee of the Communist Party of China and the State Council on Strengthening the Popularization of Science and Technology" as well as the "Outline of China's Educational Reform and Development"[82] to promote the realization of the strategic objectives of the second and third steps of China's modernization.

This National Science and Technology Conference was of such great significance to the development of science and technology and the formulation of science and technology planning in China that it enjoyed the same historical status as the 1978 National Science Conference and the 2006 National Science and Technology Conference. The convening of a national science and technology conferences in different historical periods, with the participation of the highest leaders of the party and the country, on the one hand, reflects the importance of science and technology, on the other hand, also reflects the understanding of the CPC Central Committee on the relationship between science and technology and economy in different historical periods. The 1978 National Science Conference reflected the high expectations pinned on the scientific community and science and technology during the period of restoring things to order and the Science and Technology Planning 1978–1985 was

[79]Zemin's (2006).

[80]Zemin's (2006).

[81]Zhongfa [1994] No. 11 document.

[82]Zhongfa [1993] No. 3 document.

subsequently issued. The 2006 National Conference on Science and Technology, the first science and technology conference in the 21st century, not only reflected the emphasis on science and technology by the new generation of CPC Central Committee leadership at that time but also endowed science and technology with a historical mission in the new century. In particular, it was of great significance to promote the "2006–2020 Medium and Long-Term Science and Technology Program." The 1995 National Conference on Science and Technology clearly put forward the strategy of "rejuvenating the country through science and education," which played an important guiding role in the formulation and implementation of the "National Ninth Five-Year Plan for Science and Technology Development and the Outline of the Long-Range Plan to 2010."

3. National Conference on Technological Innovation

Four years after the National Conference on Science and Technology was held, the CPC Central Committee convened the National Conference on Technological Innovation. Through the practices of reform, the CPC Central Committee had further realized the direct effect of technology on economic development. On August 20, 1999, the CPC Central Committee and the State Council made the Decision of the CPC Central Committee and the State Council on Strengthening Technological Innovation, Developing High Technology and Realizing Industrialization. The main contents of the Decision included: strengthening technological innovation, developing high technology, realizing industrialization and promoting the leapfrog development of social productive forces; deepening system reform and promoting technological innovation and commercialization and industrialization of high-tech achievements; taking effective measures to create a policy environment conducive to technological innovation, development of high technology and industrialization; strengthening the leadership of the party and the government, comprehensively promoting technological innovation, developing high technology and realizing industrialization. The National Conference on Technological Innovation was held in Beijing from August 23 to 26, 1999 on the theme of "Strengthening Technological Innovation, Developing High Technology and Realizing Industrialization." Similar to the 1995 National Conference on Science and Technology, the CPC Central Committee issued the Decision of the CPC Central Committee and the State Council on Strengthening Technological Innovation, Developing High Technology and Realizing Industrialization[83] in advance, and then convened a national science and technology conference to implement this decision. This showed that the CPC Central Committee had a further in-depth understanding of technology and technological innovation.

At that time, it was believed that the goal of the reform on the scientific and technological system was to realize technological innovation, develop high technology and realize industrialization. Despite obvious progress made in the reform of the scientific and technological system, the problem of disconnection between science and technology and economy had not been fundamentally solved. The lack of ability to transform science and technology into real productive forces and the low level of industrialization of high and new technologies were still major obstacles to the economic development of China. We should strengthen technological innovation, develop high technology and realize industrialization. This was not only a solution to the deep-seated problems facing our country's economic development but also a strategic choice to further improve the overall quality of the national economy and the overall national strength.

In fact, since the start of the 21st century, the goal of the reform on the scientific and technological system has been to establish a national innovation system. It can be seen that the CPC Central Committee has gone through a process of understanding science, technology, and innovation. As for technological innovation, the document at that time gave the following definition,

[83]Zhongfa [1994] No. 14 document.

"Technological innovation refers to the application of innovative knowledge and new technologies and processes by enterprises, the adoption of new production methods and management modes, the improvement of product quality, the development and production of new products, the provision of new services, the occupation of the market and the realization of market value. Enterprises are the main body of technological innovation. Technological innovation is an important prerequisite for developing high technology and realizing industrialization." At the beginning of the 21st century, the construction of the national innovation system was described as follows, "We should establish a market-oriented technological innovation system with enterprises as the main body, and featuring the combination of production, study and research." Obviously, the understanding of technology and technological innovation has gone through a dynamic evolution process.

Jiang Zemin pointed out at the National Conference on Technological Innovation that "the core of competition in the comprehensive national strength of all countries in the world today is knowledge innovation, technological innovation and industrialization of high and new technologies."[84] Despite its being a linear understanding of science (knowledge)-technology-economy (industrialization), this statement vested technological innovation with more missions. A deep understanding of technological innovation was not only a theoretical result after the implementation of the Ninth Five-Year Plan for Science and Technology but also had an impact on the Tenth Five-Year Plan for Science and Technology. "At present, we are studying and formulating the 10th Five-Year Plan and the medium and long-term development plan. From guidelines, strategic priorities to policy measures, we should fully consider relying on scientific and technological progress and innovation as a key factor to promoting economic and social development."[85]

Premier Zhu Rongji pointed out in his speech at the closing session of the "National Conference on Technological Innovation" that strengthening technological innovation and realizing industrialization were directly related to the process of China's socialist modernization. "A country's overall national strength and international competitiveness depends on the level of science and technology and the ability of technological innovation ... The main direction and focus of China's technological innovation in the future must be clearly defined."[86] Generally speaking, on the one hand, it is necessary to transform and improve traditional industries with high and new technologies to promote the upgrading of traditional industries; on the other hand, we should lose no time to accelerate the development of high-tech industries with market demand and prospects, drive and promote the rise of emerging industries, and take the road of technological leapfrog development with Chinese characteristics.

The convening of the National Conference on Science and Technology in 1995 and the National Conference on Technological Innovation in 1999 showed that the CPC Central Committee had further deepened its understanding of science and technology during the Ninth Five-Year Plan period, especially with regard to the tremendous role of high technology and industrialization in promoting economic development. The combination of science and technology with the economy and the combination methods were further clarified. This is the macro background for the formulation and implementation of the Ninth Five-Year Plan for Science and Technology.

III The Formulation of the "Ninth Five-Year Plan" for the National Economy and the "Ninth Five-Year Plan" for Science and Technology

Since the reform and opening up, every scientific and technological plan has been incorporated into the five-year development plan of the

[84] Zemin (1999).

[85] Zemin (1999).

[86] Rongji. (1999).

national economy in the same period, serving the economic construction as the center. The "Ninth Five-Year Plan" for Science and Technology was also an integral part of the "Ninth Five-Year Plan" for the National Economy, so it is necessary to examine the "Ninth Five-Year Plan" for the National Economy.

From September 25 to 28, 1995, the Fifth Plenary Session of the 14th CPC Central Committee was held in Beijing. The plenary session deliberated and adopted the "Proposal of the CPC Central Committee on Formulating the Ninth Five-Year Plan for National Economic and Social Development and the Long-Range Target for 2010." The full text consists of six parts: (1) An important period of China's national economic and social development; (2) Main objectives and guiding principles; (3) Main tasks and strategic layout of economic construction; (4) Main tasks and arrangements of reform and opening up; (5) Main tasks and basic policies of social development; (6) The whole Party and the people of all ethnic groups in the country unite to strive for the realization of the Ninth Five-Year Plan and the long-range goals for 2010. Li Peng, then Premier, explained the draft proposal. It was suggested that the key to achieving this goal was to implement two fundamental changes of overall significance. One was the transformation of the economic system from the traditional planned economic system to the socialist market economic system, and the other was the transformation of the mode of economic growth from extensive to intensive. These two fundamental changes were rated by *China Economic Weekly* as one of the 30 most influential events in 30 years.[87]

Li Peng pointed out in his report to the fourth session of the Eighth National People's Congress on March 5, 1996 that during the "Eighth Five-Year Plan" period, the GDP grew by an average rate of 12% a year, and the target of quadrupling the 1980 figure in 2000 had been achieved five years earlier. In the Ninth Five-Year Plan the strategy of rejuvenating the country through science and education and the strategy of sustainable development would continue to be implemented. "The implementation of these two strategies is of great significance to the development in the next 15 years and even to the realization of the whole modernization project. Economic construction must rely on science and technology, and scientific and technological work must address economic construction … We should strengthen basic scientific research and strive to make breakthroughs in advantageous fields. According to the policy of "anchoring one end securely, while letting the other roam free," we should optimize the scientific research organization structure.[88] It can be concluded from the report that the strategy that "economic construction must rely on science and technology, work of science and technology must address economic development, and efforts must be made to scale the peaks of science and technology "has become the core of the strategy of rejuvenating the country through science and education, while the core of the core is technological innovation and its industrialization.

The Proposal of the Central Committee of the Communist Party of China on Formulating the Ninth Five-Year Plan for National Economic and Social Development and the Long-Range Target for 2010, which was deliberated and adopted by the Fifth Plenary Session of the 14th CPC Central Committee. It was adopted as a national plan at the fourth session of the 8th National People's Congress in March 1996, namely the Ninth Five-Year Plan for National Economic and Social Development of the People's Republic of China and the Outline of Long-Term Goals to 2010. The Outline drawn up in accordance with the spirit of the Central Committee's Proposal will become the will of the state and the whole people after being approved by this meeting. It will unite and inspire the people of all ethnic groups in the country to work together for a better future."[89] It was proposed in the Outline that the main goal of national economic and social development during the "Ninth Five-Year Plan" period was to

[87]The outline officially released to the public consists of eleven parts.

[88]Peng (1996).

[89]People's Daily (1996).

complete the second step of strategic deployment of modernization in an all-round way and to quadruple the per capita GNP in 2000 when China's population would increase by about 300 million from 1980. Poverty would be basically eliminated and the people would live a relatively comfortable life. We will speed up the construction of a modern enterprise system and initially establish a socialist market economy system. The main goal of 2010 was to double the GNP compared with 2000, to enable people to live a better life and to form a relatively perfect socialist market economic system. Science and technology still played an important role in achieving this magnificent economic and social goal.

The Ninth Five-Year Plan of the National Economy and the Outline of the Long-range Goals for 2010 planned for social development. Under this broad framework, the Ninth Five-Year Plan and the Long-range Plan for 2010 for Science and Technology had been formulated. Therefore, the country's five-year plan is the main standard for science and technology planning. Science and technology planning cannot go beyond this big framework. The national economic plan must be added to the "science and technology planning-science and technology plan" system to form the model shown in Fig. 4.6.

From the beginning of the founding of new China to the end of the Cultural Revolution, attempts were made to incorporate science and technology planning into economic planning, but there was always overlapping in time and content. Since the reform and opening up, it has been a general feature that science and technology planning is incorporated into the national economic planning: science and technology planning serves the national economic planning.

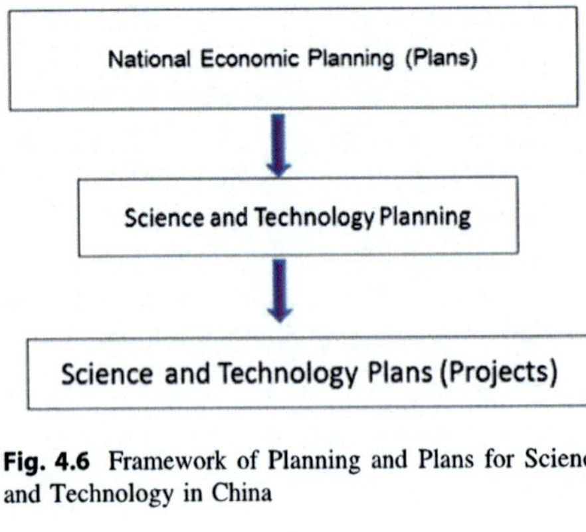

Fig. 4.6 Framework of Planning and Plans for Science and Technology in China

In accordance with the spirit of the Ninth Five-Year Plan for National Economic and Social Development and the Outline of Long-Range Goals to 2010 and the Decision of the Central Committee of the Communist Party of China and the State Council on Accelerating the Progress of Science and Technology, the State Science and Technology Commission, the State Economic and Trade Commission, the State Planning Commission and other relevant departments formed an inter-ministerial coordination group starting from the practical implementation of the idea that "science and technology are the first productive force" and the national implementation of the strategic objectives of "rejuvenating the country through science and education" and "sustainable development". After more than two years of research on the major issues involved in the preparation of the science and technology plan, the Ninth Five-Year Plan for National Science and Technology Development and the Outline of Long-Range Goals to 2010 were drafted and prepared.

In response to the national Ninth Five-Year Plan for Science and Technology, the Chinese Academy of Sciences completed the formulation of the Outline for Science and Technology Development in the Ninth Five-Year Plan and 2010 of the Chinese Academy of Sciences in August 1994. The science and technology planning of the Chinese Academy of Sciences focuses on scientific and technological work during the Ninth Five-Year Plan period, including the development of basic research on important scientific frontiers and applied basic research with major backgrounds, contributing to the technological innovation and structural adjustment of basic industries and pillar industries; vigorously developing high-tech technology; providing a scientific basis for coordinating the relationship between man and nature and promoting the sustainable development of China's social economy; providing science and technology for agricultural modernization and sustainable development.

After the Ninth Five-Year Plan for the National Economy was formulated, all trades and industries formulated their own "Ninth Five-Year Plan". For example, the State Bureau of Surveying and Mapping formulated the "Ninth Five-Year Plan" for the Development of Surveying and Mapping Science and Technology and the 2010 Long-Range Plan.[90] The guideline is: to implement the policy that "science and technology are the first productive forces," to fully implement the strategy of "developing survey through science and education and gradually realizing the industrialization of geological information "to continue to implement the guiding principle that economic construction must rely on science and technology (relying on), work of science and technology must address economic development (addressing), and efforts must be made to scale the peaks of science and technology (climbing peaks), to focus on the overall goal of surveying and mapping development, to make full use of both planning and market methods, to promote the scientific and technological progress of surveying and mapping, and to form a development pattern for the integration of science, technology and economy. The "Ninth Five-Year Plan" for Surveying and Mapping and the "Ninth Five-Year Plan" for Science and Technology overlapped to some extent, which also illustrated the influence of science and technology planning at the national level on the planning of various sub-sectors.

IV "Addressing and Relying on" Outweighs "Climbing Peaks"

On the whole, the "Ninth Five-Year Plan" for Science and Technology is not as powerful as the long-span plans for science and technology made in 1956 and 1986. Judging from the text alone, the full text of the "Ninth Five-Year Plan" for National Science and Technology Development and the Outline of Long-range Goals to 2010 (1996–2000–2010) is about 10,000 words, weaker in depth, breadth, and content than the nearly 60,000 words in the 1956 science and technology plan and the nearly 40,000 words in 2006 science and technology plan.

The full text of the "Ninth Five-Year Plan" for Science and Technology consists of seven parts: the situation and current situation; guideline and basic principles; development objectives and tasks; development priorities; deepening reform and establishing a new system for science and technology; accelerating talent training and building a high-level team of science and technology across the century; and supporting conditions and measures.

Generally speaking, the ninth five-year plan for science and technology continued the utilitarian philosophy and practical approach in science and technology planning since the transitional period. In the section "Form and Present Situation", the outline clearly pointed out that "the strategic focus of scientific and technological work has turned to national economic construction."[91] After noting that "on the whole, China's scientific and technological work has initially acquired the ability to support economic and social development and participate in international competition," it also pointed out that China's scientific and technological work still had many problems and gaps, mainly manifested in: the contribution of scientific and technological progress to economic growth was relatively low, economic construction had not yet shifted to the track of relying on scientific and technological progress and improving the quality of workers, and the national economy mainly relied on the development mode of extended reproduction. In particular, the plan pointed out that "science and technology have insufficient

[90]The Ninth Five-Year Plan for the Development of Surveying and Mapping Science and Technology and the Long-Term Plan for 2010. *China Surveying and Mapping*, 1997, (1).

[91]Ninth Five-Year Plan for National Science and Technology Development and the Outline of Long-Range Goals to 2010(1996—2000—2010)[DB/OL]. The website of the Ministry of Science and Technology of China: http://www.most.gov.cn/kjgli/lskjgh.

stamina for their own development. The research level is on a downward trend, with few major achievements in theory and breakthroughs in scientific frontiers. Outstanding academic and engineering leaders are especially scarce."[92] From this, we can see that the fact that basic science was neglected was already noticed at that time. However, it did not show the reflection or even negation of the utilitarian concept of science and technology planning, but it may be deemed as a precursor of a new awakening of consciousness.

The guideline of the outline continues to follow the guiding principle of "addressing, relying on, and climbing peaks." However, higher requirements are put forward for science and technology to support economic development and national construction. "The development of science and technology should prioritize improving China's economic competitiveness and increasing the market share of high-tech products. The development of science and technology should reflect the national industrial policy, and be conducive to the adjustment of industrial structure and the optimization of technological structure. The main task is to overcome the hot and difficult issues that need to be solved urgently in the national economic and social development.... In the development of science and technology we should set limited goals, highlight important points, and choose wisely what to do and what not to do."[93]

This clearly shows that in planning science and technology are mainly regarded as an important tool to promote economic development and national construction, and the realization of national economic interests is the main goal of science and technology planning. This is consistent with the view expressed in an article titled "Science: Endless Frontiers". However, the pursuit of economic benefits is more urgent in contemporary China, resulting in a very strong color of utilitarianism. Short-term goals overtake long-term goals in practice and planning implementation. The pursuit of materials is also the pursuit of wealth, greater than the pursuit of people, that is, the pursuit of human values. The historical merit of this utilitarian concept must be affirmed. As pointed out in this plan, while science and technology serves economic construction, its own strength has been continuously strengthened. However, in the transitional period, the problem of excessive utilitarianism in science and technology planning should be rethought sooner or later.

In "Development Goals and Tasks", the utilitarianism was also reflected in the linear understanding of the development laws of science and technology. Understanding beyond the law of development of things was written into the outline, reflecting a strong subjective volition. The strategic goal of the "Ninth Five-Year Plan" for Science and Technology and the long-term plan for 2010 was to initially build a new science and technology system by 2000 adapted to the socialist market economic system and the law of development of science and technology, so that China's overall level of science and technology would rank among the top ten science and technology powers in the world. In fact, the new scientific and technological system had not been fully realized by 2000, because on the whole, China's scientific and technological strength probably was not among the top ten in the world by 2010. Subjective good wishes overtook the law of development in science and technology. Another example was that the ratio of R&D to GDP was required to reach 1.5% by 2000, but was actually not achieved.

In the priorities of development, science and technology were required to provide more support for the national economy. It was hoped that science and technology would provide technical support in the development of the pillar industries of machinery, electronics, and construction. Worth noting was that the "Ninth Five-Year Plan" for Science and Technology, like the "Eighth Five-Year Plan" for Science and Technology, continued to vigorously promote high-tech industries and regarded high-tech industries

[92]Ibid.

[93]Ninth Five-Year Plan for National Science and Technology Development and the Outline of Long-Range Goals to 2010(1996—2000—2010) [DB/OL]. The website of the Ministry of Science and Technology of China: http://www.most.gov.cn/kjgh/lskjgh.

as the core and commanding heights of international competition. "During the Ninth Five-Year Plan period, mainly oriented by market, efforts would be focused on key areas of strategic significance to the national economy, people's livelihood and national defense construction. The implementation of a number of high-tech research and development projects would be organized in an effort to achieve major breakthroughs and innovations and occupy a place in the world's high-tech research."[94] "The sudden emergence of new and high technologies was a prominent feature of the development of science and technology in the second half of the 20th century. The integration of science and technology was most obvious in the field of high technology. The implementation of the 863 Program had enabled our country to narrow the gap from developed countries in the field of high technology. The "Eighth Five-Year Plan" and the "Ninth Five-Year Plan" for science and technology highlighted the discussion and arrangement of high-tech fields, which reflected that such plans had correctly followed the trend of this era and also showed that science and technology planning was an important means to realize core national interests.

In addition, the "Ninth Five-Year Plan" for Science and Technology did not elaborate on basic science, just like the "Eighth Five-Year Plan". In the 10,000-word text, less than 500 words were on basic science. For basic science, it was also necessary to serve the core objective of supporting economic development. "Basic research should be based on the needs of the country's economic construction and social development as well as the development trend of modern science, follow the principle of 'catching up with some but not others', prioritize national goals, aim at the international frontier and strive to scale scientific and technological peaks … Select some projects with certain advantages in China that are expected to have greater impact, to which greater investment intensity and sustained and stable support should be given. During the Ninth Five-Year Plan period, a number of major scientific projects with great scientific significance, important influence and driving effect on related disciplines were gradually established."[95] This showed that the plan for science and technology paid some attention to the free research of scientists, but this freedom must be based on the realization of national goals. On August 26, 1997, Jiang Zemin instructed in the Minutes of the Third Meeting of the National Science and Technology Leading Group that basic research was very important and basic scientific research should continue to be strengthened. Facing the 21st century, we should select some high-tech projects of strategic significance to China's economic and social development, pool necessary human, financial and material resources, and establish key bases.

Generally speaking, the "Ninth Five-Year Plan" for Science and Technology centered on the core of economic construction and the promotion of national strength, in line with the "Ninth Five-Year Plan" and supporting the "Ninth Five-Year Plan" in terms of science and technology. No matter from the textual analysis or the actual implementation of the plan, the contents about "addressing and relying on" were obviously more than those about "climbing peaks". With the former two measures, science and technology played an important role in promoting economic construction, but the latter measure was neglected. On the one hand, the overall strength of China's basic science did not grow as fast as the growth of technology categories in the 1990s. On the other hand, the utilitarian philosophy of science and technology planning was further developed and extended, covering the entire Chinese science and technology system, and exerting a direct impact on China's academic environment. This impact will be discussed later.

[94]Ibid.

[95]Ninth Five-Year Plan for National Science and Technology Development and the Outline of Long-Range Goals to 2010(1996—2000—2010)[DB/OL]. Website of the Ministry of Science and Technology of China: http://www.most.gov.cn/kjgli/lskjgh.

V. The "Ninth Five-Year Plan" for Science and Technology Promoted the National Economic Plan

1. Significant Economic Achievements

The "Ninth Five-Year Plan" for Science and Technology was a science and technology plan within the framework of the "Ninth Five-Year Plan" for National Economy. How effective was the "Ninth Five-Year Plan" for National Economy? The five years since the implementation of the "Ninth Five-Year Plan" were also the five years when China's national economy actually progressed by leaps and bounds. Some medium- and long-term targets for national economic development had been achieved ahead of schedule. After realizing the goal of quadrupling the GNP in 1995 over 1980, and through the efforts of the first two years of the Ninth Five-Year Plan, the task of quadrupling the GNP per capita over 1980 outlined in the Ninth Five-Year Plan was achieved three years in advance in 1997. In 2000, China's GDP reached RMB 8.78 trillion, exceeding USD 1 trillion converted at the exchange rate at that time. This showed that China had fully completed the second strategic deployment of modernization. Science and technology played an important supporting role. From this perspective, the "Ninth Five-Year Plan" had achieved great success.

For a scrutiny of the Ninth Five-Year Plan for Science and Technology, we only need to look at the achievements from 1996 to 2000. Although the title of the plan was the "Ninth Five-Year Plan for National Science and Technology Development and the Outline of Long-Range Goals to 2010(1996–2000–2010)," suggesting that the plan would be implemented until 2010, in fact, it was mainly implemented during the "Ninth Five-Year Plan" period because the long-term goal for 2000–2010 was only an ideal description. After entering the 21st century, the new "Tenth Five-Year Plan" would replace the implementation of the long-term goals in the "Ninth Five-Year Plan". Therefore, with an examination of the scientific and technological achievements during the period of the "Ninth Five-Year Plan," we can see the implementation effect of the "Ninth Five-Year Plan for National Science and Technology Development and the Outline of Long-Range Goals to 2010 (1996–2000–2010)."

During the Ninth Five-Year Plan period, with the enhancement of the role of science and technology in economic growth and sustainable development, governments at all levels further increased their investment in science and technology. Since the implementation of the "Ninth Five-Year Plan" for Science and Technology, 69,000 papers had been published at home and abroad and 3,697 patents had been granted. With the advancement in agricultural science and technology, more than 600 new varieties had been cultivated, with a yield increase of about 10% per unit area. With the promotion of dry cultivation and sparse planting of rice and water-saving technologies, China's grain yield increase target was realized during the Ninth Five-Year Plan period.

A number of major technological breakthroughs were made in industrial science and technology, and the technological level of key industries in China was upgraded. A series of important achievements were made in the industrialization of digital SPC exchange, oxygen-coal reinforced steel-making technology, nickel-hydrogen batteries, and amorphous materials, etc. Around the Three Gorges Project, development of a national information-based economy, integrated circuits, Qinshan Nuclear Power Station Phase II and a series of other major national construction projects, a number of key technological problems were overcome and a number of major complete sets of technology and equipment design and manufacturing technologies were mastered. The popularization and application of a number of major common technologies such as computer-aided design (CAD) and computer-integrated manufacturing system (CIMS) greatly improved the technological innovation capability of enterprises. A number of major projects such as innovative drugs and the Xia, Shang and Zhou Chronology Project contributed to social development.

In terms of high-tech research and industrialization, major breakthroughs were made in high-tech space technologies such as manned space technology, launch vehicles and satellite technology. With breakthroughs made in two-line hybrid rice, genetic engineering drugs, gene sequencing and diagnosis and treatment of major diseases, the overall level of biotechnology in China was raised close to that of developed countries. With major achievements such as high definition television, "Shenwei" computer, large-scale mono-crystalline silicon, skin stem cell regeneration technology, China leaped into the world's advanced ranks in the corresponding fields.

In the field of basic scientific research, major achievements were made in human gene sequencing, carbon nanotubes and new Nano materials, Cambrian explosion research, micro-electromechanical systems research, and South China Sea deep sea drilling. Emerging interdisciplinary fields such as surface science, non-linear science, cognitive science and earth system science developed rapidly. The construction of eight major national scientific projects, including the Chinese Continental Scientific Drilling Project and the large sky area multi-object fiber spectroscopic telescope (LAMOST) etc., created favorable conditions for basic scientific research in China.

2. Knowledge Innovation Project

In the second half of 1997, the Academy of Sciences organized forces to study China's strategic measures facing the 21st century and the era of knowledge economy, and put forward a research report entitled "Meeting the Era of Knowledge Economy and Building a National Innovation System," which was submitted to the CPC Central Committee. Suggestions on deepening the reform of science and technology system and building a national innovation system were put forward. This report consisted of four parts: knowledge economy and the opportunities and challenges faced by China; building China's national innovation system facing the 21st century; organizing the implementation of the "knowledge innovation project"; and strategic choice of the Chinese Academy of Sciences in the new era. The report put forward the idea of building China's national innovation system, including the knowledge innovation system, the technology innovation system, the knowledge dissemination system and the knowledge application system, as well as a network system composed of relevant institutions and organizations, their actors including enterprise R & D institutions, education and training institutions and government departments.[96] The formulation of the national innovation system by the Chinese Academy of Sciences was relatively close to the five major systems in the national innovation system proposed later. In the report, it was believed that organizing the implementation of the "knowledge innovation project" was a key link in the national innovation system. However, in the 11th Five-Year Plan, the construction of a "technological innovation system" was a breakthrough to build a national innovation system. It can be concluded that the Chinese Academy of Sciences and the central government had different understandings of the national innovation system. In the report it was recommended that the first phase of the "knowledge innovation project" be piloted in the Chinese Academy of Sciences. The knowledge innovation project indeed contributed significantly to the proposal and improvement of the national innovation system.

On February 4, 1998, then General Secretary of the CPC Central Committee and President Jiang Zemin gave important instructions in a report: knowledge economy and innovation consciousness are of vital importance to our development in the 21st century. It was agreed that the Chinese Academy of Sciences would carry out the pilot work. On the morning of June 9, 1998, Premier Zhu Rongji, head of the National Science and Technology Education Leading Group, held the first meeting of the Leading Group, in which the Outline for the

[96]The History of the Chinese Academy of Sciences,1997 [DB/OL]. Website of the Chinese Academy of Sciences: http://www2.cas.cn/.

Report on the Pilot Work of the "Knowledge Innovation Project" by the Chinese Academy of Sciences was deliberated and adopted in principle. Through the pilot project of knowledge innovation, the Chinese Academy of Sciences would be built into a national natural science and high-tech knowledge innovation center addressing national strategic goals and international scientific and technological frontiers, with strong and sustained innovation capability, and become a national scientific and technological knowledge base, scientific thought base and scientific and technological talent pool with international influence by 2010.

To achieve the above objectives, the pilot implementation of the Innovation Project of the Chinese Academy of Sciences was carried out in three stages: 1998–2000 was the start-up stage, when the pilot construction of eight knowledge innovation bases would be basically completed, and a modern management system for scientific research institutes be built in the pilot site units. The period from 2001 to 2005 would be an all-round promotion phase, when about 80 research institutes would be constructed as research bases, forming a new system of national scientific research institutions and a modern system of research institutes. The period from 2005 to 2010 would be the stage of optimization and perfection, when the overall objective of the National Knowledge Innovation Project in the charge of the Academy of Sciences would be realized, greatly enhancing innovation awareness and innovation capability.[97]

After several years of practice, great achievements had been made in the construction of innovative projects. For example, the overall and supporting tasks of the manned space engineering application system were successfully completed; high-performance digital TV processing chips were successfully developed; key technical support for the construction of the Qinghai-Tibet Railway was provided; 1% sequencing task of the International Human Genome Project was completed; important achievements were made in rice genome sequencing and functional research; the research on organic molecular clustering and free radical chemistry won the first prize of national natural science.

The knowledge innovation project of the Chinese Academy of Sciences improved the overall level of natural science in our country, which would be beneficial to correct the tendency towards economic construction in the formulation of scientific and technological plans in the future. The knowledge innovation project itself was also a part of science and technology planning, but this project could not change the utilitarian philosophy of science and technology planning in the whole transitional period.

3. The Reform of the Science and Technology System Was Deepened

During the "Ninth Five-Year Plan" period, breakthrough progress was made in the reform of China's scientific and technological system, and the periodical goals set by the state for scientific and technological system reform were basically achieved. The strategic focus of scientific and technological work was turning to the main battlefield of national economic construction, and the scientific and technological strength of enterprises was further strengthened. On September 15, 1996, the State Council made the Decision on Deepening the Reform of Science and Technology System during the Ninth Five-Year Plan, pointing out that the main objectives of deepening the reform of science and technology system during the Ninth Five-Year Plan period were: to fully implement the idea that science and technology are the first productive forces, to further implement the strategy that economic construction must rely on science and technology, and the scientific and technological work must address economic construction and strive to climb the peaks of science and technology.

In February 1999, the General Office of the State Council forwarded the Opinions on the Reform of the Management System of Scientific

[97]The History of the Chinese Academy of Sciences,1997 [DB/OL]. Website of the Chinese Academy of Sciences: http://www2.cas.cn/.

Research Institutions under the 10 National Bureaus Managed by the State Economic and Trade Commission by six departments including the Ministry of Science and Technology, and the State Economic and Trade Commission etc.,[98] explicitly proposing the restructuring of 242 national technological development research institutes under the 10 National Bureaus. Based on the actual situation, these research institutions chose their own reform methods, including transforming into enterprises, entering enterprises in whole or in part, transforming into intermediary organizations, etc. After the transformation of scientific research institutions, legal person registration was handled in accordance with relevant regulations. The whole reform was completed in the first half of 1999. The conversion plan of 242 scientific research institutions was as follows: 131 institutes joined enterprises; 40 enterprises were transformed into scientific and technological enterprises subject to local management; 18 institutes retained the nature of public institutions and were transformed into intermediary institutions; 24 institutes joined universities or were merged; 29 institutes were transformed into large-scale scientific and technological enterprises directly under the central government.[99] Most of them were converted to enterprises or enterprises-related institutions. After the reform and opening up, the reform of the scientific and technological system began, thus advancing to a situation more in line with the market economy. However, the transformation of many basic research and public welfare research institutions was not smooth, leading to a decline in the level of scientific research.

4.6 A Summary Review and Enlightenment

This chapter discusses China's science and technology planning during the period of economic transition. The transitional period in this book refers to the period from the end of the "Cultural Revolution" to the end of the 20th century, that is, the period from 1978 to 2000, which is defined as a period of transition, because after the "Cultural Revolution", China's development was transformed from planned economy to market economy. Since the start of the 21st century, new developments have taken place in China's development, resulting in changes of corresponding science and technology planning concepts. Therefore, the transitional period is defined as the period after the "Cultural Revolution" and before the beginning of the new century.

This chapter studies the "1978–1985 Science and Technology Program," the "1986–2000 National Medium-and Long-Term Science and Technology Development Program," the "1990–2000–2020 National Medium-and Long-Term Science and Technology Development Program," the "Ten-Year Science and Technology Development Plan" and the "Outline of the Eighth Five-Year Plan (1991–1995–2000)," as well as the "Ninth Five-Year Plan" for National Economic and Social Development and the Outline of Long-Range Goals to 2010. The scientific and technological program of 1978 was somewhat rash because it was hoped that the scientific losses of the "Cultural Revolution" be made up for as soon as possible. Later, it was changed to a program to tackle scientific and technological problems. Guided by this program, China's scientific and technological development gradually changed from relying solely on science and technology planning to a scientific and technological system of "science and technology planning-scientific and technological plan."

The scientific and technological program for 1986–2000 was relatively mature and well demonstrated. Up to now, the academic circles generally believe that three scientific and technological plans are better plans, that is, the 1956–1967 science and technology program, the 1986–2000 science and technology program, and the 2006–2020 science and technology program. These programs are better because besides sufficient argumentation and good implementation effect, they have also pioneered some new situations or traditions. The 1956–1967 science and

[98]Guo Ban Fa [1999] No. 18 document.

[99]Shiguo (2002).

technology program laid the foundation for China's big science system and the top-level concept tradition of the three big programs. The science and technology program formulated for the period from 2006 to 2020 was being implemented. It is difficult to evaluate its effect, but it is expected to be a good science and technology plan. The 1986–2000 science and technology program was a relatively mature one in the transitional period. China changed from a social pattern of consolidating political power and class struggle to a social pattern of settling down to economic construction and improving people's livelihood. Everything was centered on economic construction, and almost all social sectors served this purpose. Therefore, the science and technology planning from 1986 to 2000 was consistent with the utilitarian planning philosophy.

The "National Medium-and Long-Term Science and Technology Development Program for 1990–2000–2020" has not been implemented since its formulation, but the utilitarian philosophy continues at the textual level. The "Eighth Five-Year Plan" and the "Ninth Five-Year Plan" were matched with the five-year plan of the national economy to better serve the center of economic construction.

It was exactly 22 years before the "Cultural Revolution" from the first science and technology planning in 1956 to the end of the "Cultural Revolution" and the beginning of the reform in 1978. The transitional period also lasted 22 years from 1978 to 2000. Two different historical periods corresponded to different science and technology planning philosophy. Science and technology planning before the "Cultural Revolution" had three top-level concepts, which had already been discussed before, namely, "planning science and technology, technology first," "the supremacy of the national core interests" and "strengthening political functions". After the Soviet Union stopped rendering aid, "self-reliance" became the fourth largest planning concept. After entering the transitional period, the fourth concept was not inherited due to the high expectations of the reform and opening-up for the introduction of foreign technologies, until the 21st century when the national innovation system was built. During the transitional period, the first three concepts changed into "planning science and technology, technology first," "utilitarian tendency" and "strengthening economic function".

Technology planning played a very important role in the whole science and technology planning during the transitional period, whether from the perspective of text analysis or concrete implementation. In pursuit of economic benefits and GDP growth rate, technology played a greater and more direct role than basic science. Therefore, in science and technology planning, we will continue to implement the concept of technology first. In some periods of time, "technology first" almost became "technology alone". Therefore, during the period from 1980s to 1990s, China's basic science showed an overall downward trend. In several scientific and technological plans during the transitional period, high and new technologies were, especially stressed to "develop high technologies and realize industrialization." This further highlighted the position of technology in science and technology planning. The planning philosophy of "planning science and technology, technology first" is very clear.

This philosophy has ensured the rapid promotion of the overall scale of science and technology in China. China has earned the name "World Factory", which is somehow related to its emphasis on technological development. The rapid development of technology has made our country leap forward in many fields, and greatly promoted economic efficiency and the improvement of the country's overall strength. However, it must be pointed out that there are many deficiencies in our country's promotion of the concept "technology first". Apart from the result that the development status of basic science is not commensurate with technology, with regard to the technology itself, introduction is stressed while digestion and absorption are belittled. Therefore, indigenous innovation has been, especially emphasized in the science and technology planning at the beginning of the 21st century.

The philosophy of "the supremacy of national core interests" before the "Cultural Revolution" was changed into the philosophy of "utilitarian tendency". A country's interests have many aspects, but not all interests are core interests. For example, the national territorial sovereignty is not the same with the interest status of inland river shipping, as the former is a core interest while the latter is a general national interest. Before the "Cultural Revolution", for the new regime, consolidating the regime, maintaining national unity and developing protective strategic high-tech weapons were core interests of the country. The development of "atomic and hydrogen bombs and man-made satellites" in our impoverished country embodied the planning concept of "the supremacy of national core interests." After the start of the transitional period, everything was centered on economic construction. Although economic development also involved national core interests, economic development covers many aspects, and not all contents can be deemed as the country's core interests. What's more, in economic development, interests in different fields often conflict with each other. Economic construction is set as the core, as a result, the function of science and technology changed from supporting and consolidating the political power to supporting economic development, and the development policies of "addressing, relying on and climbing peaks" was gradually enhanced. In the policy, "climbing peaks" was not as effective as the first two policies during the transitional period. The idea behind these three development policies is a utilitarian tendency to pursue short-term economic benefits. Therefore, these scientific and technological plans all revolve around how to serve the economic construction concretely. Pursuing short-term benefits and consciously or unconsciously ignoring long-term development and comprehensive balance are not only the embodiment of the national economic plan but also the embodiment of science and technology planning. All pursuit of economic utility reflects the science and technology planning philosophy of "utilitarian tendency".

The positive effect of this philosophy of "utilitarian tendency" must be affirmed; otherwise, the country cannot develop very fast, and many problems may arise. China's huge population pressure is beyond the imagination of any country. Without pursuing economic benefits and faster development speed, it would be difficult to reflect the superiority of socialism and national rejuvenation would be impossible. Therefore, "utilitarian tendency" is a reasonable choice.

Another planning philosophy consistent with "utilitarian tendency" is "strengthening economic function". Before the "Cultural Revolution", it was "to strengthen political functions," because planning for science and technology was an extension of political activities, not an agreement among scientific communities. The wavering policies for intellectuals also show that in the scientific communities, political factors are more important than scientific and technological factors. After entering the transitional period, science and technology were regarded as productivity and then became the primary productive force. Intellectuals naturally became masters of the country and part of the ruling class. The political function of science and technology planning still exists and continues to exist, but it is not as prominent as before. Instead, the economic function of science and technology planning is strikingly prominent. Science and technology are not only for the development of science and technology itself but also for the development of economy. Science and technology can promote social development, which is the basic principle of sociology of science. The promotion functions include economic function, cultural function, political function, educational function and other social functions. Among these functions, the economic function is especially prominent, above everything else.

"Strengthening economic function" makes the development of science and technology in China deliberately inclined to the economic aspect. Great achievements have been made and the country's strength has been greatly improved. At the same time, the scale and conditions of science and technology have also been greatly improved.

"Science and technology planning, technology first," "utilitarianism tendency" and "strengthening economic function" are three parallel and consistent philosophy of planning for science and technology that complement each other to jointly guide the formulation and implementation of science and technology planning during the transitional period, and play a huge guiding role in promoting the development of science and technology and the economic development of our country. As China's S&T system has been transformed into a system of "S&T planning guiding S&T plans," and S&T planning occupies a central position in the whole S&T system of the country, the three major concepts of S&T planning in the transitional period have also been naturally extended to the whole S&T system. Although the whole S&T system in China does not all contain these three concepts, their influence is certainly felt throughout the system.

References

Department of Development Planning of the Ministry of Science and Technology, Science and Technology Development Plans and Programs of the People's Republic of China (1949–2005), 2008.

Editorial, Weathering the Storms of the Century and Striving for Prosperity. People's Daily, March 05, 1996.

Liang, Ma. 2001. Two Voices about "Science and Technology Is the First Productive Force". Journal of Hangzhou Normal University (Humanities and Social Sciences Edition), (1).

Lianzhong, Cui, et al. 2000. *History of the World (Contemporary Times)*. Beijing: People's Publishing House.

Mingyu, Wu. 2008. Recalling the National Science Conference in 1978, Caijing, (11).

Muoruo, Guo. 1978. The Spring of Science, Speech at the Closing Ceremony of the National Science Conference in 1978. People's Daily, April 1, 1978.

Peng, Li. 1996. Report on the Ninth Five-Year Plan for National Economic and Social Development and Outline of Long-Term Goals for 2010. People's Daily, March 19, 1996.

Research group of the History of Chinese Academy of Sciences, National Science Conference: Making the National S&T Plan, Science Times, March 13, 2008.

Review of the seven S&T plans in the People's Republic of China. Science and Technology Daily, March 21, 2006.

Rongji, Zhu. 1999. Speech at the Closing Session of the National Conference on Technological Innovation. People's Daily, August 27, 1999.

Science and Technology Development Plans and Programs of the People's Republic of China (1949–2005),compiled by the Department of Development Planning of the Ministry of Science and Technology, (internal report), 2008.

Selected Works of Deng Xiaoping (Vol. 2). Beijing: People's Publishing House, 1994.

Shiguo, Tang et al. 2002. Research Report on the Reform and Development of China's Science and Technology System.

State Science and Technology Commission, Guidebook to China's Science and Technology Policies—Science and Technology White Paper No.1.Beijing: Scientific and Technical Documentation Press, 1986.

The 6th Five-year Plan for National Economic and Social Development. Bulletin of the State Council, 1983, (9).

The Six Points here are Extracted from the State Council's notice on Issuing the Guideline for National Medium/long-Term Science and Technology Development, Bulletin of the State Council, 1992, (7).

Xiaoping, Deng. 1978. Speech at the Opening Ceremony of the National Science Conference. Guangxi Agricultural Science, (4).

Xin, Li. 2003. *Collection of Current Science and Technology Laws and Regulations in the People's Republic of China*. Hefei: Anhui Audio & Video Press.

Zemin, Jiang. 2006. *Selected Works of Jiang Zemin (Volume 1)*. Beijing: People's Publishing House.

Zemin, Jiang. 1997. Raise the Great Banner of Deng Xiaoping Theory and Push the Cause of Building Socialism with Chinese Characteristics into the 21st Century, The Fifteenth CPC National Congress, September 12, 1997.

Zemin, Jiang. 1999. Speech at the National Conference on Technological Innovation. People's Daily, August 24, 1999.

Ziyang, Zhao. 1986. Report on the 7th Five-year Plan, Bulletin of the State Council of the People's Republic of China, (10).

5 Reflections on S&T Programs and Planning Philosophy in the Twenty-First Century

Abstract

This chapter describes the development and changes of China's Medium and Long-Term Science and Technology Program and philosophy in the twenty-first century. The Chinese government has a deeper understanding of the nature of science and technology, the advice from academic circle and the awareness of policy makers, which made basic research getting more attention in science and technology planning. Medium and Long-Term Science and Technology Program in the new century have three planning philosophy: "technology comes first in S&T plan", "people centrism" and "indigenous innovation".

Keywords

High-Tech · Innovative country · Debate about China's science · People centrism · Indigenous innovation

5.1 Special Plan for the Development of Science, Technology and Education Under the 10th Five-Year Plan (Science and Technology Development Plan) (2001–2005)

I. Political and economic changes in the new century

Since the beginning of the twenty-first century, China has been playing an increasingly important role on the world stage with its growing national strength, but it's also faced with more complicated national conditions and world environment, which poses a higher demand on science, technology (S&T) and their planning. On October 21, 2001, the 9th informal meeting of APEC leaders, a large-scale, high-level international meeting, was held in Shanghai.

The Community Party of China (CPC), the ruling party of China, has to press ahead and keep abreast with the times in the new historical period. In February 2000, Jiang Zemin put forth the important thought of "Three Represents" when inspecting Guangdong. He stressed that the Party must always represent the requirements of the development of China's advanced productive forces, the orientation of the development of China's advanced culture, and the fundamental interests of the overwhelming majority of the

Z. Chen, *China's Medium and Long-Term Science and Technology Program*,
https://doi.org/10.1007/978-3-030-75146-3_5

Chinese people. This was later generalized as the important thought of "Three Represents" that became the "essence of the Party, foundation of its rule and source of its power." At the 16th CPC National Congress held on November 8–14, 2002, the resolution on the *Constitution of the Communist Party of China* (amendment) was passed, upholding "Three Represents" as a guiding thought that the Party must adhere to for a long time to come. It was officially put into the *Constitution of the People's Republic of China* in 2004 when the National People's Congress (NPC) approved the constitutional revisions. The initiation of the "Three Represents" reflected the CPC's deep understanding of its mission as well as new requirements of the times, especially representing the requirements of advanced productive forces. This meant that the CPC has had new perceptions of S&T, and S&T and its planning were therefore endowed with more meanings. "Science and technology are primary productive force and the concentrated reflection and main symbol of advanced productive forces. We must acutely grasp this objective trend… and strive to make substantial progress in our productive force development. This is an important duty that the CPC has to fulfill to represent the requirements of the development of China's advanced productive forces."[1] Ranking first of the "Three Represents", "representing the requirements of the development of advanced productive forces" provided the theoretical and legal basis for further raising the social status of S&T and intellectuals. On the one hand, S&T and their plans were put in a higher position in China; on the other hand, S&T, as the direct embodiment of productive forces, was fortified and legalized as a ruling instrument. Their practical value was expressed in theories: S&T not only generates the knowledge for understanding the world and the tools for reforming it but also becomes the ruling party's instrument and approach for state governance.

The political atmosphere in China was more democratic and relaxing. On December 26, 2002, the political bureau of CPC Central Committee held the first collective study session, the content of which was "earnestly implementing the *Constitution* and building a moderately prosperous society in all respects." It decided to step up the study efforts, not only encouraging self-study but also organizing collective study and making it a consistent system, which waged a new study wave across the Party. In 2004, the 4th plenary session of the 16th CPC Central Committee adopted the *Decision on Strengthening the CPC's Governance Capability*, which proposed to "build a learning Party" through a plenary session decision for the first time. From the 16th CPC National Congress to September 28, 2007, the political bureau of the CPC Central Committee organized 44 collective study sessions. During the 5 years, 89 experts and scholars were invited to the Huairentang of Zhongnanhai to discuss major issues in economy, politics, law, culture, S&T, history, and other aspects, and two of those discussions involved the formulation of medium/long-term programs.[2] The earnest attitude toward the study of the new CPC Central Committee demonstrated the Party's maturity in state governance and gave a strong push to fostering a clean political environment and encouraging free thinking, which was two of the prerequisites for healthy S&T development. An atmosphere of rational reflection on S&T was taking form. Though the utilitarian concept was still present in S&T plans, the general perception was much more rational than before.

In terms of economic background, China has come to the "third step" of its "three-step" strategy of modernization in the twenty-first century, when economic construction and national development both have entered a new historical period. On January 24, 2000, the then Premier of the State Council Zhu Rongji expatiated on the "Western Development" strategy at a State Council meeting about that topic. On May 18 that year, the National Planning Commission announced ten major projects in the western region, lifting the curtain on the massive development campaign.

On November 10, 2001, the 4th WTO Ministerial Conference held in Qatar's capital city Doha reviewed and unanimously passed the

[1]Zemin (2006).

[2]Mingzheng (2013).

decision to admit China into the WTO, putting an end to China's 15-year-long efforts to resume its status as a GATT contracting party and to join the WTO. "Joining the WTO is a strategic decision made by the CPC Central Committee and the State Council and a historic milestone in our journey of reform and opening up. It marks the beginning of a new stage in our opening up."[3] After we joined the WTO, the industry structure adjustment and technology transfer in the global scope provided new opportunities for our economic construction. Meanwhile, the economic and technological integration worldwide and the achievement of a number of technological breakthroughs created the external conditions for us to make big technological strides, but they also put new restrictions on our S&T and its planning.

II. Special S&T plan in the 10th five-year national economic plan

1. 10th five-year national economic plan

As all the S&T programs after the reform and opening up were launched hand in hand with national five-year plans, it is necessary to first take a look at the 10th five-year economic plan. The 5th plenary session of the 15th CPC Central Committee was held in Beijing on October 9–11, 2000, which reviewed and passed the *Suggestion on Formulating the 10th Five-year Plan for National Economic and Social Development.* The plenary session concluded that in the new century, China will enter the development stage of building a moderately prosperous society in all respects and promoting modernization at a faster pace. In view of the overall outlook and deployments made by the 15th CPC National Congress for China's modernization drive in the new century, the 5th plenary session put forth the general arrangements in the 10th five-year period. Regarding the role of science, technology and S&T plans in the 10th five-year plan, the plenary session pointed out, "We must rely on institutional and technological innovation to promote economic development and structural adjustment...Technological progress is the decisive factor to enhance the overall national strength. We should continue to implement the strategy of strengthening the nation through science and education, actively advance high-tech research with strategic significance, develop high-tech industries, and improve the capability of consistent technological innovation."[4] The 10th five-year economic plan associated S&T with technological innovation, which "is playing an increasingly important role in economic and social development and has become the key factor of developing productive forces."[5] Later the *Suggestion* was adopted at the NPC and became a national will.

From March 5 to 15, 2001, the 4th session of the 9th NPC was held in Beijing, at which Zhu Rongji made a report on the outline of the 10th five-year plan for national economic and social development. The report consisted of ten parts: (1) Review of economic and social development in the 9th five-year period; (2) Objectives and guidelines for the 10th five-year period; (3) Cement the fundamental position of agriculture and work hard to increase farmers' income; (4) Energetically improve and upgrade the industrial structure; (5) Implement the Western Development strategy to promote balanced regional development; (6) Implement the strategy of strengthening the nation through science and education, and strongly develop talents; (7) Further deepen the reform and expand opening up; (8) Continuously improve people's lives and the social security system; (9) Continue with the sustainable development strategy, and (10) Step up the efforts for cultural development, democracy, the rule of law, and national defense.[6] The report set the main economic and social objectives for the next five years: to keep the national

[3]People's Daily, November 11, 2001.

[4]Bulletin of the 5th plenary session of the 15th CPC Central Committee, *Party Building 11* (2000).

[5]"Editorial: A Great Campaign in the New Century," *People's Daily*, October 12, 2000.

[6]"Outline of the 10th Five-year National Economic and Social Development Plan," *Bulletin of the State Council of PRC (12)*, 2001.

economy growing at a fairly high speed; to deliver obvious results in the strategic adjustment of economic structure; to considerably improve the quality and benefit of economic growth; to lay a solid foundation for doubling the 2000 GDP by 2010, and so on. The meeting adopted the resolution on the outline of the 10th five-year plan for national economic and social development and the report on it.

The report put S&T in a position as important as reform and opening up, and recognized reform, opening up and technological progress as the drivers of social development and national rejuvenation. Reform and opening up is the only way to build a strong nation with affluent people, technological progress and innovation are critical for enhancing the overall national strength, and economic development and structural adjustment, therefore, have to depend on institutional and technological innovation. The third chapter of the report on the 10th five-year plan elaborated on science, technology, education and talent, including a specific part dedicated to S&T. It's worth noting that "national innovation system" was mentioned in a national economic plan for the first time, although the five innovation systems that were later widely known among the people, namely knowledge innovation, technological innovation, national defense innovation, scientific and technological intermediary, and regional innovation, were not specified yet. The 10th five-year plan set the general framework for the 10th five-year S&T plan.

2. Formation of special S&T plan

The *10th Five-year Plan for National Economic and Social Development—Special Plan for the Development of Science, Technology and Education (S&T Development Plan)* was an important part of the 10th five-year plan and an elaboration and crystallization of the outline. In October 1998, the National Leading Group for Science, Technology and Education held the 2nd meeting, at which it was decided to compile the 10th five-year S&T plan. At the end of that year, the Ministry of Science and Technology (MOST) kicked off the research on the S&T plan, and formed a general strategic research team composed of experts in science and technology, economy and management to specifically study the domestic and external environment for S&T development, scientific and technological needs, and the guideline for S&T work. In the meantime, several hundred experts were divided into 11 research groups, respectively focusing on agriculture, industry, social development, basic research, high-tech research, international S&T cooperation, S&T input and supportive conditions, S&T team building, regional S&T development, S&T popularization, and development of national high-tech zones. A primary research report of nearly two million characters was formed in the end.

In early 2000, the leading group for the drafting of 10th five-year S&T plan was formed, which consisted of responsible officials from the National Planning Commission, MOST, Ministry of Education, etc., and a work group was also set up, which had to complete the draft plan in June 2000. In this period, General Secretary Jiang Zemin and Vice Premier Li Lanqing debriefed the drafting group several times. The draft plan was passed in principle at the 6th meeting of National Leading Group for Science, Technology and Education. Upon the approval of the State Council, the National Planning Commission and MOST officially promulgated the *10th Five-year Plan for National Economic and Social Development—Special Plan for the Development of Science, Technology and Education (S&T Development Plan)* in May 2001. It differed from previous S&T plans in that it was directly affiliated to the 10th five-year national economic plan.

3. Special educational plan

Unlike previous S&T programs, the 10th five-year S&T plan was titled *Special Plan for the Development of Science, Technology and Education*. On the one hand, it meant science, technology and education were given equal importance and the strategy of "strengthening the nation through science and education" included an S&T strategy and an education strategy. On the other hand, it meant there was also a special 10th five-year plan for education. Indeed the nation also formulated the *10th Five-year Plan for National Economic and Social Development—Special Plan for the Development of Science, Technology and Education (Educational Development Plan)*, in which utilitarian objectives, such as vitalizing the economy and boosting rapid development, were taken as the guiding thought for educational development. "Efforts must be made to make education a central factor for economic prosperity and national rejuvenation...We should truly make education a pioneering and fundamental knowledge industry with overall significance, and operate it as an investment in the future of the nation, the society and the individuals...Great efforts must be made to make sure that a group of advantageous disciplines will occupy a commanding height on the forefront of world technology, or at least catch up with it... Priority will be given to the construction of about 500 key disciplines to sharpen the capability of fostering high-caliber innovative talents and resolving the nation's major issues concerning economic construction, technological innovation, and social development. We will work very so hard that about 100 key disciplines will reach or approach the world's advanced level by 2005."[7] This statement well illustrated the direct connection between educational plan and S&T plan.

III. The guideline of *innovation and industrialization*

The guideline for the 10th five-year S&T plan was different from that of the 9th. "During the 10th five-year period and for the time beyond, our S&T work will focus on economic construction and structural adjustment, make overall progress with clear priorities, develop high technology and realize technological industrialization, improve the capability of continuous technological innovation, and achieve leapfrog technological development."[8] This was later referred to as the guideline of "innovation and industrialization". The 9th five-year S&T plan followed the guideline that "economic construction has to depend on science and technology, scientific and technological work must help economic construction, and new heights have to be scaled," but the 10th one shifted to the guideline of "innovation and industrialization", placing more emphasis on the industrialization of S&T. However, as science couldn't be industrialized directly, it mostly referred to technology, especially high technology.

To transform the guideline into reality, the 10th five-year S&T plan made concrete arrangements on the following principles[9]:

1. Highlight national objectives. Key areas of S&T development were determined according to the nation's major demands, including strategic adjustment of economic structure, sustainable development and national security, and S&T and economy were integrated more closely.
2. Be market-oriented. The market's basic role in resource allocation was given full play, and market competition and cooperation were introduced to make S&T activities more efficient and beneficial, so as to push S&T development on all fronts.
3. Make technological breakthroughs. Strengthen the alliance of industry, universities and research institutes in key S&T areas where we had comparative advantages or that were of strategic significance, and strive for substantial breakthroughs in those areas.
4. Intensify indigenous innovation. Combine independent R&D with introduction, digestion and absorption, reinforce basic research and emphasize original innovation, increase S&T input by a large margin, and constantly improve the capability of indigenous innovation.

[7] "Notice of National Planning Committee on Issuing the 10th Five-year Plan for National Economic and Social Development—Special Plan for the Development of Science, Technology and Education (Educational Development Plan)," *Bulletin of the State Council of PRC (12)*, 2002.

[8] "Notice of National Planning Committee and Ministry of Science and Technology on Issuing the 10th Five-year Plan for National Economic and Social Development—Special Plan for the Development of Science, Technology and Education (S&T Development Plan)," *Bulletin of the State Council of PRC (11)*, 2002.

[9] Ibid.

5. Strengthen military-civilian integration. Under the guideline of "integrating military and civilian sectors and peacetime and wartime activities," energetically develop dual-purpose technologies and promote two-way technology transfer between the two sectors.
6. Reflect people-centrism. Give full play to the central role of innovative talents in S&T development.
7. Promote coordinated S&T development among regions. Taking the opportunity of the Western Development, promote S&T cooperation and coordinated development among regions.

These principles showed that the idea of building a national innovation system was in the making, and S&T plan would make a turn from reforming the S&T system to establishing the national innovation system. "People centrism" appeared in S&T plan for the first time, indicating that the CPC had reflected on the purpose of economic development. When this reflection was projected to the S&T field, it triggered reflection on the nature of S&T development and human development. However, the utilitarian philosophy of S&T plan had a strong inertia and the 10th five-year S&T plan didn't specify what to reflect on and how.

It must be pointed out that the utilitarian philosophy still had a strong presence despite the initial signs of reflection, and there were obvious utilitarian practices too. Regarding basic research, for example, "We will concentrate resources on making major breakthroughs in important areas that play a critical role in economic and social development as well as in national security, or where China has an edge. By 2005, we will strive to approach or reach world cutting-edge level in several important scientific fields and strategic high-tech areas, register a batch of ground-breaking research results with international influence, and lay a solid S&T foundation for our long-term economic and social development."[10] Basic science needed long-term accumulation, so did major breakthroughs. The thought and act of specifying, by way of a government directive, a period or even a date to deliver an S&T result showed the linear understanding of science, and it was utilitarian to think that increased S&T input would naturally lead to more results. It was hard to reach world cutting-edge level in certain areas within five years, or four years to be more specific because the 10th five-year S&T plan was officially issued in May 2001. Besides, given the actual implementation, major basic innovations had to be fortified, so it was more appropriate to understand the objectives as a reflection of the dire yearning for major results in basic science.

The utilitarian planning philosophy could be seen in the plan. The 10th five-year S&T plan proposed that "by 2005, the ratio of total R&D expenditure to GDP will be raised above 1.5%," but the ratio was only 1.34% in 2005.[11] It was very hard to raise that indicator, and the inconsistency between the target and actual result indicated the lack of coordination in input between the S&T plan and national economic plan.

The plan also said that during the 10th five-year period, our S&T work must revolve around the central task of the strategic adjustment of economic structure. Under the guideline of "innovation and industrialization", strategic deployments were made at two levels—"promoting the upgrade of industrial technology" and "improving the capability of continuous technological innovation." The first level meant that enterprises, as the main entities of technological innovation, had to grasp key, high and new technologies, and use them to promote the technological upgrade and structural adjustment of industries. The second level meant that universities and research institutes should be brought into full play, and the capability of continuous S&T innovation must be enhanced in strategic high-tech research and original basic research, aiming for leapfrog technological progress in key areas where we had comparative advantages. The

[10] Ibid.

[11] 2005 China Statistical Report on Science and Technology, http://www.sts.org.cn.

plan didn't mention leapfrog progress in science because that was extremely difficult.

The 10th five-year S&T plan also made deployments for Western Development for the first time by actively implementing the "S&T Action for Western Development." More deployments were made for basic science than in the 9th five-year S&T plan, the contents accounting for 10% or so of the total. Regarding S&T system reform, a very important subject in previous S&T plans, the 10th five-year S&T plan proposed to establish a national innovation system. In the 10th five-year period, the general goal of S&T system reform was further emancipating the S&T productive force and primarily establishing a vibrant national innovation system that met the requirements of socialist market economy and conformed with the laws of S&T development, so as to lay a sound institutional foundation for greater development of S&T productive force. However, the plan didn't give a systematic account of how to establish the national innovation system, a task that would be accomplished under the *2006–2020 Medium/Long-term Science and Technology Development Program*. The 10th five-year S&T plan would step up the construction of scientific research system, technological development system, and S&T service system, strengthen the industry-university-research institute alliance, and form an innovative network and operating mechanism with enterprises as the main entities, and research institutes, colleges, service agencies and government organizations interacting with each other. This was somewhat different from the notion of technological innovation system that we are familiar with today, meaning that the Party's and the government's perception and understanding of national innovation system have deepened over time, as clearly reflected by the S&T plan.

IV. Effect of implementation: leapfrog development in high technology

The 10th five-year S&T plan was well executed and many S&T indicators displayed the tendency of faster improvement.

1. Remarkable effect in general

During the 10th five-year period, the number of projects under various S&T programs kept increasing, up to 15,549 in 2004, an increment of 1347 from the previous year. 86.9% of those projects went well, of which 82.7% were on schedule and 4.2% ahead of schedule. There were 6245 projects under three main programs (basic research program, Program 863 and National S&T Challenge Tackling Program), 92.1% of which went well, including 3.7% ahead of schedule; of the 9304 projects under three industrial programs (Torch Program, Spark Program and program of promoting S&T results), 83.4% went well, including 4.5% ahead of schedule.[12] In 2005, the total output value of the Chinese high-tech industry reached RMB3436.7 billion, up 23.8% from the previous year, and its R&D intensity (the ratio of R&D expenditure to total industrial output value) was 1.05% as opposed to 0.54% in the manufacturing sector that year. In comparison, R&D intensity was above 10% in the US, Canada, UK, and Japan, around 8% in Germany and France, and over 4.5% in ROK and Italy. The high-tech R&D input in China was obviously not enough.

In 2004, altogether 265,000 technological contracts were signed across the country, with a total value of RMB133.44 billion. Compared with 2003, the number of signed technology contracts was down 1.25 percentage points, but the transaction amount was increased significantly by 23.0%,[13] as shown in Fig. 5.1.

Regarding the category of S&T results, 2129 basic theoretical results were registered nationwide in 2005, down 25.0% from the previous year and accounting for 6.6% of the total registration; 28,559 applied technological results were registered, up 4.4% from a year ago and accounting for 88.3% of the total; while 1671 soft scientific results were registered, up 10.1% from a year ago and taking up 5.2% of the total. Most of the registered S&T results were focused

[12] Ibid.

[13] 2006 China Statistical Report on Science and Technology, http://www.sts.org.cn.

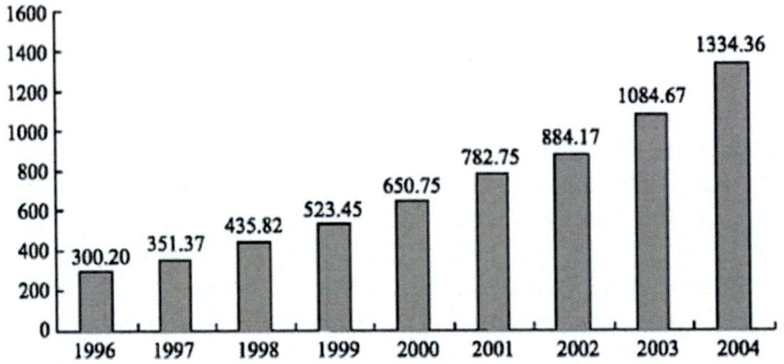

Fig. 5.1 Transaction amount of technology contracts nationwide (1996–2004)[14] (Ordinate represents money, Abscissa represents time)

on applied technologies, while the number of basic theoretical results shrank rapidly.

2. Huge success of Shenzhou spaceship series

On March 25, 2002, the Shenzhou III spaceship developed by China was successfully sent to the pre-set orbit from the Jiuquan Satellite Launch Center.[15] Jiang Zemin gave a speech at the launch site, saying that the manned space program was a strategic decision made by the CPC Central Committee according to the general trend of world S&T development and China's development of S&T undertakings and modernization. In the past decade, substantial progress was made in the engineering and construction of this field in the spirit of "two bombs and one satellite", a manned space system with Chinese characteristics has taken primary shape, and China's manned space technology has reached a new height.[16]

Building on the "two bombs and one satellite," China's manned space program has made remarkable achievements, which not only demonstrated our increased national strength but also showed that the management mode of major scientific projects remained effective for S&T development. The manned space program was under the unified management of the government-backed Office of China Manned Space Program, which was responsible for planning, technology, and funding. We can see that the planning mode of "big scientific projects" determined in the 1956 S&T program and the "national system" had a persistent bearing on China's S&T cause, and they will be continued in S&T plans of the twenty-first century.

3. National Top Science and Technology Award

After entering the twenty-first century, the national award system was reshuffled, with a National Top Science and Technology Award established. The award in a way reflected the results of previous S&T plans and scientific programs. In 1999, the State Council launched a comprehensive reform of national S&T awards and enacted and implemented the new *Regulation on National Science and Technology*

[14]Based on 2005 China Statistical Report on Science and Technology, http://www.sts.org.cn.

[15]In 1992, the Chinese government, making overall considerations for political, economic, S&T and national defense factors, held that it was necessary to develop manned space flight immediately, and decided to start with manned spacecraft and carry out manned space programs. It determined the three-step strategy for manned space development. First, launch manned spacecraft, build primary experimental manned spacecraft facilities, and conduct space application experiment; second, after successfully launching the first manned spacecraft, make breakthrough in the rendezvous and docking technology between manned spacecraft and space vehicle, modify and launch a space lab with manned spacecraft technology, so as to pave the way for moderate-scale space applications with short-term human care; third, build space station and solve large-scale space applications with long-term human care.

[16]China Manned Space, http://www.cmse.gov.cn/cha_fxrw/show.php?itemid=102.

Awards, which listed five national awards—National Top Science and Technology Award, National Award for Natural Sciences, National Award for Technological Invention, National Award for Science and Technology Progress, and China International Science and Technology Cooperation Award.

On February 19, 2001, the CPC Central Committee and State Council held a national science and technology award conference in Beijing, at which Jiang Zemin conferred the 2000 National Top Science and Technology Award to Wu Wenjun and Yuan Longping. This was the highest honor conferred by the nation to scientists who have made outstanding contributions to S&T development, starting in 2000 according to the decision made by the CPC Central Committee and State Council. It's worth noting that Wu Wenjun and Yuan Longping were awarded for their achievements made in the 1970s.[17] Since then, a national S&T award conference has been held every year, demonstrating how much importance the Chinese government attaches to science and technology. From 2000 to 2014, twenty-five prestigious S&T experts won this honor, with an absence in 2004 (see Table 5.1).

The National Top Science and Technology Award is equivalent to that of the Nobel Prize in terms of the amount of money, but it's different in three important ways. First, National Top Science and Technology Award is an award conferred by the nation rather than by a scientific community. Second, basic science, especially pure basic science, takes up a small proportion. Third, the award is based on general and comprehensive superiority rather than prominent innovation of individuals. "In the review and selection of National Top Science and Technology Award winners, the criterion is the overall assessment of the candidates' S&T contributions during their whole career, or in other words, the accumulation of their advantages in scientific research."[18] While this shows the unprecedented importance that China attaches to S&T and relevant experts, it also reflects the utilitarian side of the award. The Nobel Prize is inclined toward pure basic research, whereas our top S&T award has a preference for applied basic research and applied research, namely results in the Pasteur's Quadrant.

Besides, only 1.5 awards or 30% of the five national awards (National Top Science and Technology Award, National Award for Natural Sciences, National Award for Technological Invention, National Award for Science and Technology Progress, and China International Science and Technology Cooperation Award) are for basic sciences, while technological award stake up 70%. This seems to say that it's reasonable to give more awards to technologies that can boost economic and national development in a short span of time, but it also reflects a utilitarian view about science and technology.

5.2 National Medium- and Long-Term Program for Science and Technology Development (2006–2020)

I. Background changes in the new century

1. Building an innovative nation

The 2006–2020 national medium/long-term S&T development program has historic significance for several reasons—it is China's first long-term S&T program in the twenty-first century, the first S&T program after China established the socialist market economy and joined the WTO, and the first one to be formulated and implemented by the new Chinese government after the core of the CPC's 4th-generation leadership was established. This program displayed some new features.

In the twenty-first century, the new technological revolution has arrived with a swifter and stronger momentum than the one in the twentieth century, harboring new and major breakthroughs.

[17] Winners of 2000 National Top Science and Technology Award, website of MOST: http://www.most.gov.cn/bstd/cx/kjjlcx/kjjl2000/200802/t20080214_59059.htm.

[18] Yan (2007).

Table 5.1 Winners of National Top Science and Technology Award (2000–2014)

2000	Wu Wenjun	Born in 1919, mathematician
	Yuan Longping	Born in 1930, father of hybrid rice
2001	Wang Xuan	Born in 1937, founder of laser phototypesetting system of Chinese characters
	Huang Kun	Born in 1919, physicist
2002	Jin Yilian	Born in 1929, prestigious expert on high-performance computer
2003	Liu Dongsheng	Born in 1917, earth environment scientist
	Wang Yongzhi	Born in 1932, space technology expert
2005	Ye Duzheng	Born in 1916, meteorologist
	Wu Mengchao	Born in 1922, hepatic surgery scientist
2006	Li Zhensheng	Born in 1931, geneticist, founder of distant hybridization of wheat
2007	Min Enze	Born in 1924, petrochemical catalyst expert
	Wu Zhengyi	Born in 1919, botanist
2008	Wang Zhongcheng	Born in 1925, neurosurgery expert
	Xu Guangxian	Born in 1920, chemist
2009	Gu Chaohao	Born in 1926, mathematician
	Sun Jiadong	Born in 1929, rocket and satellite expert
2010	Shi Changxu	Born in 1920, metallography and material expert
	Wang Zhenyi	Born in 1924, hematologist
2011	Wu Liangyong	Born in 1922, expert on architecture and city planning
	Xie Jialin	Born in 1920, physicist
2012	Zheng Zhemin	Born in 1924, dynamicist
	Wang Xiaomo	Born in 1938, radar engineering expert
2013	Zhang Cunhao	Born in 1928, physical chemist
	Cheng Kaijia	Born in 1918, nuclear weapon expert
2014	Yu Min	Born in 1926, nuclear physicist

People generally believe that information technology or IT will be the engine that pushes economic growth and knowledge dissemination and application in the twenty-first century, while life sciences and bio-technologies will improve the living quality of mankind. Moreover, energy technology, nanotechnology, space technology and others will bring about profound technological transformation.

Major breakthroughs in basic research will broaden the horizon for the humankind to understand objective laws and promote technological and economic development. Since knowledge plays a bigger role in economic and social development amid the new technological revolution worldwide, the increase of national wealth and the improvement of human life have come to depend more on knowledge accumulation and innovation. S&T competition, or knowledge competition, becomes the focus in the international competition of overall national strength. Political thinkers hold that in today's world, he who has an edge in knowledge and S&T innovation will grasp the initiative in development. All countries in the world, especially developed ones, have made pushing S&T progress and innovation a national strategy and increased S&T input by a large margin. The political, economic and military competition in the world has to a large extent been translated into S&T competition. Therefore, China's S&T development and planning are faced with bigger challenges and higher requirements in the new century.

On January 9–11, 2006, the CPC Central Committee and State Council held the first national science and technology conference in the twenty-first century to make arrangements for the implementation of the *Outline of National Medium/Long-term Science and Technology Development Program (2006–2020)*. It could be viewed as a mobilization conference for the implementation of the plan, at which Hu Jintao said in his speech, "We have more urgent demand for a solid scientific foundation and strong technological support than ever before."[19] This indicated that the CPC and the Chinese government had a deeper understanding of S&T and a stronger desire to boost economic development through S&T development. What's the direction of our S&T development in the new century? Hu Jintao pointed out that the current status of China's S&T development didn't adapt to our urgent need for adjusting economic structure and transforming economic growth mode, for steering the economic and social development to the path of people-centered, comprehensive, coordinated, and sustainable development, and for building a moderately prosperous society in all respects and continuously improving people's living standards. To shift China's economic growth from being resource-dependent to being innovation-driven, we have to earnestly steer economic and social development to the path of scientific development. This implied that our S&T development was gradually moving from the sole pursuit of aggregate national wealth to advancing the development of individuals. On January 26, the CPC Central Committee and State Council made the *Decision on Implementing the Outline of Science and Technology Plan and Fortifying the Capability of Indigenous innovation.*

The purpose of center of the 2006–2020 medium/long-term S&T program was to realize an innovative country. "To build an innovative country, the key is making the enhancement of indigenous innovation capability the strategic pivot of S&T development and promoting its leapfrog progress. The fortification of indigenous innovation capability should be made the central link in adjusting industrial structure and transforming growth mode, so as to build a resource-saving and environment-friendly society. It should also be regarded as a national strategy and implemented in every aspect of the modernization drive."[20] The S&T development strategy began to change as the nation's strategic intention changed.

Regarding the guideline of S&T development, Hu Jintao made the following remarks: "To take the path of indigenous innovation with Chinese characteristics, the key lies in upholding the guideline of 'indigenous innovation, making breakthroughs in priority areas, supporting development and leading the future'. Indigenous innovation means that we should work harder on original innovation, integrated innovation, and re-innovation based on introduction, digestion and absorption, with the aim of strengthening the nation's innovation capability. That is a summary of our half-century-long S&T development experience and an important choice we have made to realize the great rejuvenation of the Chinese nation. It must be implemented in the whole process of our S&T development."[21]

As a matter of fact, China adopted different S&T development guidelines in different stages of economic development. There were arguments over the guideline of "indigenous innovation, making breakthroughs in priority areas, supporting development and leading the future" before and after it was put forth, and the summary could be seen as an important understanding of S&T from the political perspective. It also meant that S&T and political development were closely related in China. Hu's remarks indicated that the CPC began to reflect on past S&T planning concepts, and a new planning concept was primarily formed.

[19]Jintao (2006).

[20]Ibid.

[21]Ibid.

2. The scientific outlook on development

The 6th plenary session of the 16th CPC Central Committee was held on October 8–11, 2006, which reviewed and passed the *Decision on Several Major Issues Concerning the Building of a Socialist Harmonious Society*. The session put forth the idea of harmonious society and held that social harmony is the intrinsic nature of socialism with Chinese characteristics and an important guarantee for the country's affluence and strength, the nation's rejuvenation and the people's happiness. Hu Jintao pointed out that China has come to a new historical starting point, when various deep-rooted contradictions gradually come to the fore, so we should put building a socialist harmonious society in a more prominent position. The initiation of harmonious society further manifested the central government's understanding of social development and its vigilance against the difficulties in achieving the third-step strategic goal. It has come to realize the two-sidedness of science, especially technology development, or in other words, "science and technology is a double-edged sword." They can either advance or undermine social harmony—it all depends on how the policy makers view and use them. Therefore, the notion of harmonious society posed higher requirements on S&T plan makers, who should take into account both the upside and downside of science and technology.

On June 25, 2007, Hu Jintao gave a speech at the class of provincial and ministerial cadres in the Central Party School. He said the primary goal of Scientific Outlook on Development was the development and the center was people centrism, which brought out the relation between S&T development and human development. S&T plan was a guiding blueprint of S&T development intended to better boost it, while S&T development was aimed at more freedom for mankind, but planned S&T development was bound to limit some of the human freedoms because freedom couldn't be planned, or else it couldn't be called so. How to address the contradiction between planning and freedom then? The idea of "people centrism" was an attempt to answer this question and has gradually become the new S&T planning philosophy.

On October 15–21, 2007, the 17th CPC National Congress was held in Beijing, at which the Scientific Outlook on Development was written into the *CPC Constitution*. "Scientific Outlook on Development is a scientific theory that builds on Marxism, Leninism, Mao Zedong Thought, Deng Xiaoping Theory, and the important thought of Three Represents and presses ahead with the times. It is an important guideline for economic and social development in China and a significant strategic thought that must be resolutely implemented in building socialism with Chinese characteristics."[22] General Secretary Hu Jintao elaborated on this theory in his report to the 17th CPC National Congress. It was proposed based on China's basic national condition of being in the early stage of socialism and intended to meet the new development demands. In the new century, China's development displayed a series of new features. On the one hand, immense achievements were made on all fronts, but on the other hand, deeper contradictions continued to exist, as reflected in the following areas. (1) Economic strength was notably enhanced but structural contradictions and the extensive growth mode didn't change fundamentally; (2) The socialist market economy was primarily established, but social stratification continued to widen; (3) Coordinated development saw obvious progress, but urban-rural and regional differences were more conspicuous; (4) Democracy and the rule of law made headway, but people had higher requirements; (5) The society was more vibrant and dynamic, but social management faced new challenges; (6) Overall national strength was improved, but the pressure from developed countries' economic and technological superiority continued to exist. "China's development features in the current stage are a representation of our basic national condition of being in the early stage of socialism in the new century and new development stage."[23]

It must be noted that the "science" as in "Scientific Outlook on Development" didn't mean the same as in "science and technology plan". They were two different concepts, the

[22] *Constitution of the Communist Party of China* (General Outline) (Beijing: People's Publishing House, 2007).

[23] Jintao (2007).

former involving more connotations outside of the scope of science. However, this in a way implied that in contemporary China, science was not only applied in social system and human activities but also could be sublimated to the ideological level and as the state's governing concept. Apparently, science, technology, and S&T plan should all serve the Scientific Outlook on Development. It took a process for science to become an ideology, and the proposal of Scientific Outlook on Development could be seen as a symbol.

This theory was also a theoretical reflection on China's sole pursuit of GDP growth in the twentieth century. The center of Scientific Outlook on Development was people centrism, which meant promoting all-round human development, and development being for the people, dependent on the people, and shared by the people. This outlook had a little direct effect on the formulation of the 2006 medium/long-term S&T program because it only became the Party's guiding thought after the program was released, but in the long run, it constituted the macropolitical background of its implementation, which must be influenced by it. Scientific Outlook on Development was one of the restrictive standards for the formulation of the 11th and 12th five-year S&T plans, and the "people-centric" philosophy was sure to be reflected in future S&T programs.

3. Entering a new stage of economic development

Economically speaking, China has set the goal of building a moderately prosperous society in all respects. By 2020, such a society will have been built to benefit over a billion people and the GDP of 2000 will be quadrupled. Since China set the reform goal of establishing socialist market economy in 1992, its economic system reform has seen substantial progress both in theory and practice. Against the backdrop of market economy, the structure and approach of our S&T activities underwent fundamental changes compared with the period of the planned economy. Economic and technological globalization became the new trend, and China has been merging into the international community on all fronts, especially after it joined the WTO. It had to seek its own innovations and breakthroughs on the basis of drawing on advanced science and technology from developed countries.

On June 25, 2007, the National Bureau of Statistics issued the *Review Reports on Economic and Social Development from the 16th to the 17th CPC National Congress*, which stated that since the 16th CPC National Congress, national economy had increased quickly, steadily and consistently. From 2003 to 2006, China's GDP increased by 10.4% on annual average from RMB12033.3 billion in 2002 to RMB21087.1 billion in 2006, jumping from the 6th to the 4th position in the world. In 2002, our per capita GDP exceeded USD1, 000 for the first time, and it exceeded USD2, 000 four years later in 2006, jumping from the 132nd to the 129th position. According to World Bank's standard, China had stepped from a low-income country to a middle-income country, which proved the effects of our S&T plans while posing higher demands for future ones. The academia generally believes that the stage when per capita GDP is in the range of USD1, 000-3, 000 is critical for development with many contradictions.

II. **Arguments over S&T programs**

1. Argument between the economic circle and S&T circle

There were arguments over certain issues of S&T plan when the first plan was being made, and future S&T plans always came with arguments that, however, only happened in a small scope. But S&T programs in the twenty-first century have encountered serious arguments. When the outline of the 2006 medium/long-term S&T program was being formulated, the planning office held several symposiums of experts, entrepreneurs, and other related parties, solicited opinions from central departments, local governments, CAS, CASS and other organizations, and handed out questionnaires to 500 enterprises. People from 600 enterprises participated in the planning. Premier Wen Jiabao chaired seven

State Council meetings to listen to relevant reports, and State Councilor Chen Zhili debriefed the MOST and planning office several times, chaired planning symposiums, and attended international forums in that theme. Going through five stages—preliminary preparations, framework design, task listing and policy sorting, formation of draft, and solicitation of opinions, the *Outline of National Medium/long-term Science and Technology Development Program (2006–2020)* was revised 12 times and sent to 124 local governments and departments for opinions during the drafting period, receiving more than 500 revision suggestions, including disagreements.

Major arguments appeared in the early research stage. Around June 2003, the leading group of national S&T planning held the first strategic seminar at Jingxi Hotel, Beijing. It invited three economists—Jiang Xiaojuan, head of National Academy of Economic Strategy of CASS, Lin Yifu, director of China Center for Economic Research of Peking University, and Hu Angang, director of Institute for Contemporary China Studies of Tsinghua University—to give lectures on the future development strategy of Chinese economy, China's national development strategy in the next 20 years, and some major issues concerning S&T plan. Their lectures triggered intense arguments among economists and people in the S&T circle.

Economists tended to look at things according to economic laws, especially cost and profit analysis. In Jiang Xiaojuan's opinion, "in the context of globalization, China, when developing its industrial technologies and upgrading the industrial structure, can make full use of the opportunities presented by economic globalization, and upgrade the industrial structure by encouraging outbound direct investment by multinationals and introducing advanced technologies from foreign countries."[24]

Applying the theory of comparative advantage, Lin Yifu pointed out in his report that "a developing country like China should fully utilize its technological distance from developed countries and push its industrial technology upgrade by introducing foreign technologies. It is a feasible, low-cost strategic choice of good benefits."[25] When explaining his view, Lin said the China currently had a similar situation to that in Japan when it decided to adopt the "introduction" path. Japan began to rise based on technological introduction and only shifted to innovation later, so China should follow suit, he suggested. He emphasized that there were two ways of technological change—independent invention and introduction, the latter with relatively low input and little risk. For a developing country, the key in whether it could boost economic development by using the technological gap lied in its ability to promote quick technological innovation based on the introduction of foreign technologies.

Hu Angang held that technological introduction and indigenous innovation were not in an "either-or" relation but were complementary to each other, and China should transform from imitative growth to innovative growth in the long term. However, as an economist, he pointed out that the choice between independent R&D and introduction of foreign technology mainly depended on cost and the market.[26]

But people in the S&T circle were strongly opposed to the economists' view that China's S&T development should focus on "introduction". Ma Junru, former director of State Administration of Foreign Experts Affairs, insisted that introduction could not solve our problems. State-owned large and medium-sized enterprises, such as those engaged in energy, electric power and mining, used the most capital and materials of the nation, and had spent hundreds of billions on several rounds of technological renovation. If we emphasized technological introduction over innovation, our large SOEs would still have to buy foreign technologies for upgrade in a few decades.

In fact, a lot of technologies couldn't be introduced. The mathematical model we used for weather forecast was imported from Europe, but Europe made it clear that the model it exported to

[24]Feng (2004).

[25]Ibid.

[26]Ibid.

us must be ten years behind, not to mention defense technologies concerning national security, such as satellite gyroscope and early warning plane. "We have to know that lots of technologies cannot be bought with money. First of all, the essence of market economy is competition, and enterprises would never sell their core technologies to competitors. Second, developed countries in the west have explicit restriction on us because China is a socialist country. Anyone in the science and technology circle and business circle would remember the Coordinating Committee for Export to Communist Countries,[27] which set numerous restrictions on export to China. Third, without independent R&D, we will be in a passive position even when introducing technologies," said Kong Deyong.[28]

Officials at MOST said that in China's technological introduction, complete equipment sets (hardware) accounted for the majority, while technological license and consulting service (software) only took up less than 20% each, indicating Chinese enterprises' general reliance on foreign technology. At present, the introduced technologies were not sufficiently and efficiently digested and absorbed, and the investment in digestion was only 1/3 of that in introduction. As a result, although some industries looked flourishing on the surface, they had actually fallen into the trap of "introduction—backwardness—introduction again—still backwardness."

Those experts that were opposed to introduction mostly had worked for S&T management in the government, and their personal experience in China's S&T development gave them a profound understanding of technological introduction, which alone was unable to improve China's S&T innovation capability. Although the economic theory of comparative advantage made some sense, S&T development couldn't completely rely on economic laws; we also had to consider and refer to the laws governing S&T development and social development, the international environment, background of the time, and so on.

For instance, the "market for technology" approach adopted by the Chinese auto industry basically proved a failure. We let out our market, but technology, especially core high technology, was still controlled by others. In August 2005, Long Yongtu, secretary general of Boao Forum for Asia, and He Guangyuan, former minister of machinery industry, argued about the topic of independent auto brand at Guangzhou Auto Forum. Mr. Long believed "in the era of economic globalization, the auto industry will be an international industry. If the last five or six large automakers can land in China, I'll count it as a victory for the Chinese auto industry. We don't necessarily have to create a Chinese auto brand with sole Chinese investment."[29]

Then he added, "don't try to create an independent brand for its own sake." That didn't mean we didn't encourage domestic auto industry to work on innovation and develop independent brands, but the timing wasn't ripe yet for domestic automakers to develop their own brands.[30] Long also put forth the ideas of "market for jobs" and "market for fiscal revenues", which were similar to "market for technology".

But his views were opposed and criticized by He Guangyuan, former minister of machinery industry, who believed brand stood for intellectual property (IP), and IP stood for the owner's control of knowledge, which was its interest. Therefore, it was unreasonable to say that we didn't have to develop our own brands.

This argument became heated with the media's hype-up, and set people thinking deeply about the "Latin American mode that opens the gate wide open for multinationals and ignores the development of national brands." Both sides had some points and they once again called people's attention to independent brands and innovation,

[27]The Coordinating Committee for Export to Communist Countries was secretly formed at America's initiation in November 1949. Headquartered in Paris, it was an unofficial international organization formed by western developed industrialized countries in the field of international trade after WWII, and its tenet was restricting its member states in exporting strategic supplies and high technologies to socialist countries.

[28]Feng (2004).

[29]Shen and Yongtu (2005).

[30]Ibid.

which was supported by quite a number of people in the auto industry.

Japan became an innovative country based on the technological introduction, but that didn't necessarily suit our conditions. Japan didn't succeed by introducing technologies, but through innovation based on introduction, which was well evidenced by the amount of money it spent on secondary innovation. "From the 1950 to 1970s, Japan's introduction spending increased by 14 times, but the R&D spending on digestion, absorption and innovation increased by 73 times."[31] Besides, the time background was different too. Back then knowledge economy didn't appear, and Japan had America's support and unilateral access to the American market after WWII for a variety of reasons, but none of those conditions existed today. Therefore, China had to take the path of indigenous innovation.

After this round of arguments, the economic circle gradually became inclined to indigenous innovation. In an article of 2005, Lin Yifu said "while exerting comparative advantages, we should also attach importance to technological innovation…When we have comparative advantages in a certain industry but countries more developed than us don't, we have to carry out independent R&D."[32] Jiang Xiaojuan admitted in an article of 2008 that "the growth mode driven by low-cost labor and consistent increase of factor input won't sustain and has to be changed."[33] This meant that the argument had helped the economic and S&T circles to reach an agreement, and indigenous innovation not only became one of the S&T planning concepts in the new century but also a governance concept. Of course, there wasn't a universal conclusion as to technological introduction was good or bad, and certain technologies still had to be imported, but we had to take the path of indigenous innovation regarding strategic high technologies and core technologies.

[31]Xia (2004).

[32]Lin (2005).

[33]Xiaojuan (2008).

1 Arguments within the S&T circle

The drafting of the 2006–2020 S&T program wasn't a plain sailing. Apart from the argument between economic circle and S&T circle, the S&T circle had arguments internally as well. This program was discussed for nearly three years and revised tens of times; at first the S&T circle had intense internal arguments, and some scholars took the new plan as putting too much stress on big projects—all these factors slowed down the drafting process. When the work group began to finalize the program in the autumn of 2004, *Nature* magazine published a few commentaries, including criticisms of the S&T plan. They said China's S&T fund management wasn't transparent enough, peer review was weak, and scientific research was of low credibility.

In March 2006, America's *Science* magazine published articles by two Chinese reporters that recorded some arguments in Chinese S&T circle about our medium/long-term S&T program. Qiu Chengtong believed that basic research hadn't occupied a central position on the government's agenda, and scientific activities were closely connected with economic development just like earlier. Although the Chinese government tried to encourage foreign and overseas Chinese scholars to work in China, those specializing in basic research were far from enough. It was a bad thing that funding was persistently committed only to those areas that were currently considered important. The S&T plan ignored some important areas such as math.[34] The articles published by the famous science magazine *Nature* about China's science development will be discussed later.

According to American neurobiologist Rao Yi, the most innovative ideas usually came from a very small number of creative scientists at extremely rare moments, but planning large projects required the consensus of many scientists, so it was unrealistic to expect highly innovative scientific projects from the plan.

Tsinghua University, said "I strongly oppose the system of a chief scientist controlling tens of millions of research fund." In August 2004, eleven Chinese biologists in America jointly

[34]Xin and Yidong (2006).

submitted an open letter to Premier Wen Jiabao, in which they expressed their concerns over the ultra-large biological project in the draft S&T plan.[35] They were all Chinese scientists who had lived overseas for many years and enjoyed great influence in the biological field.

The letter read, "In the past century, results in biomedicine research have played an important role in improving human health and promoting socioeconomic development, but hardly any of those significant discoveries was achieved through anticipated or planned scientific projects. Quite on the contrary, they were achieved by scientists in small labs through creative research using their own ways and methods."[36] "Many major economic and health issues that China is faced with today have to be resolved through biomedicine research, but we cannot expect the pre-planned 'big science' projects to do that, otherwise we may get just the opposite. We suggest that in the medium/long-term S&T program, the nation should control the investment in major areas of biomedicine that are closely related with national welfare and people's livelihood, create a reasonable fund distribution system, and encourage a large group of small labs to compete for the research fund fairly, so as to fully unleash the scientists' creativity and initiative."[37] The fundamental reason why the eleven scholars questioned the large-scale biological project was that technological progress isn't achieved through big projects, especially not in biomedicine. Engineering projects like railway and bridge can be broken into small projects, but scientific project is an integral whole and scientific discovery is a matter of chance. Even if thousands of labs are working on a scientific project, the effect may be the same as one lab doing the same thing.

Some scholars suggested bringing in overseas scholars to join the review. Shi Yigong held that the government must establish a more open system to allow the participation of overseas scholars. Some critics were worried the fund might be wasted, and the expensive new equipment might be left to waste due to the lack of talents who grasped the new technologies. Others thought the plan might have a strategic mistake—limiting a 15-year-long plan to a few key topics would make it hard to change the research direction in the future. Certain scholars even predicted the drain of talents in unpopular fields. Deng Xingwang, head of Beijing-based National Institute of Biological Sciences and an agrobiologist, said "researchers will shift from their original directions to areas supported by the plan, otherwise they won't be able to obtain research fund."[38]

The other side of the argument—scientists working in China—were generally supportive of the plan. Lu Yongxiang, president of CAS, held that new scientific knowledge and inventions had to be transformed and industrialized, and the plan was an important platform to change China from the largest developing country to a global source of drive. It was regarded as a valuable attempt at turning a sporadic scientific research system into a harmonious academia that could compete with the west and make constant innovations.

Scholars that participated in formulating the plan for up-growth, reproduction and biological research believed they would create a review mechanism based on the academic capability to allocate the fund. Zhang Yaping from the Kunming Institute of Zoology of CAS said the ultra-large biological project would promote exchanges between genetics, developmental biology and evolutionary studies, which was a very promising direction.

Some supporters of the program held that it wasn't easy for domestic scientists to conduct scientific research with limited funding. RMB2.5 billion was the money for a long period of time and had to be distributed to each and every specific project, which didn't get much individually. Zhu Qingshi, an academician of CAS, said China had to carry out "big science" projects. Some biologists in the country also said there was no big project in life sciences in the past; now that the nation began to pay attention to this field, they wouldn't miss this opportunity.

[35]Yuan (2004).

[36]Ibid.

[37]Ibid.

[38]Xin and Yidong (2006).

Weather in China or the US, the approval of large projects was neither easy nor simple but required a long period of demonstration. The larger the project and the more extensively it involved, the more arguments. The process of demonstration and decision making should be improved to be more open and democratic. Scientific or engineering projects usually reflected the proposing scientists' preference for the areas they specialized in and their perception of the importance of those areas. The "big science" projects should not be linked with the interests of certain individuals or groups, and the government must have the final say considering China's limited funding. China had its special conditions and reasons to launch "big science" projects given the specific background.

An expert in the basic science team who participated in drafting the national medium/long-term S&T program pointed out that they launched the "big science" projects in hopes of pulling together the whole nation's forces to push China to the international arena of certain scientific fields at an early date. In comparison with foreign countries, the scientific research in China was of a low level, but we had abundant human resources and the advantage of teamwork, so the "big science" mode, to some extent, could overcome the disadvantage of low research level. At present, China hoped to develop life sciences within a limited period of time and soon secure a footing on the world stage of this field,[39] for which, according to him, the "big science" mode was a good approach. Small-scale research was too slow for the complex life systems, and "big science" mode could speed up the research by utilizing the multitudes of data generated in successful human genome studies. Protein was the most basic unit of genes, and it was necessary to digest and transform these data through the "big science" mode, as workshop-style research couldn't meet the demand of gene study.

Manned space flight, trunk airliner, large-piece equipment, large marine transportation vessel, IC and computer operation system were all strategic industries, the development of which reflected a nation's will and spirit. The national medium/long-term S&T development program was a comprehensive, strategic and forward-looking top-level design of the priorities and resources of China's S&T development. The outline of the plan was a government action that had to reflect national will rather than become a coverall summary. China's S&T progress had to be firmly based on indigenous innovation. Instead of debating whether the "big science" or "small science" mode was good for China, we should blaze a path of indigenous innovation with Chinese characteristics according to our special conditions and strategic needs.

In twenty-first century S&T programs, China must take the path of indigenous innovation. Facts have proved over and again that the real core technology cannot be imported. Only by acquiring the ability of independent development will China be able to compete in the international market without being controlled by others. Stored Program Control (SPC) exchange, for example, cost nearly USD500 when China imported it from France in the 1980s, but after China successfully developed it on its own its price quickly fell and was less than one US dollar in the end.

Besides, the National Natural Science Foundation, the main source of fund for basic research in China, was increased dramatically, giving the scientists more freedom in choosing topics. This also proved the special conditions in China. The critics offered reasonable advice, but it didn't necessarily suit China.

Those that were obsessed with globalization stressed using foreign capital and technologies to make up for our "comparative disadvantages", but technological introduction and foreign direct investment (FDI) were of little help for enhancing China's S&T competitiveness. Foreign enterprises barely opened up their technology in the host country, and they were controlling technological transfer and diffusion to China more strictly, so their technological overspill had little direct effect on enhancing our technological innovation capability. On the contrary, technological introduction led to "path dependence" and suppressed the improvement of indigenous

[39]Yuan (2004).

innovation capability. The inertia of "path dependence" has caused the failure of the "market for technology" path that Chinese auto industry has taken in the past 20-plus years.

Indigenous innovation reflects national will—it is the strategic consideration made by the nation to solve substantial difficulties in national economy, such as irrational economic structure and extensive growth mode. It not only represents a strategic shift in S&T development path, but is also a major breakthrough in economic development strategy and policy, and a strategic arrangement to raise the low technological level of Chinese industry and free us from foreign control in key technologies.

At last, the national S&T leading group ended the argument by shifting the focus to how to formulate the specific S&T plan. The argument deepened the public's understanding of this S&T plan, and extensive public participation was a feature of the plan. It revealed the differences and conflicts between the popular science mode in the west and the big science mode with Chinese characteristics, which promoted the implementation of the "indigenous innovation" philosophy in new S&T programs.

III. S&T program of the new century receiving close attention from the government

At the 16th CPC National Congress in November 2002, Jiang Zemin called for strong development of education and science and proposed to "formulate a long-term development plan for science and technology." On January 6, 2003, the then premier of the State Council Zhu Rongji held the 12th meeting of the National Leading Group of Science, Technology and Education, at which Xu Guanhua, Minister of Science and Technology, made a report on China's S&T development strategy. The meeting decided that MOST would work with relevant departments to study and formulate the long-term S&T plan of the new century.

The first plenary session of the new State Council held on March 22, 2003 decided to study and formulate the national medium/long-term S&T development program, and this strategic decision that concerned the nation's long-term interests was officially put on the new administration's work agenda. On May 30 that year, Premier Wen Jiabao chaired a meeting of the National Leading Group of Science, Technology and Education for the first time at Zhongnanhai, at which they reviewed and passed on principle the *Work Plan for Formulating National Medium/long-Term Science and Technology Development Program* submitted by MOST. It was pointed out at this meeting that the first two decades of the twenty-first century were a strategic period of opportunities for China's scientific, social and S&T development, during which we should give full play to the role of national medium/long-term S&T program and exert S&T as the primary productive force. The plan had to make overall and holistic arrangements from a comprehensive, strategic and forward-looking perspective; set clear goals and priorities; and put strategic research in a very important position while putting heads together in a scientific and democratic spirit.

On June 6, 2003, the State Council formed the leading group of national medium/long-term S&T development program involving 23 departments to make decisions on major issues that appeared during the plan-making process. Premier Wen Jiabao and State Councilor Chen Zhili were head and deputy head of the leading group, and 24 ministerial-level leaders were group members, including CAS president Lu Yongxiang, Chinese Academy of Engineering (CAE) president Xu Kuangdi, director of the General Equipment Department Li Jinai, and Minister of Science and Technology Xu Guanhua. The office of the leading group that comprised 11 departments was responsible for organizing the planning work. It was set in MOST, with Minister Xu Guanhua as director and leaders of NDRC (National Development and Reform Commission) and MOF (Ministry of Finance) as deputy directors.

With the approval of the State Council, an expert panel focusing on the overall strategy of national medium/long-term S&T development program was formed, with Zhou Guangzhao, Song Jian and Zhu Guangya as conveners and 21 senior scientists including Wang Xuan as members. The expert panel's duty was to provide advisory opinions on the direction of strategic research and on major issues, tasks and strategic goals of S&T development, and provide advisory opinions and suggestions on major disputes that appeared in the process of strategic research and plan formulation. Later 20 strategic research groups were formed, which comprised about 2000 experts from the S&T circle, social science circle and management circle, who engaged in the strategic research on 20 specific topics. Such a senior leadership and rigorous organization was rarely seen in previous S&T programs. Attention on the political level had something to do with the quality of S&T program.

The leading group worked out a "three-step" timetable for formulating the plan, which was divided into three stages—the first stage (June–December 2003) was for strategic research (Table 5.2); the second stage (January–June 2004) was for drafting the *Outline of National Medium/long-term Science and Technology Development Program (2006–2020)*; and the third stage (July–September 2004) was to review and adopt the *Outline* and the *2006–2010 National Science and Technology Development Plan* (the 11th five-year S&T plan).

More than 1000 leading researchers from various circles were gathered in the strategic research period, and 1000-plus non-leading researchers participated in the study of specific subjects. Based on the strategic research, more than 2000 experts from relevant departments, CAS, CAE and some universities, research institutes and enterprises jointly drafted the outline, which was revised 12 times and went through five stages—preliminary preparations, framework design, task listing and policy sorting, formation of draft, and solicitation of opinions. The draft outline was reported to the leading group multiple times, extensively solicited opinions from departments and local governments several times, was revised and finalized, and reviewed and adopted by the executive meeting of the State Council. On December 30, 2005, the *Outline of National Medium/long-term Science and Technology Development Program (2006–2020)* was officially issued by the State Council.

This S&T program was formulated in a much different way from previous ones. For instance, it was the first time in China's S&T planning history that CAS, CAE and CASS all provided advisory services, the first time that a large-scale planning strategy forum was held, and the first time that social science experts were engaged. Three major mechanisms were used in the process—communication and coordination, strategic consulting and public participation. Ever since the planning office started working, it held several high-quality consultative meetings with overseas experts and S&T luminaries. On November 13, 2003, the "*Chinese Science and Technology Toward 2020—International Forum on National Medium/long-term Science and Technology Development Program*" was held, which was attended by 17 famous S&T strategy experts, scientists and government S&T officials from France, America, Canada, Japan, Russia and relevant UN bodies, as well as 20 overseas Chinese experts and scholars (Fig. 5.2).

The principle of "public participation" was reflected when formulating the national medium/long-term S&T development program, which never happened before. Some national newspapers opened columns to publish the opinions of experts in various sectors on major issues in strategic research. In May 2003, MOST launched a website on "national medium/long-term S&T program," on which it posted the major strategic issues related with the plan to interact with the public. The website included over 10 million characters and was viewed more than 100,000 times, collecting nearly 1000 opinions and suggestions from the public.

Also different from previous S&T programs was the fact that voices from enterprises, including private ones, were given much weight during strategic research. The planning office held several symposiums of enterprises and handed out 3000 questionnaires among the public. China Mobile, Datang Telecom Technology, SAIC Group and China Aerospace Science and Technology Corporation (CASC) all

Table 5.2 2006–2020 National Medium/long-term Science and Technology Development Program—strategic research[40]

No.	Subjects of strategic research
Subject 1	Comparison of S&T development in China and abroad; strategic study of China's medium/long-term S&T development
Subject 2	S&T system reform and study of national innovation system
Subject 3	Study of integration of military and civilian innovation systems
Subject 4	Study of regional S&T development
Subject 5	Major S&T needs and S&T development for building a moderately prosperous society in all respects and taking the path of new-type industrialization
Subject 6	Study of "agriculture, farmers and the countryside" in China
Subject 7	Population, health, and public security
Subject 8	Study of issues concerning sustainable development
Subject 9	Study of issues concerning manufacturing development
Subject 10	Study of issues concerning the industrialization of high and new technologies
Subject 11	Study of issues concerning modern service industry
Subject 12	Study of issues concerning national defense technology
Subject 13	Study of issues in cutting-edge science
Subject 14	Study of strategic high technologies
Subject 15	Study of S&T HR
Subject 16	Study of S&T popularization
Subject 17	Study of issues concerning S&T input
Subject 18	Study of S&T conditions and platforms
Subject 19	Study of basic, public-welfare S&T work
Subject 20	Study of S&T laws and policy environment

sent representatives to express their views at the symposiums.

Members of the leading group also visited and consulted with senior S&T experts, many of whom commented that the new-generation Party leadership attached great importance to S&T planning and attentively listened to the scientists' opinions, and that political leaders and scientists jointly thought about the nation's future and made scientific judgments. Only when political leaders and scientists worked closely together could we formulate an S&T plan that would deliver tangible and substantial results.

[40]Office of national medium/long-term science and technology development plan, "Report on the Strategic Research on National Medium/long-term Science and Technology Development Program," 2006.

The "scientist + politician" combination was perhaps the common feature of all S&T programs formulated since the People's Republic of China was founded, only with different degree and rate of combination in different periods and each specific plan. Generally speaking, the more supportive political leaders are, the better the S&T program and the bigger its role will be. Of the ten MLST programs so far, the ones of 1956–1967, 1986–2000 of 2006–2020 were acknowledged as rather important in history, all of which featured the effective teamwork of political leaders and scientists. "We firmly believe that with the concerted efforts of the S&T circle and the whole society, this medium/long-term program will be a significant milestone and

Fig. 5.2 Wen Jiabao meets with some foreign experts attending the "Chinese Science and Technology Toward 2020—International Forum on National Medium/long-term Science and Technology Development Program"[41]

leave a deep footprint[41]in China's S&T development history."[42] All S&T programs after the PRC was founded were made within the political framework with the united efforts of the scientific and political circles. The formulation and implementation of national S&T development programs was a government action with clear-cut goals.

Besides, this S&T program also displayed a deeper understanding of the difference between science and technology. The full name of this program was *Outline of National Medium/long-term Science and Technology Development Program (2006–2020)*, separating science and technology for the first time instead of putting them together like before as if they were one and the same, which meant the Party and the government better understand their differences. This was a positive sign. In view of the process of making this S&T plan, the new government obviously hoped it could be mentioned in the same breath as the 1956 one and make historic contributions to China's scientific, technological and social development like that one did in the early years of the PRC. Therefore, it hoped to roll out S&T tasks and projects as momentous as the "two bombs and one satellite".

IV. The much expected S&T program of the new century

1. Priorities

As the *Outline of National Medium/Long-term Science and Technology Development Program (2006–2020)* is being implemented, it's more appropriate to analyze its contents. The plan is quite long and consists of ten parts: (1) Prelude; (2) Guideline, objectives and general arrangements; (3) Key areas and priority subjects; (4) Major projects; (5) Cutting-edge technologies; (6) Basic research; (7) S&T system reform and establishment of a national innovation system; (8) Several important policies and measures; (9) S&T input and basic conditions and platforms; and (10) Cultivation of talents. This S&T program differs from previous ones in many ways. The prelude stresses that "China is a big economy, but not a strong one, and a

[41]Xinhua News Agency: http://www.xinhuanet.com.

[42]Guanhua (2004).

fundamental reason is the weak innovation capability."[43] The S&T program has to answer how to turn a big economy into a strong one.

The *Outline* generalizes the guideline of the medium/long-term S&T program with 16 characters and clearly explains its connotations, "The guideline of S&T work in the next 15 years is indigenous innovation, breakthroughs in key areas, supporting the development and leading the future." Indigenous innovation means intensifying original innovation, integrated innovation and innovation based on introduction, digestion and absorption in order to enhance the nation's innovation capability. Breakthroughs in key areas mean focusing the resources on making breakthroughs in key areas that have certain foundation and advantages and concern national economy and security as well as people's livelihood, and achieving leapfrog development. Supporting development means specifically tackling major, key and common technologies in view of the urgent real demand to support sustained and coordinated economic and social development. Leading the future means to making deployments in cutting-edge technologies and basic research from a futuristic perspective, create new market demand, foster emerging industries and lead the future economic and social development. This guideline summarizes China's S&T development experience in the past half-century and more and is an important choice we've made to realize the great rejuvenation of the Chinese nation."[44]

Before and after the guideline was put forth, there were many arguments and the guideline was revised over and again, which reflected people's rational reflection on S&T development. The plan explicitly responded to the aforementioned arguments over technological introduction. During the 20-plus years of reform and opening up, China imported a lot of technologies and equipment, which played an important role in raising the technological level of our industries and boosting economic development. However, we must be clearly aware that the sole reliance on technological introduction and neglect of digestion, absorption and re-innovation would eventually weaken our capability of independent R&D and make us fall farther behind the world advanced level. Facts showed that core technologies in key areas that concerned the economic lifeline and national security couldn't be bought. To grasp the initiative in fierce international competition, we had to improve our indigenous innovation capability, control a batch of core technologies in several important fields, possess a batch of IPs, and foster a number of internationally competitive enterprises. In short, improving the indigenous innovation capability must be made a national strategy and implemented in all aspects of the modernization drive and all industries, sectors and regions, so as to raise our nation's competitiveness by a large margin. This statement clearly explained the importance of indigenous innovation and its critical role in promoting national economy, indicating a change in the general philosophy of S&T program in the new century. The planning philosophy of "indigenous innovation" was applied to the new S&T program.

The plan set the following goals: by 2020, the total social R&D input will account for more than 2.5% of GDP, the contribution to economic growth made by S&T progress will exceed 60%, reliance on foreign technologies will be lowered to <30%, and the annual grant of Chinese invention patents and the number of scientific papers quoted will both rank among the top five in the world. Those are the signs of China becoming an innovative nation. Although in China, the word "innovation" has different and more extensive meanings than overseas, the somewhat ideological publicity of it has fostered a consensus among all Chinese people on innovation and the national innovation system, and indigenous innovation, therefore, became one of the main philosophy of S&T programs in the twenty-first century.

[43] "Outline of National Medium/long-term Science and Technology Development Program (2006–2020)," *Bulletin of the State Council of PRC (2)*, 2006.

[44] Ibid.

The program proposed 11 key areas and 62 priority subjects (see Table 5.3).[45]

Table 5.3 shows that the new S&T program makes all-round deployments in key areas that have substantial impacts on China's economic and social development in the twenty-first century. The 62 priority subjects involve all aspects but are mostly related with technology, with the focus on cultivating indigenous core technologies. This reflects not only the idea that "technology comes first in S&T plan" but also an emphasis on "indigenous innovation". Such a detailed technological plan is of much greater depth and breadth than previous ones. In this sense, only the S&T program of 1956 can be compared with that of 2006. In fact, the latter was just intended to make the same important contributions as the former.

2. Major projects and cutting-edge technologies

The 1956 S&T program made history because it laid a solid foundation for China's scientific, technological and educational systems and for the realization of a string of major S&T projects such as the "two bombs and one satellite". In China's history of S&T programs, the implementation of several major projects, represented by manned space flight and hybrid rice, played a significant role in enhancing our overall national strength. In the twenty-first century, Western developed countries such as the US and European countries, as well as ROK all carried out major programs toward national goals, which was an important way to improve national competitiveness. For instance, the US had the Manhattan Project, Project Apollo, Star Wars Program, National Information Infrastructure (NII) program, the National Nanotechnology Initiative (NNI) and the Hydrogen Program in recent years; Europe has launched the EURECA, FP7, Airbus and GALILEO in succession; and ROK has carried out the Leading Technology R&D Program and alternative energy program.

The 2006 S&T program highlights a number of major projects in the strategic research stage,[46] with the aim of driving leapfrog development in productive forces through local S&T advancement and filling strategic blanks. The *Outline of National Medium/Long-term Science and Technology Development Program (2006–2020)* finalizes 16 major projects, including core electronic devices, high-end general chips and basic software, ultra-large-scale IC manufacturing technology and complete process, new-generation broadband wireless mobile communication, large plane, manned space flight and lunar exploration program. These projects cover strategic industries such as information and biology, substantial and pressing issues concerning energy, resource, environment and human health, and technologies of military-civilian integration and national defense.

These are "big science" projects that cannot be accomplished through the market mechanism alone, but have to rely on the centralized distribution of resources and concentrated efforts by the nation. These projects further manifest China's institutional superiority as a socialist country in its ability to pull together resources for massive undertakings, which is an important reason why S&T program have been formed consistently. There have been so many arguments because the government hasn't really stepped in. For instance, the "big plane project" that will be discussed later as a case study started at the time of Mao Zedong and made satisfactory progress, but it has been up and down several times for various reasons, an important one being that it wasn't implemented as a national strategic task like the "two bombs and one satellite", and a big science project like that couldn't be accomplished simply by market forces and relevant departments. That's also why our S&T programs have to reflect the national will.

[45] Based on "Outline of National Medium/Long-term Science and Technology Development Program (2006–2020)".

[46] Major project refers to major strategic project, key and common technology and major project accomplished within a specified period of time through breakthrough in core technology and resource integration in order to achieve national goals. It's the priority of priorities in China's S&T development.

Table 5.3 The 11 key areas and 62 priority subjects

Key areas	Priority subjects
1. Energy	(1) Industrial energy conservation; (2) Clean, efficient exploitation and utilization of coal, liquefaction and multi-generation; (3) Prospecting, exploitation and utilization of oil and gas resources in complex geology; (4) Low-cost, large-scale exploitation and utilization of renewable energy; (5) Ultra-large-scale power transmission and distribution and grid security guarantee
2. Water & mineral resources	(6) Optimized allocation and comprehensive exploitation and utilization of water resources; (7) Overall water conservation; (8) Seawater desalination; (9) Resource prospecting to increase reserve; (10) Efficient exploitation and utilization of mineral resources; (11) Efficient exploitation and utilization of marine resources; (12) Comprehensive resource zoning
3. Environment	(13) Comprehensive pollution control and waste recycling; (14) Restoration of ecological functions in ecologically vulnerable regions; (15) Marine ecology and environmental protection; (16) Global environmental change monitoring and countermeasures
4. Agriculture	(17) Discovery, storage and innovation of germplasm and directed breeding of new species; (18) Healthy breeding and epidemic control of livestock, poultry and aquatic products; (19) Deep processing and modern storage and transportation of agricultural products; (20) Comprehensive exploitation and utilization of agroforestry biomass; (21) Ecological security of agroforestry and modern forestry; (22) Development of environment-friendly fertilizers and pesticides and eco-friendly agriculture; (23) Multi-function agricultural equipment and facilities; (24) Agricultural precise operation and informatization; (25) Modern dairy industry
5. Manufacturing	(26) Basic and general parts; (27) Digital and intelligent design & manufacturing; (28) Green and automated development of process industry and its equipment; (29) Recyclable iron & steel process and equipment; (30) Large marine engineering technology & equipment; (31) Basic raw materials; (32) New-generation information materials and devices; (33) Key materials and engineering of military industry
6. Transportation	(34) Transportation infrastructure construction and maintenance technology and equipment; (35) High-speed railway system; (36) Low-energy-consuming and new-energy automobiles; (37) Efficient transportation technology and equipment; (38) Intelligent transport management system; (39) Transportation security and emergency guarantee
7. Information industry and modern service industry	(40) Information technology and large application software supporting modern service industry; (41) Key technology and service for next-generation network; (42) High-efficiency trusted computer; (43) Sensor network and intelligent information processing; (44) Digital media content platform; (45) HD large-screen display; (46) Information security of core applications
8. Population & health	(47) Safe contraception and prevention and treatment of birth defect; (48) Prevention and treatment of major non-contagious diseases such as cardio-cerebrovascular disease and tumor; (49) Prevention and treatment of common and frequent diseases in urban and rural communities; (50) Inheritance and innovative development of traditional Chinese medicine (TCM); (51) Advanced medical equipment and bio-medical materials
9. Urbanization & urban development	(52) Urban zoning and dynamic monitoring; (53) Urban function improvement and intensive use of space; (54) Architectural energy conservation and green building; (55) Quality assurance of ecological and living environment in cities; (56) Urban information platforms

(continued)

Table 5.3 (continued)

Key areas	Priority subjects
10. Public security	(57) National public security emergency information platform; (58) Early warning and rescue for major production accidents; (59) Food safety and entry/exit inspection and quarantine; (60) Prevention and fast disposition of public emergency; (61) Bio-security guarantee; (62) Major natural disaster monitoring and prevention
11. National defense	

The program makes forward-looking deployments for cutting-edge technologies,[47] which, as they represent the direction of the most advanced technologies in the world, will lead the formation and development of emerging industries in the future and will assist in the upgrade of industrial technologies. The S&T program lists cutting-edge technologies in eight aspects, as shown in Table 5.4.[48]

Table 5.4 shows a spate of strategic high technologies that are critical for economic development and national security. Technological plan is always given priority in the Chinese history of S&T plans, exactly as in "technology comes first in S&T plan." Technology takes up a large proportion in every S&T plan partly because it's easier to predict and make research arrangements for technology than for science, and partly because technology can yield results in a short period of time, so that national will and people's interests can be reflected quickly. Therefore, "technology comes first in S&T plan" is a common concept shared by all S&T programs in China.

3. Basic research deployments in advance

"Technology coming first" doesn't mean the medium/longer S&T program neglects science, especially basic science. Instead, it overcomes this tendency that appeared in the previous three S&T programs (Table 5.5) and puts basic science in a prominent position, which is its important feature. The *Outline* holds that "in developing basic research, we must combine serving national goals with encouraging free exploration, follow the laws of scientific development, value scientists' exploring spirit, and emphasize science's long-term value. We should give it steady support, make deployments in advance, and adjust them dynamically according to new situations in science development."[49] The statement of combining serving national goals with free exploration indicates that the Party and the government have had a deeper understanding of the characteristics of free research and basic science. Basic science must serve national goals, especially for a fast-developing country like China as we need to resolve key and difficult issues in future development through basic research. The principles for selecting research directions are as follows: the subjects or research fields should be of strategic, overall and long-term significance for economic and social development and national security; be of critical importance for development despite current weakness; or be able to drive the integration of basic science and technological science and lead future high-tech development.

It is better to conduct basic scientific research in multiple directions without setting specific goals. Basic science not only serves the country but also the whole human race. If a specific direction or goal is set for it, it will become applied basic scientific research in the Pasteur's

[47]Cutting-edge technology refers to forward-looking, pioneering and exploratory high technology. It is an important basis for future technological upgrade and the development of emerging industries, and a comprehensive reflection of a nation's high-tech innovation capability.

[48]Based on "Outline of National Medium/Long-term Science and Technology Development Plan (2006–2020)".

[49]"Outline of National Medium/long-term Science and Technology Development Plan (2006–2020)," *Bulletin of the State Council of PRC (2)*, 2006.

Table 5.4 Cutting-edge technologies in the 2006–2020 national medium/long-term S&T outline

Cutting-edge technology	Specific subjects
1. Bio-technology	(1) Target discovery; (2) Animal/plant variety and pharmaceutical molecule design; (3) Gene manipulation & protein engineering; (4) Human tissue engineering based on stem cells; (5) New-generation industrial bio-technology
2. Information technology	(6) Intelligent sensing; (7) Self-organized network; (8) VR
3. New material technology	(9) Intelligent materials and structure; (10) High-temperature superconducting; (11) High-efficiency energy material
4. Advanced manufacturing technology	(12) Extreme manufacturing; (13) Intelligent service robot; (14) Major product & facility life forecast
5. Advanced energy technology	(15) Hydrogen energy and fuel cell; (16) Distributed energy supply; (17) Fast neutron reactor; (18) Magnetic confinement fusion
6. Marine technology	(19) 3D monitoring of marine environment; (20) Multi-parameter fast detection of ocean floor; (21) Natural gas hydrate development; (22) Deep-sea operation
7. Laser technology	
8. Space technology	

Table 5.5 The 2006 S&T program makes basic research deployments in four aspects[50]

Basic research	Specific subjects
1. Subject development	(1) Basic subjects; (2) Inter-disciplines & emerging ones
2. Cutting-edge science	(1) Quantitative research and systematic integration of life process; (2) Condensed matter & new effect; (3) Deep-level structure of matters and large-scale physical laws of the universe; (4) Core math and its application in crossover fields; (5) Earth system processes and their effects on resource, environment and disaster; (6) Chemical process of the creation and transformation of new matters; (7) Brain science & cognitive science; (8) Scientific experiment and innovation in observing method, technology and equipment
3. Basic research oriented toward the nation's major strategic demands	(1) Biological basis for human health and diseases; (2) Scientific issues regarding biological inheritance and modification in agriculture and agricultural sustainability; (3) Mechanisms of how human activities affect earth system; (4) Global change and regional response; (5) Complex systems, disaster formation and their forecast and control; (6) Key scientific issues in energy sustainability; (7) New principles and methods of material design and preparation; (8) Scientific basis for manufacturing in extreme environmental conditions; (9) Major dynamics issues in aviation and spaceflight; (10) Scientific basis that supports IT development
4. Major scientific research programs	(1) Protein study; (2) Quantum regulation & control; (3) Nano study; (4) Developmental & reproductive study

Quadrant rather than pure free research. In China's S&T programs, both technology and basic science serve national interests. Some of the arguments in 2003 and 2004 over S&T plan and issues discussed in the supplements to *Nature*, which will be mentioned later, stem from the understanding of basic science. To respect the laws governing subject development while serving specific national goals—such a "basic scientific research mode with Chinese characteristics" has a strong national orientation.

[50]Based on "Outline of National Medium/Long-term Science and Technology Development Plan 2006–2020".

4. New S&T planning philosophy

The S&T program also makes deployments for S&T system reform and talent strategy, the former aiming to ultimately create a national innovation system. As our perception and understanding of this system have deepened over time, the *Outline* defines and elaborates on it in great detail.

"The purpose of deepening S&T system reform is to advance the construction of a national innovation system, which is a social system where the government takes the lead, the market plays a basic role in resource allocation, and various S&T innovation entities interact closely and efficiently. In the current stage, the construction of a national innovation system with Chinese characteristics has the following focuses. First, build a technological innovation system with enterprises as the principal entities and integrating industry, colleges and research institutes, and take it as the entry point to comprehensively build a national innovation system. Second, build a knowledge innovation system integrating scientific research with higher education. Third, build a national defense S&T innovation system integrating the military and civilian sectors. Fourth, build regional innovation systems with respective characteristics and advantages. Fifth, build a social, networked S&T intermediary service system."[51]

The proposal of the five innovation systems indicates the maturity of the innovation system theory with Chinese characteristics and demonstrates the formation and perfection of an important philosophy in China's S&T programs—"indigenous innovation". The proposal for a national innovation system also reflects the CPC's deeper and more mature understanding of S&T.

Besides, the 2006 medium/long-term S&T program reflects a new philosophy—people centrism, which is the center of the Scientific Outlook on Development and is fully manifested throughout the plan. For instance, China has the second-largest primary energy consumption in the world and is facing serious energy problems, such as sharp conflict between energy supply and demand, unreasonable energy structure, low use efficiency and pollution caused by coal burning. If we don't diversify the energy portfolio and implement energy conservation through S&T innovation, China's reliance on petroleum import and natural gas import will exceed 55% and hit 25%–40% respectively in 2020, when the people not only have to bear the rising energy price but may also face petroleum and power shortage. The new S&T program lists industrial energy conservation, clean and efficient development and utilization of coal, and low-cost and large-scale exploitation and utilization of renewable energies as priority subjects, and lists the development of large oil and gas field and coal-bed methane as well as new-type nuclear power stations in the 16 major projects. They will provide S&T support for dealing with the possible energy crisis in the future and guarantee basic energy supply in daily life. The program also gives priority to efficient transportation system, low-energy-consuming and new-energy vehicle, intelligent transportation management system, and so on, which are targeted at three hotspot issues concerned by the public—transportation energy consumption and pollution, traffic safety, and traffic congestion in big cities.

Population and health are a cornerstone for social stability and economic development. Two of the 16 major projects are about health—invention of substantial new medicine, and prevention and treatment of major infectious diseases such as AIDS and viral hepatitis. All these tell us that the S&T program attaches great importance to the quantity, structure and quality of population, highlights people's livelihood, and follows the concept of "people centrism".

The Party and the government paid close attention to the S&T program. To prevent it from staying on the paper, the government later rolled out many implementing rules and set about formulating the 11th five-year S&T plan, so that the medium/long-term S&T program could be implemented through every five-year and annual S&T plans. In 2006, the State Council issued the "*Notice on Several Auxiliary Policies for*

[51]"Outline of National Medium/long-term Science and Technology Development Program (2006–2020)," *Bulletin of the State Council of PRC (2)*, 2006.

Implementing the Outline of National Medium/Long-term Science and Technology Development Program (2006–2020)",[52] whereby it rolled out 60 implementing rules covering ten aspects, including S&T input, tax incentive, financial support, government procurement, innovation based on the introduction, digestion and absorption, IP creation and protection, talent team, education and science popularization, S&T innovation base and platform, and better coordination and overall arrangement.[53] Departments and local areas all worked out their own plans according to the *2006–2020 Medium/Long-term S&T Program*, such as the *Outline of the Action Plan for Improving the Scientific Competence of all Chinese (2006–2010–2020).*

5.3 Nature's Comment on China's S&T Development and Plan and Reflection

China formulated ten medium/long-term S&T programs from the founding of the PRC to 2020, including the 2006–2020 program. How did the first eight S&T program turn out? It's not the job of this book to make a comprehensive, quantitative assessment on S&T programs, but we can make a qualitative comment.

As mentioned earlier, the *Nature* magazine weighed in on S&T program in 2006. This world-top science magazine issued supplements in 2003 and 2004 (Figs. 5.3 and 5.4) in Chinese under the title of "China voices I & II," which had never happened before and constituted a milestone in China's S&T history. *Nature* had published special issues on China before,[54] but the supplements, which focused on criticism and advice, were more influential, and many articles therein talked about China's backward management of and innovation in S&T programs. In 2003–2004, the 2006–2020 S&T program was being formulated and was definitely influenced by *Nature*'s "China voices", so we'd better elaborate on this here.

Nature and *Science* are acknowledged as the best science magazines in the world. Having papers published in them is considered an honor for scholars and scientists, and many Nobel prizes have come out of their articles. Many universities in China offer bounties to authors whose papers are published by those two magazines, and this also shows their high position in the Chinese academia. Therefore, *Nature*'s China supplements can be viewed as a helpful comment on China's S&T development and problems in the past few decades. Since the medium/long-term S&T program is in a central position in China's S&T development system, an overall grasp of our S&T development would help us better understand the historical evolution and changing philosophy of S&T programs, and it's

[52]Document titled Guo Fa (2006), No. 6.

[53]Reply of the General Office of State Council on the approval to formulate implementing rules for "Several Auxiliary Policies for Implementing the Outline of National Medium/long-term Science and Technology Development Plan," *Bulletin of the State Council of PRC (17)*, 2006.

[54]In the autumn of 2001, *Nature* issued a supplement titled *Rise of the Dragon* (Readings from *Nature* on the Chinese Fossil Record). The Chinese and English versions were published simultaneously in China and Britain. Jiang Zemin, the then chairman of the People's Republic of China, wrote the title for the Chinese version, quoted from Ji Qiang, Institute of Geology, Chinese Academy of Geological Sciences, Beijing. *Working for the Rise of Palaeontology in China—Congratulations on the Issuance of the "Rise of Dragon," Supplement to Nature*, quoted from *Geological Review (1)* 2002 (p. 8). I contacted the publishing house of *Nature* in Japan and was presented the supplement *China voices* of 2003 and 2004, but not *Rise of the Dragon*.Besides, *Nature* held a premiere of its supplement *Focus on Shenzhen* at the China Hi-tech Fair in Shenzhen on October 12, 2007, the first time it dedicated a supplement to a city in Chinese mainland. The issue on September 26 that year also published a long commentary titled "Enter the dragon" (*Nature 449*, 502–504, 26 September 2007), which introduced the history, present and future of Shenzhen from the perspective of its S&T development. But I think this supplement was more for the publicity of *Nature* and wasn't nearly as important as *China voices*.

Fig. 5.3 Cover of supplement to Nature 2003

convincing to draw on the comments from *Nature*.

I. **Supplement to *Nature* 2003: problems in China's science**

Why did a world-top magazine issue a supplement on China's science problems? Doctor Philip Campbell, editor-in-chief of the 2003 supplement and also of *Nature*, explained this in his foreword to the supplement.

He began the foreword with this:

"We at *Nature* feel proud of our history of keeping in touch with the Chinese scientists. More importantly, *Nature* has published excellent scientific research results from China at an increasingly higher frequency. But what you are reading now is unprecedented. It's a supplement

Fig. 5.4 Cover of supplement to Nature 2004 and some authors

to *Nature* especially compiled for Chinese researchers and published in Chinese."[55]

Then he kind of asked and answered a question all by himself (the bold-face is noted by the author of this book).

"Why would we do that? The main motivation for us to do this shows that we've realized some positive trends. **China is rapidly growing into a primary economy in the Asia Pacific and the whole world**, partly because of their inborn ambition and partly because of **their persistent pursuit of new technologies**. Equally important is the fact that China has never given up the efforts to become a world-class scientific power all these years. But these trends have a downside. **Despite its continuous development and the Chinese government's huge input in scientific research, China hasn't fully exerted its scientific potential** for reasons that are discussed in the articles of this supplement. This has not only resulted in its weak presence on the world science stage but also cost it the opportunity to fully utilize the scientific and technological results in other countries, as well as the opportunity to develop its own economy and improve the quality of life for its people. This is a thin supplement, but I hope **it will be an important attempt at helping to change this very abnormal phenomenon in international scientific research**. I invited a group of experts specializing in different fields to provide constructive opinions and advice on China's scientific and technological development. My original thought was that these authors must be senior researchers in the Western scientific and cultural environment, but it's also very important that they have first-hand experience of the scientific environment and opportunities in China."[56]

1. Inefficient indigenous innovation and inadequate academic exchange

The 2003 supplement had eight articles, which expounded the authors' understanding of China's S&T development in different disciplines. For the sake of demonstration, this book quotes these authors extensively. The first article was *Building Scientific Research Institutes for China—Cultural Reflection* by Mu-ming Poo,[57] who had a Chinese cultural background and had conducted scientific research for a long time both in China and the US. His article offered many insights. Based on his experience in building some research institutes in China in the past 20 years, the author increasingly realized that the obstacles to Chinese institutes standing out in the world might come from cultural rather than economic factors.

According to Mu-ming Poo, "homage to authority and existing rules is a main obstacle to scientific breakthrough and technological innovation…Any challenge to current evidence, hypothesis and concept, no matter how naive or weak it may seem, must be encouraged and taken seriously…Chinese research projects are often reputed as 'filling a scientific blank in China'. If research always follows the existing thoughts and Western modes, China will come to disadvantage in competition."[58] He encouraged critical academic exchanges. "A universal problem in Chinese academia is that scientists don't have public, critical academic exchanges. All in all, a positive attitude to academic criticism will help foster such a scientific culture in which scientific truth is much more important than 'saving face'."[59]

Speaking of non-science obstacles, Mu-ming Poo said "complaints about administrative restrictions and trivialities are universal around the world, but there are other obstacles for Chinese scientists…The government fund is allocated in a top-down approach, which gives the administrative departments of various levels a lot of responsibilities and power. As a result, administrative personnel in research institutes usually function as 'commanders' who control the fund and resources, give orders instead of

[55]Campbell (2003).

[56]Ibid.

[57]Poo (2003). He is a professor of molecular and cell biology at University of California, Berkeley, director of the department of neurobiology, and head of Institute of Neuroscience, CAS.

[58]Ibid.

[59]Ibid.

providing services for scientists."[60] Regarding long-term goals of scientific development, the author asked "regarding the unsolved major scientific issues, do we have enough resources and determination to compete with the powerful Western research institutes?" He said "organizing and mobilizing the scientists to work on a specific common goal doesn't necessarily lead to any result. The main job of the research institute is to provide a stable and favorable environment where scientists can freely explore subjects that interest them, and to provide institutional incentives to guide them to explore major scientific issues from different directions."[61] Mu-ming Poo was actually criticizing the big science projects in China's S&T plans.

At last, the author reminded that it was a bad idea and probably a waste of resources to compete with mainstream Western labs along their current directions. It took time to deviate from the mainstream influence and find one's own research direction, which required patience and perseverance from the scientists and administrators. In sum, science, like the evolution of species, might also develop through "splendid isolation".

Mu-ming Poo was to the point about the problems in China's S&T development and scientific community, for which there were many reasons. However, the author of this book holds that China's scientific undertaking isn't just of the scientific community but also of the nation. S&T plan is the nation's approach to science execution and behind it is the concept that the nation hopes to apply to S&T development.

Chien and Chien co-authored *The Silk Road of Molecular Medicine in China*.[62] They held that "for a long time, China has been in the dark about the latest scientific development in the west due to its separation from Western culture by the Pacific Ocean and the Western boundary. Moreover, the Chinese scientific circle advocates the principle of 'experience and authenticity' but lacks adventurous reform and exploration. These have seriously impeded the development of modern science in China. Therefore, the most challenging objective for Chinese scientists may be to get familiar with global scientific and technological results through extensive exchanges, for which they have to cross the boundary in both geography and the way of thinking."[63] Innovation, the only way to success, usually happens at the intersection of different fields and cultures.

Xiangzhong Yang wrote China's Opportunities and Challenges in the twenty-first Century: R&D of Therapeutic Cloning, Human Embryonic Stem Cells and Embryo Biotechnology.[64] He believed China had natural advantages in embryo cloning. "To be objective, China's embryo biotechnology research doesn't fall far behind world advanced level in certain fields and is in a leading position in specific directions. But the competition has just begun and China has a long way to go before it becomes a world leader. Take genetically modified animal technology for example. China and Western countries started almost at the same time in the 1980s, and China was technically capable of producing genetically modified animal and cloned animal then. But today, when some European and American countries have begun phase-3 clinical experiment on genetically modified drug-protein, China's genetic modification industry is still nascent, not even beyond the stage of protein expression."[65]

As to the reasons, Xiangzhong Yang said, "the scientific research fund in China, especially the input in life science, is far from enough. The Chinese government's financial support for scientific research doesn't consist with its status as the world's 6th largest economy."[66] Regarding the evaluation system, he stressed that "apart from funding and talent, we also need a favorable

[60]Ibid.

[61]Ibid.

[62]Chien and Chien (2003). Chien, guest professor of Medical schools of California. University of California, San Diego and Salk Institute; Chien, senior researcher at Dupont, Wilmington, Delaware.

[63]Ibid.

[64]Yang (2003). He is belong of the Center for Regenerative Biology, University of Connecticut.

[65]Ibid.

[66]Ibid.

environment and managing and operating system. Here I want to especially point out that for the research on embryo biotechnology, a set of review system (including recruitment and promotion) has to be established in China using internationally accepted academic indicators such as the quantity of papers published, their quality (publication, quote rate) and reputation in the international academia. These, and not the result appraisal meeting or news report, should be the standards especially for the distribution of research fund…There is no denying that there isn't a strong atmosphere of research cooperation in China yet. It's not uncommon to see labs in different universities doing similar or even identical research, and experimental apparatus is bought over and over, but these labs work on their own and hardly ever cooperate or communicate even in the same department. This at least caused the repetition of research and waste of money, and sometimes bad competition or even vicious mutual attack…In the past, many international cooperation projects only wanted to capitalize on the fame of the scientists and labs, making the cooperation superficial…We must advocate initiative while encouraging cooperation."[67] S&T plans, according to Yang, were partly to blame because planned S&T development and planned thinking might block the mind and suffocate the innovative spirit.

These articles talked about inadequate indigenous innovation and academic exchanges. Scientific research needed full academic exchange and debate, which Chinese scientists obviously didn't do as well as their Western counterparts. The fact that S&T became a means to vying for resources instead of satisfying curiosity resulted in the formation and expansion of China's utilitarian philosophy of S&T development. This was especially obvious when China was in the transition period, when our S&T programs displayed a strong "utilitarian tendency". When studying overseas, I (the author of the book) was deeply impressed by the pure academic aspirations of Western students and scholars. China's inadequacy in indigenous innovation was attributed to not well understanding this issue. After the Soviet Union stopped providing assistance for us, China had the planning philosophy of "independent development based on domestic conditions," but after China launched the reform and opening up, Chinese had such high expectations for technological introduction that who neglected indigenous innovation. When the failure of "market for technology" finally woke us up, China took a long and hard look in the twenty-first century and came up with the planning philosophy of "indigenous innovation".

2. Different perspectives of developing science

Tso wrote *Scientific Agriculture—New Vision for 2050.*[68] He voiced his views from an agricultural perspective. "China should take immediate actions to change its rural development path. It is estimated that there are 150,000 fine Chinese scientists working overseas. The Chinese government should lure them back and work for its scientific and technological development."[69] As to the Western Development Program, Tso said America only had 120 years of experience and achievements in developing its Western region, and Xinjiang might become the California in China. He stressed the importance of S&T development and basic science. "A nation's scientific and technological capability is built on basic science and advanced technology, both of which are indispensable. It's wrong for some universities to sacrifice regular biological subjects for bio-engineering…China should make utmost efforts in GMO, but it cannot neglect basic science either."[70] S&T plans in the transition period did neglect basic science.

Chung-I Wu, Suhua Shi and Ya-ping Zhang co-authored *Ecology, Evolution and Bio-*

[67]Ibid.

[68]Tso (2003). He is a senior scientist of United States Department of Agriculture, honorary chairman of Institute of International Development and Education in Agriculture and Life Sciences.

[69]Ibid.

[70]Ibid.

diversity.[71] In their opinion, "In the past 50 years, China has made remarkable achievements in fauna and flora catalogue, ecological investigation, system growth and biogeographical analysis."[72] After introducing the basic principles of biodiversity studies, they concluded that it was particularly difficult to consistently maintain and preserve biodiversity. The sixth article *Nanotechnology and Biotechnology Companies—Case Studies* was written by James C. Hsiao and Kenneth Fong,[73] who said China's "Nano Science Program" (1990–1999) was much earlier than the National Nanotechnology Initiative (NNI) launched by President Clinton in 2000. But they mentioned market risks, "scientists and technological experts usually overestimate their inventions and underestimate the difficulty in translating them into commodities…Nanotechnology and biotechnology that are drawing increasing attention will be the next favorites."[74]

The eighth article, *Follow Your Heart*, was written by Alice Shih-hou Huang,[75] who shared her opinions about how young students should choose their research directions. "In many Chinese families today, science and technology have replaced the traditional idea of imperial examination…Friends who admire Einstein see fame and fortune in him. An analysis of the lives of successful scientists today would show that causes are international, rewards are different, and approaches are often unpredictable…Only when study is applied to meet new challenges will current methods be useful."[76]

These articles discussed from different perspectives how to develop S&T and make it generate real benefits. China attached too much importance to the social, especially economic functions of science, technology and S&T plans, and the opinions from overseas Chinese scholars shed light on how to bring those functions into better play.

3. Funding for scientific research and review mechanism

The seventh article, *Improving the Yield of Scientific Research in China is Challenged*, was written by Ray Wu,[77] founder of the Society of Chinese Bio-scientists in America (SCBA), which was an influential society with 3000 members in America, all being senior experts in the biological field. He will be quoted extensively here. Ray Wu pointed out in the article that, "In the past 20 years, especially the recent 5 years, the Chinese mainland has made substantial progress in basic scientific research, which, however, still has a very low yield as evidenced by the small number of high-quality and influential papers published every year. I think there are a few reasons for this: the small number of prolific scientists; seriously insufficient fund for basic research; and unsatisfactory research project review system and fund-distributing system—all of them have hindered Chinese scientists in carrying out advanced and innovative research. Another key reason is that most universities don't encourage students to have critical and innovative thinking."[78]

The way to solve this problem is increasing the fund for basic research, establishing a fairer and more transparent research plan review and

[71] Wu et al. (2003). Chung-I Wu is a professor at the Department of Ecology and Evolution of the University of Chicago; Shi Suhua is a professor at the national key lab of biological prevention and control, School of Life Sciences, Sun Yat-sen University; Zhang Yaping is a professor at Kunming Institute of Zoology, CAS.

[72] Ibid.

[73] Hsiao and Fong (2003). Hsiao is chairman of Shanghai Inframat Nano Materials Co., Ltd.; Kenneth Fong is chairman of Hi-Shine Investment Holding Co., Limited.

[74] Ibid.

[75] Huang (2003). She is a Senior Councilor for External Relations and Faculty Associate in Biology and an academician of Academia Sinica in Taiwan. She used to be a professor of microbiology and molecular genetics at Harvard Medical School, and president of New York University Polytechnic School of Engineering.

[76] Ibid.

[77] Wu (2003). Ray Wu is a professor at the Department of Molecular Biology and Genetics, Cornell University. He's been engaged in the basic research of bio-chemistry and molecular biology in the US for more than 50 years and published several papers on prestigious academic magazines.

[78] Ibid.

fund distribution system, and reforming the basic structure and administrative system of universities and research institutes. Ray Wu believed "China allocated too little fund for basic scientific research, which accounted for only 0.02% of GDP in 2000. Although its annual R&D expenditure accounts for 1.1% of GDP now, most of the money goes to development projects. In comparison, the US government allocates about USD30 billion, or 0.3% of GDP, to basic research on life sciences every year, plus another USD30 billion from the industrial sector and various foundations, which is basically zero in China...The small input in basic scientific research reflects the amount of attention paid by the nation."[79]

Regarding the review and selection of research project, Ray Wu said "the research plan review and fund distribution system in China needs substantial improvement. Basically, the majority of its basic research fund is distributed through the following three channels: CAS, MOST (especially Program 973), and Natural Science Foundation of China (NSFC). Several scientists who can be trusted told me that big projects (key and major ones) of NSFC and some projects of CAS and MOST often don't go through rigorous peer review, and the distribution of funds is usually affected by senior officials. To sum up, these three channels of distributing basic research funds all have problems in big project review. Some say that 'strict review for small fund, loose review for moderate fund, and no review for large fund'. Reform of this system is imperative. The way to reduce these problems is inviting a group of qualified reviewers from abroad to join the jury, and they should make up over half of the jury."[80]

Wu also said reward for researchers must be based on their performance in the research project rather than on their seniority or administrative performance. It was also necessary to encourage outstanding scientists to stay in research positions instead of transferring them to administrative ones. He was obviously confident in the future of Chinese science.[81]

China's scientific research fund was insufficient, and its distribution was sometimes irrational too, implying both resource shortage and waste. The reasons were very complicated. China's S&T programs are in effect important methods to distribute and deploy its S&T resources. How to make those resources serve national goals and scientists more effectively is a question faced by future S&T programs. The science review mechanism in China has many problems, which concern the S&T system reform, but it seems rapid improvements have been made there according to a recent MOST document. However, it's unrealistic to put foreign experts on over half of the jury seats because should that happen, we wouldn't be able to guarantee that the jury would endorse national will when there was a conflict between national goal and the goal of the scientific community. This is what this book has been saying over and over again. China's S&T program isn't an agreement or research outline of the scientific community, or the research plan after such agreement or outline is approved by the central government. It is a government action and a top-down approach to achieving national goals. That's where S&T plan with Chinese characteristics differs from any foreign disciplinary plan that is independently made, and it is exactly these Chinese characteristics that have ensured China's economic takeoff and massive enhancement of national strength, which, nonetheless, come with many problems as well.

II. **2004 supplement to *Nature*: reflection on and expectation for S&T program**

Authors of the articles in the 2003 supplement to *Nature* were all scholars who had worked in the US for many years. Their opinions were valuable but people would doubt how much they knew about the special conditions in China because immense changes were taking place in the country every year. Then *Nature* issued

[79]Ibid.

[80]Ibid.

[81]Ibid.

Fig. 5.5 Dr. Philip Campbell, editor-in-chief of Nature, is giving a speech at the Nature China Voices Forum[82]

Fig. 5.6 Nature China voices forum[83]

another supplement in 2004, and the authors of articles in it were all outstanding scholars who had worked in the Chinese mainland for many years. The magazine also hosted two "Nature China Voices Forum" to celebrate the Chinese supplement published on November 18, 2004 under the title of "China voices II: Keeping pace with change"—a very Chinese and very popular phrase these days. Both forums were directed at Chinese scientific community (Figs. 5.5 and 5.6). The first one was held in Shanghai on November 22 in conjunction with Shanghai Institute for Biological Sciences, CAS, followed by the second one in Shenzhen co-hosted with Tsinghua Shenzhen International Graduate School and the University Town of Shenzhen on November 23.

Doctor David Cyranoski, editor-in-chief of the 2004 supplement, asked a question in the preface: "China is making continuous progress (in science[84]), but which way should it go?" He wrote,"China is marching forward like a bull in scientific and economic fields, but high speed has brought imbalance. It is both a developing and a developed country, and its people are facing problems that exist in both poor and rich countries. **China has the best scientists in the world, but most of them have had no standard training or advanced equipment. Huge and imbalanced—China doesn't have an existing mode to follow** (the bold-face is noted by the author of this book). Therefore, discussing this issue became very important, especially in the scientific field. As a platform for discussions, *Nature* issued the first 'China voices' in January 2004, the first independent foreign-language supplement of the magazine ever. The English version that came later let more readers know about this. 'China voices II' moved the platform to Chinese mainland. The authors last year mostly lived overseas, but the authors in this supplement mostly live in China or have worked with scientists in Chinese mainland for many years."[85] After summarizing the "views of Chinese scientists" in the 2004 supplement, Dr. David Cyranoski made a positive conclusion:

> These are all open questions. These commentators should be the first group to openly discuss them, not the last. I would like to thank them for having this discussion. Their contributions to this supplement showcase their care for China's scientific future, a future that definitely will be bright and thriving.[86]

[82]"*Nature* China Voices Forum held at Tsinghua Shenzhen International Graduate School," Tsinghua news: http://news.tsinghua.edu.cn.

[83]Ibid.

[84]The word "science" is added by the author.

[85]Preface to the Chinese version of the 2004 supplement to *Nature*.

[86]Ibid.

1. Reflection on S&T programs

The nine articles in the 2004 supplement expressed the authors' views on the foundation, mechanism and future of Chinese science development from different angles. For demonstration's sake, the main points of their views will be quoted here.

The article co-authored by Yi Rao, Bai Lu and Chen-Lu Tsou was *The Fundamental Change Needed in Chinese Science and Technology: from Traditional Rule of Man to the System of Competition—Will the Medium/long-term S&T Program Leave Excellent Legacy or Miss a Good Opportunity?*[87] All the three authors were renowned scholars, and the article, including "medium/long-term program" in its title, was met with vehement responses. Earlier in the book, we mentioned some of its viewpoints when talking about the argument over S&T plan in the new century. The authors pointed out boldly and straightforwardly that the fundamental problem with the current S&T system in China was the rule of man, and expressed their worries over the 2006–2020 medium and long-term S&T program. "We believe that the fundamental institutional problems with China's science and technology development haven't been solved yet. Its science and technology management stays in the stage of 'rule of man', interpersonal politics are playing an important or dominant role on multiple levels of the social and technological circles, while professional superiority or inferiority cannot play a due role in the current system."[88] They proposed a "competitive science and technology system" that separated administration from scientific and technological work. "Politics and science and technology are seriously misplaced at least in three aspects. First, on the national level, leaders are sometimes assigned to take a direct part in selecting specific S&T topics. Second, experts are put on occasions that facilitate academic politics rather than professional review. Third, middle and lower-level officials in various departments have too much power… 'Separating politics from science and technology' should start with defining the role of political leaders, experts and administrative personnel. Three important principles have to be observed in the reform to improve S&T efficiency in China. For one thing, political leaders can decide the total S&T fund in light of social needs with reference to the opinions of S&T leaders and experts. For another, other decisions should be made by experts in each specific sector, but the expert jury only reviews topics on which the experts have relative advantages. Thirdly, specific topics have to be selected by experts in the corresponding field. Administrative personnel in various departments are not allowed to control or sway the decisions of sponsorship, but should stay in their service-providing positions."[89] Sharp as they were, their opinions laid bare the imperfections in our S&T plan regarding S&T system reform.

The three authors were worried that big subjects in the medium/long-term program would magnify existing problems. "Technologically developed countries have big subjects too, but with different origins and approaches. Chinese departments make new plans every few years that usually contain big subjects, but these plans are mostly made under the lead of officials and administrative personnel."[90] Regarding the process of how the subjects came into being, the authors said what's in the medium/long-term program about the formation and selection of big subjects indicated the existing defects in S&T management.

In the end they concluded that the historical legacy of medium/long-term program was an important choice. "The success of the medium/long-term progam and relevant big subjects may become an important legacy of this administration. A more significant contribution of the 1956 program might be the increase of

[87]Yi et al. (2004). Yi Rao is a professor at Northwestern University, deputy director of its neuroscience institute, and co-director of Shanghai Institute for Advanced Studies, CAS; Bai Lu is a director at National Institutes of Health and science advisor to MOST; Chen-Lu Tsou is a CAS Academician.

[88]Ibid.

[89]Ibid.

[90]Ibid.

total S&T funding, establishment of technological institutions, and the creation of colleges and new departments…For state leaders, creating a sound S&T system and enabling excellent S&T talents in China to exert their creativity and focus on what they are good at is a topic more important than specific S&T subjects…The Chinese S&T management system lags far behind its economy: when planned economy is virtually replaced by the competitive market economy, if the S&T management system that is an outcome of the age of planned economy isn't reformed thoroughly, it will impede China's S&T progress…A sound system cannot be established by Chinese S&T professionals on their own…When conditions are right, systematic transformation will be a historic feat of far-reaching influences."[91]

This article was critical of and worried about China's S&T planning, especially that in the new century. It might be a little exaggerated and some of its points didn't suit China's reality, but it had reference values for China's S&T plans in the past few decades. The defects and deficiencies had something to do with the concept behind S&T plans. To reflect on them, we must reflect on the concepts behind their formulation and implementation as well.

Mu-ming Poo was the only scholar whose articles were included in both supplements. This time he explained his understanding of S&T plan and Chinese science from the perspective of the scale of science. In his *Big Science and Small Science*,[92] Mu-ming Poo said "in China, 'big science' suits the nation's usual approach of planning and running scientific research in a 'top-down' manner…The operation of 'big science' is like 'planned economy' while 'small science' is more like 'market economy'."[93] He gave an example of "big science". In 1971, the Nixon administration announced the "war on cancer" program, vowing to defeat cancer by 1976, which failed. 30 years later, cancer remained a grave threat to human life and the human genome program was highly successful, which, unlike "war on cancer", was technology-oriented and had an achievable goal. When answering why we had to conduct "small science", Poo said "history has told us that most leading breakthroughs in the field of life sciences came from 'small science' institutes pushed by individual scientists…The development of science is hard to predict and needs multiple labs to do explorations in different ways…More importantly, small labs provide the close teacher-student relationship that is needed to cultivate young scientists…To put it simply, 'small science' labs are the main locations where major scientific research progress is achieved and the best environment to foster next-generation scientists."[94]

Regarding the risk of "big science", Mu-ming Poo admitted that big programs were risky in themselves, so there should be a detailed plan and a correct estimation whether there were enough well-trained researchers in a specific field to conduct the programs. With limited resources, the allocation of lots of them to big programs would squeeze the resources for small ones, causing serious negative impacts on the entire scientific circle…Big scientific programs usually stressed the importance of developing large platforms of biotechnology, but China needed to have powerful basic biological labs before any sophisticated technology could be developed; otherwise, we would have to spend huge amounts of money and import platforms from foreign countries. However, equipment and platforms should only be imported when there were a lot of labs that were able to effectively use them, otherwise resources would be wasted… Those fund-driven programs barely operated effectively…As a matter of fact, participating in big programs like 863 or 973 used to be considered by Chinese scientists as an opportunity to obtain a large sum of research fund without a rigorous peer review. Speaking of the waste in Chinese scientific sector, Poo said "many foreign scientists were impressed by the latest instruments in Chinese institutes, but rarely expressed

[91]Ibid.

[92]Poo (2004).

[93]Ibid.

[94]Ibid.

appreciation of how efficiently they were used. Some Chinese scientists…never think that the money spent on the instruments that sit idle in their lab is enough to sustain a whole poor village for several years…The annual budget of National Natural Science Foundation of China (NSFC) is about RMB2 billion, but a gigantic medium/long-term research program still under discussion may draw in RMB10 billion. Based on the nation's total investment in natural scientific research now, it's irrational to spend most of the fund on a few big programs."[95]

Poo's worry that the major projects in S&T plans would impede the development of basic sciences made perfect sense. However, one of the philosophy guiding China's S&T program was "technology comes first", and most of those major projects were technological sure-wins that would largely boost economic development and national strength, so it was reasonable to invest in basic sciences later after a certain amount of wealth had been accumulated. Mu-ming Poo's worry was more about "technology should not be equaled with science." Since the domestic scientific community was rather weak and wasn't very eager or willing to "speak the truth", opinions from prestigious overseas Chinese scientists could help us reflect on our S&T programs and the philosophy behind them.

As to basic science, applied science and technology, Poo pointed out that Chinese science policy makers and managers generally took science and technology (S&T) as the same thing and planned and managed scientific plans and engineering plans in the same way without any difference. Important breakthroughs were only possible when scientists were devoted to their research whole-heartedly, not joining an existing project half-heartedly just for the fund. In view of inadequate high-quality researchers, organizing large projects would lead to the waste of resources. Therefore, a more suitable strategy for basic science was to divide the fund among many small projects while stressing the quality and originality of research. British scientist Max Perutz said "scientific innovation cannot be organized. Top-down instructions will kill creativity." Scientific development had to depend on government support, but the direction and content of scientific research were hard to plan, for which Max Perutz was a good example. The Cambridge molecular biology lab that he created and led cultivated 12 Nobel Prize winners, laying the foundation for modern molecular biology and today's biotechnological industry.

At last, Mu-ming Poo emphasized the need to enhance basic scientific research. "What's surprising and baffling is that only one of the 20 subjects listed in the national medium and long-term S&T program (2006–2020) is about basic science, which directly reflects the government's lack of attention to it. The main bottleneck of China's future technological development is exactly the weak basic science, especially in the field of bio-medicine and biotechnology…In an age when originality and IP are given more weight, it will be extremely costly and inefficient to share basic scientific research results with developed countries…Basic science is the source of scientific culture."[96]

The position of basic science in China was indeed inconsistent with the nation's economic strength. Although we often claimed that substantial results had been achieved in certain areas of basic science, there were actually very few if we ignored the inciting publicity, lagging far behind developed western countries where basic science had witnessed continuous and substantial breakthroughs. It's been more than 60 years since the PRC was founded, but in basic science, China haven't had lots of natural science Nobel Prize of our own yet, which was partly due to our S&T programs and philosophy. Reflecting on them could be helpful for the formulation of future S&T programs and S&T development.

Zhu Chen wrote *Historical Mission of Young Science Managers—A Rational View of Young Leaders in Chinese Scientific Research Institutes*.[97] As a leader in CAS, he gave his opinions of scientific research management from an administrative perspective. By comparing the age

[95]Ibid.

[96]Ibid.

[97]Chen (2004).

structure of leaders at CAS institutes and other world-renowned institutes, he found that those in China were ten years younger than those in America on average. Chen explained why so many young scientists were in administrative positions. "There are both historical reasons and realistic needs. The expanding scale of scientific research and the ongoing reform of research institutes have given unprecedented opportunities for young scientists. Traditional ideas and personal aspirations also play a part…Many leaders in administrative departments tend to show their attention to talents through administrative promotion, which is without any doubt related with the traditional Chinese culture of 'official standard'."[98]

Regarding the pros and cons of institute leaders getting younger, Zhu Chen said, "generally speaking, this phenomenon is very positive for China's scientific research development…A successful young research manager is the best publicity to attract overseas Chinese students to come back."[99] But he also expressed his worry about the lack of scientific masters in China. The problem was that a well-developed scientific research management system hasn't been established yet, so leaders were often trapped in the trivia of daily management but didn't have enough time for thinking and research. A basic solution was to establish a reasonable management mode as soon as possible, whereby scientists play a leading role supported by a professional management team…The practice of linking the leadership of scientific research institute with administrative rank might encourage academic corruption.

These articles reflected on S&T plans and talked about their goals and S&T system reform. Scientists were quite concerned about political and administrative impacts on science and technology, which was a characteristic of Chinese S&T plans. We should look at this in a positive light. Without the positive political impact, S&T development in China wouldn't have obtained resources so easily and rapid development would be empty talk. As to their underpinning concepts, their existence spoke for their rationality. The purpose of reflecting on S&T plan wasn't to negate it, but to understand China's national conditions and "science conditions" more deeply, so that future S&T plans could play a better role.

[98]Ibid.

[99]Ibid.

2. Future prospects of science in China

In his *Is China Ready to Deal with Major Microbial Threats?*,[100] Ho offered his opinions about the damage of potential epidemic in China, and asked if China was fully prepared for other major microbial diseases that might appear or spread in the future, if it was able to effectively cope with the next global flue or bio-terrorist attack. He suggested, "To prepare China for major microbial threats, we should solve two basic issues. The first is the lack of fund in relevant basic research fields; the second is the poor health care infrastructure and the lack of attention to it."[101]

Shigang He wrote *A Fine Fowl Perches only on a Fine Tree—On Talent Policy and Environment in China's Scientific Research*,[102] in which he said the three talent programs that China has implemented since the mid-1990s (CAS' "100 Talent Program", NSFC's "National Outstanding Youth Research Program," and Ministry of Education's (MOE) "Chang Jiang Scholars Program") have effectively mitigated and reversed the talent outflow and were beginning to take good effect. However, some of those talents were "phantom personnel" (personnel who were not really in position). At the end of 2002, Shigang He and 25 other scientists submitted a letter exposing "phantom personnel" to MOST, MOE, CAS, and NSFC,[103] drawing their attention to the improvement of the management of those programs.

[100]Ho (2004). He works at the Aaron Diamond AIDS Research Center of Rockefeller University.

[101]Ibid.

[102]He (2004). He is a researcher at the Institute of Biophysics, CAS.

[103]http://www.xys.org/xys/ebooks/others/science/report/gongkaixin.txt.

He asked a question, "What improvements should China make to better attract talents? Creating a fair, just and transparent research fund distribution mechanism and minimizing the interference by administrative personnel in the selection of research programs—this is the most important factor to attract research personnel… Another factor that affects China's scientific research environment is the annoying red tapes needed to import equipment and reagent. Some ridiculous rules are still in effect."[104] He said it was against the laws of scientific research to manage basic research as an engineering project. Another factor that was indispensable for a dynamic and competitive talent environment was also lacking in China, namely the auxiliary policies to facilitate talent flow…America's NIH expressly stated that when the leader of a foundation-funded project was transferred to another organization, he or she could take away all the project funds and the equipment bought with them. Such a policy was very attractive for researchers and protected national resources.

Shanchao Wu, Hui Meng and Cunhao Zhang co-authored *Academic Misconduct in Chinese Science and Technology Community*,[105] according to which it was a consensus in Chinese S&T circle that scientific ethics was a more serious issue in China than overseas. Meanwhile, China lagged far behind other countries in terms of public understanding of scientific ethics, the importance attached by the S&T circle to this issue, disclosure of contraventions in that regard, and relevant department's prevention of scientific misconduct. The article offered suggestions on China's "long and hard journey of scientific ethics construction."

Ruiping Xiao and Heping Cheng wrote *Will China Merge into the Mainstream of Molecular Medicine?*,[106] which believed China had unique resource advantages in the post-genome age when molecular medicine became the mainstream of bio-medicine. "China clearly realizes that molecular medicine needs the new mode of 'coordinated big science'," but serious improvements had to be made in several directions. A new operation mechanism had to be set up for research institutes to avoid bureaucracy and adapt to large-scale scientific coordination and integration, the peer review system of research fund had to be further perfected, and IP regulations protecting the output of bio-medicine research had to be enacted and implemented as soon as possible.

The eighth article—*Overview of Primate Research in China: History and Outlook*—was written by Weizhi Ji and others.[107] It read, "the Chinese government has energetically intensified international cooperation this year and profoundly realized its importance…In contrast, cooperation and communication of domestic scientists is much less and more difficult than international cooperation."[108] Ji Weizhi said China had rich resources in primate research but had to properly and carefully coordinate, organize and manage the relation between scientific research and ethics.

The last article in this supplement was *Changes in China's Environmental Policies—The Past, Present and Future of Chinese Dams*[109] co-authored by Wenwei Ren, John Meligrana and others. The article compared the different conclusions drawn from the advice of building dams on the Yangtze River and the Jinsha River, analyzed the shelving of the latter, and concluded that the Chinese government attached more importance to environmental protection than before. More importantly, scientists played an important role when passing these views to the government. In the end, it pointed out that the decision of building a dam, which was made by the government and a few administrative bodies in the past, has now become a public process involving several interest groups.

These articles provided their views and suggestions on some problems in China's S&T

[104] http://www.xys.org/xys/ebooks/others/science/report/gongkaixin.txt.

[105] Wu et al. (2004). They all engage in international scientific ethics.

[106] Xiao et al. (2004).

[107] Ji et al. (2004).

[108] Ibid.

[109] Ren et al. (2004).

development in different angles, but all displayed confidence in China's scientific future and its robust prospects. It's foreseeable that voices from the Chinese scientific community (including Chinese-born scientists) will be given more value in the formulation and implementation of future S&T plans. The comments and reflection on S&T plans and their concepts can help us understand the decades-long history of S&T plans and the evolution of planning concepts.

3. Problems generalized by the two supplements

Through the two "China voices" supplements to *Nature*, first-rate scientists in China and abroad gave comprehensive comments on China's scientific panorama since the PRC was founded, especially in recent years. They affirmed the tremendous achievements that China had made, especially technological and economic achievements, and pointed out many problems. The articles could be categorized as follows.

First, some articles were on the cultural and ideological level, such as the one by Mu-ming Poo who had a very strategic vision. Traditional culture both promotes and restricts modern science greatly, and how to make good use of historical and cultural resources is a new topic in S&T planning.

Second, most articles focused on S&T system. At least 9 of the 17 articles mentioned the various problems in China's S&T system, including (1) not enough attention was paid to basic science; (2) the fund application and evaluation system were not very reasonable; (3) the importance and complexity of IP was not fully realized; (4) administrative interference was strong; (5) there wasn't sufficient exchanges among scholars, especially domestic ones; and (6) science and technology were not differentiated clearly enough. Although the articles criticized these problems, their opinions and suggestions were helpful.

Third, some articles discussed the potential of scientific development in China. Generally speaking, science in China has developed at a fast pace in the past few decades, with immense quantitative progress and varied speed of qualitative improvement. In fields where China had unique resources, it had the potential to reach the world's advanced level, such as biological resources, embryonic study, and medical causes of diseases.

As Dr. Philip Campbell, editor-in-chief of the first supplement, said, although the Chinese government had invested hugely in scientific research, it hadn't fully exerted its scientific potential. This has not only resulted in its weak presence on the world science stage but also cost it the opportunity to fully utilize the scientific and technological results in other countries, as well as the opportunity to develop its own economy and improve its own people's quality of life. Why hasn't China fully released its scientific potential? What kind of role have the S&T plans since 1949 played in China's S&T development? These questions have been discussed earlier in the book. We need to look at the two *Nature* supplements in a more positive light in order to better understand the development of S&T plans and the evolution of planning concepts.

III. *Nature*'s affirmation of scientific development in China

Nature magazine seemed to be always concerned about the development status of science in China. A few years later, it issued another supplement in July 2008 to introduce and comment on the scientific situation here (Fig. 5.7).

The cover of this supplement was the Bird's Nest, China's national stadium, and the feature picture was a soaring red dragon. It included 16 papers in 8 columns such as editorial, news, and commentary, which covered a range of aspects from China's economic, S&T, environmental to population policies, giving a panoramic portrayal of and comment on the country.

The supplement published an editorial titled *China's Challenges*, saying that China's development was remarkable, but the reality behind the brilliant data was more complicated. Several papers in the supplement mentioned China's scientific development. For instance, in a news feature titled *China, a Great Competitor*, the author said Chinese economy had developed

Fig. 5.7 "China's Challenges", 2008 supplement to Nature

rapidly since the reform and opening-up was launched in 1978, lifting tens of millions of people out of poverty. In the past 20 years, China's input in scientific research had increased at the annual rate of nearly 20%, jumping from USD7.5 billion in 1991 to almost USD90 billion in 2006, but the input in basic research only took up 5%, much lower than the 20% in developed countries. However, China had realized this and was actively advancing basic research and indigenous innovation.[110] The number of Chinese students obtaining scientific and engineering degrees was rising year by year from 115,000 in 1995 to 672,000 in 2004, more than that in the US and Japan. In 2007, 32,000 of the 142,000 foreign students that received doctor's degree in the US were Chinese, second only to India. The global proportion of papers published by Chinese scholars increased from 0.2% in 1980 to 7.4% in 2006, surpassing Japan for the first time. Nonetheless, despite the huge amount of papers, their "quality isn't that good—for decades, the score of quotation and influence of Chinese scientific research results has lingered around 0.35, which, however, has risen to 0.73 since 1995, even approaching 1.0 in material science and nanotechnology."[111]

Another article titled *End of a Scientific Giant* pointed out that China had shown the signs of becoming an important scientific force. The number of papers published by Chinese scholars and the number of doctor's degree granted every year was ranking high in the world. Moreover, more and more overseas Chinese scholars had come back in recent years to contribute to China's scientific development. However, the economy, form of capital, communication, organizational structure and specialization have gone through fundamental changes in the past few decades, which meant the US wouldn't easily and willingly pass the scientific crown to any possible successor, like China. A more possible situation was that the multi-polar global economy that was in the making would lead to multiple scientific centers, including the US, EU, Japan, China, Russia, and India.[112]

This supplement was much different from the ones in 2003 and 2004 and showed more affirmation and appreciation of China's scientific progress. Science and technology in China had indeed undergone profound changes. Scientists' doubts, criticisms and reflection on China's S&T problems, though making perfect sense, were based on the scientists' point of view, but China's S&T and S&T plans were not just about the scientists but were the combination of science and politics. The politicians had insight and vision that scientists didn't, so we couldn't analyze China's S&T panorama from a single perspective. That is the purpose of dedicating such a large part of this chapter to the papers in *Nature*, and also the reason why the planning concepts behind Chinese S&T plans are more like political than scientific concepts.

That a world's top publication has paid consistent attention to China and issued several supplements to introduce and discuss China's scientific status tells us that China's S&T development has drawn close attention in the world. Although some of the viewpoints were biased due to the limited knowledge of China's special national conditions and scientific conditions (a generalization of the environment and

[110]Nature, Vol. 454, Issue no. 7203, 24 July 2008.
[111]Ibid.
[112]Ibid.

status of S&T development in a specific time and space), the issuance of the supplements in a way implied that the S&T plans made after the PRC was founded have pushed our S&T development and largely boosted domestic economic strength. S&T plans have indeed accomplished and are continuing to fulfill significant historical missions.

5.4 "11th Five-Year" Science and Technology Development Plan

The "11th five-year" S&T plan was actually a continuation of the *2006–2020 National Medium and Long-term Science and Technology Development Program*, as the latter made the former's formulation a task in the third stage when conducting strategic studies. The outline of the medium/long-term S&T program specifically mentioned this. "This outline should be better aligned with the 11th five-year national economic and social development plan. To make the outline more operable, we need to align its contents with the 11th five-year national economic and social development plan based on priority, including priority subjects, major projects, cutting-edge technology, basic research, establishment of basic conditions and platforms, and S&T system reform. We will select from them the key tasks that have to kick off immediately or have to be solved in the 11th five-year period, and make detailed arrangements and deployments in the 11th five-year national economic and social development plan."[113]

I. Formulation of the "11th five-year" S&T plan

The formulation of the 11th five-year plan began in 2004. It had to fully reflect the strategic thought of the national medium/long-term S&T development plan and create a good beginning for its successful implementation. It also should be well connected with the 10th five-year plan so that all key works could proceed forward steadily.[114]

In October 8–11, 2005, the 5th plenary session of the 16th CPC Central Committee was held in Beijing, at which the delegates listened to and discussed the work report of the political bureau of CPC Central Committee, and reviewed and passed the *Suggestion of CPC Central Committee on Formulating the 11th Five-year Plan for National Economic and Social Development*. The plenary session proposed the main targets of economic and social development in the 11th five-year period: to double the 2000 per capita GDP in 2010 on the basis of perfecting structure, increasing benefits and reducing resource consumption; to notably raise the resource utilization efficiency and lower energy consumption per unit GDP by about 20% from the end of the 10th five-year period; to foster a batch of enterprises with unique advantages, independent IP, famous brands and strong international competitiveness; and to ensure that socialist market economy system is fairly well developed, open economy reaches a new level, and international payment is in basic balance. The session believed that developing science, technology and education and strengthening the talent team was the decisive factor for enhancing national competitiveness, so we had to earnestly carry out the strategies of "strengthening the nation through science, technology and education" and "strengthening the nation through talents." In developing science and technology, we must stick to the principle of "indigenous innovation, breakthroughs in key areas, supporting development and leading the future," keep improving enterprises' innovation capability, and build the national innovation system more quickly.[115] The suggestion stated that the first twenty years of the twenty-first century were an important period of strategic opportunities for China, and the 11th five-year period was

[113]"Outline of the National Medium/long-term Science and Technology Development Program (2006–2020)," *Bulletin of the State Council of PRC 2* (2006).

[114]Guanhua (2004).

[115]Bulletin of the 5th plenary session of the 16th CPC Central Committee. Database of CPC national congresses: http://cpc.people.com.cn/GBL.

particularly crucial. The 11th five-year plan set out a series of important tasks, such as building the socialist new countryside, perfecting and upgrading the industrial structure, promoting coordinated regional development, building a resource-saving and environment-friendly society, and advancing the construction of a socialist harmonious society. They constituted the background for the formulation of the 11th five-year S&T plan.

Making more efforts to implement these two strategies indicated the central government's deeper understanding of the relation between S&T and S&T workers. As the talent strategy was elevated to be a national strategy equally important as the first one, intellectuals were placed in a more prominent position. The transition from emphasizing science and technology on the material level to placing equal emphasis on scientists as well as a significant progress.

The 11th five-year S&T plan was formulated according to the 11th five-year national economic and social development plan and the 2006 medium/long-term S&T program. Since the 10th five-year S&T plan was put into practice, it has advanced S&T work following the guideline of "innovation and industrialization" and improved the capability of sustained S&T innovation. The three strategies of "talent, patent and technical standard" took obvious effects, 12 major projects were successfully carried out around national strategic goals, and substantial S&T achievements were scored represented by the Shenzhou-VI manned spacecraft, super hybrid rice, high-performance computer, SARS vaccine, and quantum entanglement. In 2005, China's R&D input accounted for 1.3% of GDP, R&D personnel reached 1.2 million man-year (full-time equivalent), the number of S&T papers ranked 5th in the world, and the number of applications for patent for invention from China reached 93,000, of which 21,000 were granted. All these meant that the tasks in the 10th five-year S&T plan were basically accomplished.

The 11th five-year S&T plan upheld the guideline of "indigenous innovation, breakthroughs in key areas, supporting development and leading the future," and organized major projects with indigenous innovation as the centerline. It vowed to comprehensively build the national innovation system and create a favorable environment for indigenous innovation, so as to provide powerful S&T support for building a moderately prosperous society in all respects and a socialist harmonious society.

To achieve the medium/long-term goal of S&T development, namely "becoming an innovative country," the 11th five-year S&T plan pledged to lay a solid foundation in three areas.

First, further improve the national innovation system with Chinese characteristics, and put in place the necessary S&T institution for building an innovative country. With the establishment of the technological innovation system, which features enterprises as the main entities and the integration of industry, universities and research institutes, as the breakthrough point, we will take faster steps to build a national innovation system in which technological innovation, knowledge innovation, defense technology innovation, regional innovation and technological intermediary service are integrated in mutual promotion.

Second, primarily complete the S&T infrastructure and platforms that meet the needs for S&T innovation, and provide the necessary S&T conditions for building an innovative country. We will build a batch of world-class basic, public and strategic scientific research bases.

Third, foster a large, high-caliber team of innovative talents, and ensure the necessary S&T talents for building an innovative country. We will increase the total number of S&T professionals to more than 50 million, considerably raise the ratio of scientists and engineers, and increase the full-time equivalent of those engaged in R&D per 10,000 working people to 14 man-years.[116]

II. The planning philosophy of "indigenous innovation"

"Indigenous innovation", as one of the main philosophy in S&T plans of the twenty-first

[116]"National 11th five-year science and technology development program (2006–2010)," *China Science and Technology Information 22* (2006).

Table 5.6 Main indicators of 11th five-year S&T plan[117]

Indicators	2010 goals
Total R&D input/GDP	2%
Foreign technology dependence	<40%
Number of cites of international scientific papers	World's top 10
Annual grant of invention patents by Chinese	World's top 15
Contribution made by S&T progress to economic growth	>45%
Added value of high-tech industry/added value of manufacturing	*18%*
Total S&T human resources	50 million
Total S&T personnel	7 million
Number of scientists and engineers engaged in R&D (full-time equivalent)	1.3 million man-year

century, was continued in the 11th five-year S&T plan, which was an intermediate stage on our way to becoming an innovative country, and its indicators were all targeted at that goal (Table 5.6).

The 11th five-year S&T plan made detailed arrangements for major projects and cutting-edge technologies in 2006–2010, and major projects were launched one after another as long as conditions were ripe. In developing basic science, "serving national goals" and "encouraging free exploration" were combined, and the disciplinary layout was bettered with the focus on talent, environment and key breakthroughs, in the endeavor to score substantial progress in emerging inter-disciplines and scientific forefront. In mainstream directions of world scientific development and basic research fields where we had advantages and unique features, we strove to register a batch of original innovations with great international influence, so as to significantly raise our position in the international scientific circle. Four major scientific research programs would be implemented at a faster pace, namely protein, quantum control, Nano science and technology, and development and reproduction, in order to enhance China's original innovation capability.

The 11th five-year S&T plan highlighted the principal position of enterprises and boosted the construction of national innovation system with Chinese characteristics across the board. It elaborated on five innovation systems (technological innovation system, knowledge innovation system, defense innovation system, regional innovation system, and intermediary service system), with a special emphasis on the first one. The plan vowed to give full play to the guiding role of economic and S&T policies, and encourage and guide enterprises to truly become the principal entities that make R&D input, carry out technological innovation activities, and use innovative results. The national S&T plan implementation mechanism was adjusted and more support was given to technological innovations by enterprises.

In sum, the 2006–2020 medium/long-term S&T program and 11th five-year S&T plan were more detailed, solid, and appropriate than previous ones, and their discussions and deployments of basic science were in greater depth. Such progress reflected the government's new planning philosophy, namely "people centrism" and "indigenous innovation". The 11th five-year S&T plan proved a great success. Stiglitz, a winner of the Nobel Economics Prize, expressed his appreciation of China's 11th five-year plan when giving a speech at Tsinghua University (Fig. 5.8).

According to relevant statistics, the 11th five-year S&T plan was implemented smoothly. In 2008, the R&D expenditure stood at RMB457 billion, up 23.2% from the previous year and taking up 1.52% of GDP, of which RMB20 billion went to basic research. The nation

[117] Ibid.

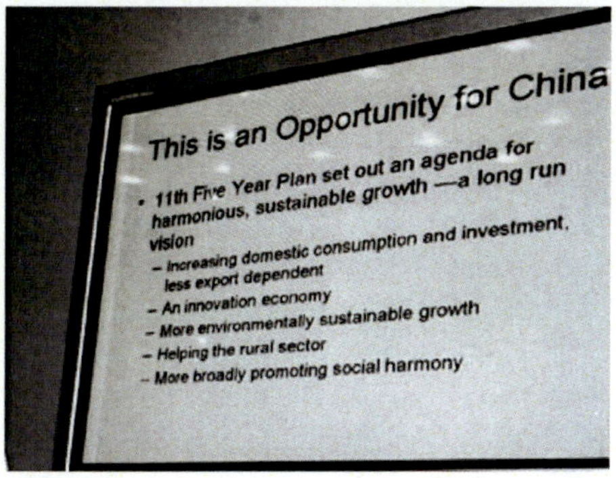

Fig. 5.8 Nobel Economics Prize winner Stiglitz's comments on the 11th five-year plan during his speech at Tsinghua University[118]

arranged 922 subjects of S&T programs and 1205 subjects under the "863" program throughout the year. It granted 412,000 patents, including 352,000 domestic ones (85.5%); 94,000 of those were invention patents, including 47,000 domestic ones (49.7%). Also this year, 226,000 technology contracts were signed to the tune of RMB266.5 billion, up 19.7% from the previous year.[119] These figures proved the successful, even accelerated, implementation of the 11th five-year S&T plan, but we must keep in mind that our basic research remained weak and needed fortification.

In 2007, China's basic research expenditure accounted for 4.70% of its total R&D expenditure, the lowest among the 24 countries that revealed this data. This ratio was around 20% in most developed countries, above 10% in Japan and close to 15% in Russia, but it has stayed around 5% for many years in China according to historical data.[120] The new S&T plan paid close attention to the weak basic research that had been a weak link in our S&T development by largely increasing the fund for R&D and basic research, but more time was needed to change its weak status owing to inertia (Fig. 5.9).

It's worth noting that from 2007 to 2009, the most serious economic crisis swept across the world since the Depression of the 1930s, and all countries were adjusting their internal and external policies. China took active steps to cope and hoped to defuse the financial crisis through science and technology. This new external condition might be reflected in future S&T plans.

III. A glimpse of the 12th five-year S&T plan

The 12th five-year S&T plan was an important stage that pushed China's S&T development to a new level, and a new step to further advance the strategic deployments in the *Outline of National Medium/long-term Science and Technology Development Program (2006–2020)*. It drew the overall blueprint for our S&T development through 2015. It's not the right time to give it a comprehensive evaluation in this book, so we'll only briefly introduce it on the strategic level. Consisting of 12 aspects, the 12th five-year S&T plan proposes to deeply implement the *Outline of National Medium and Long-term Science and Technology Development Program (2006–2020)* (2006), promote the perfection and upgrade of industrial structure, serve major S&T needs for improving livelihood, foster innovative talents and teams, boost the construction of national innovation system, and lay a solid foundation for achieving strategic S&T leaps and building an innovative country. Unlike its predecessors, this plan put forth the indicators of S&T development (Table 5.7).

These indicators would give an impetus both to technological and economic functions and to original innovation. The plan arranged 11 major scduientific projects, including core electronic devices, high-end universal chips and basic software products, ultra-large IC manufacturing equipment and complete sets of process, new-generation broadband wireless communication network, and premium numerically-controlled machine tool and basic manufacturing equipment. It also proposed to strongly foster 11

[118]Science Net. Liu Li's blog: http://www.sciencenet.cn/m/user_content.aspx?id=221587.

[119]2008 Statistical Bulletin of National Economic and Social Development of the PRC (2009).

[120]China science and technology statistics: http://www.sts.org.cn.

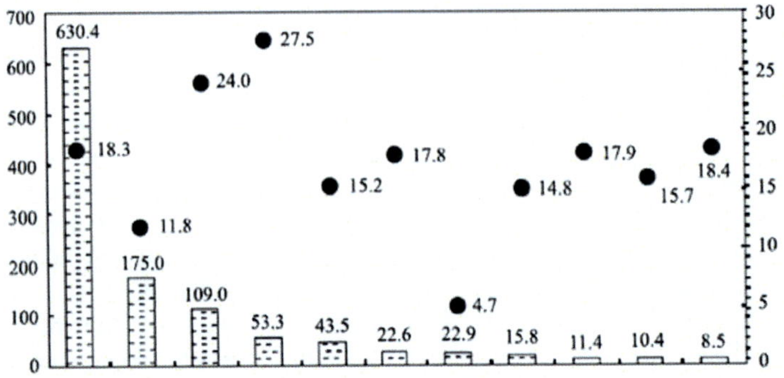

Fig. 5.9 Basic research expenditure and its percentage in total R&D fund in some countries in 2007[121] (Ordinate in the left representation Basic research expenditure (USD100 million), Ordinate in the right representation Ratio of basic research expenditure (%). Abscissa from left to right, US; Japan; France; Italy; ROK; Spain; China; Russia; Denmark; Israel; Norway)

Table 5.7 Main indicators of S&T development in the 12th five-year period

Indicators	2010	2015
Percentage of R&D fund in GDP (%)	1.75	2.2
R&D personnel per 10,000 working people (man-year)	33	43
World ranking of cites of international scientific papers (position)	8	5
Possession of invention patent per 10,000 people (piece)	1.7	3.3
Application of invention patent by R&D personnel (piece/100 man-year)	10	12
National total volume of technology contracts (RMB 100 million)	3906	8000
Percentage of added value of high-tech industry in added value of manufacturing (%)	13	18
Percentage of citizens with basic scientific quality (%)	3.27	5

strategic emerging industries, including energy conservation and environmental protection, new-generation IT and bio-industry, and highlighted them in blocks, which never happened before.

The plan also made deployments in other sectors, such as pushing breakthroughs in core and key technologies of major areas, livelihood technology, basic research and S&T system reform. It was quite detailed and much anticipated. The plan made "promoting scientific development and indigenous innovation" the main tasks, with a special stress on tackling climate change and benefiting the people through science and technology. It made overall deployments for S&T development, emphasized the market's basic role in resource allocation, and exerted its own role in guiding the distribution of S&T resources and steering social S&T resources. Oriented by the nation's substantive needs, it vowed to steadfastly push for collective technological breakthroughs in key areas, so as to boost the general technological jump of important strategic industries. To build a national innovation system, the plan put its focus on building a technological innovation system with enterprises as the main entities, and diverted innovative resources toward them.

Besides, the 12th five-year S&T plan provided important indicators and definitions in the appendix, which was rarely seen in previous plans, and it continued the planning philosophy of "people centrism" and "indigenous

[121]Department of Development Planning of MOST (2008).

innovation". China will continue to formulate S&T plans in the future and they will play an important role in economic and social development. In this process, the S&T planning philosophy will keep evolving along with the time, national conditions, and scientific conditions.

5.5 Review and Elevation

This chapter discusses S&T programs in the new century, including the 10th and 11th five-year S&T plans as well as the 2006–2020 S&T plan, and our new planning philosophy.

As a special plan, the 10th five-year S&T plan had the full name of *10th Five-year Plan for National Economic and Social Development—Special Plan for the Development of Science, Technology and Education*. Being implemented to assist the 10th five-year national economic plan, it wasn't as important as previous five-year S&T plans. It followed the guideline of "innovation and industrialization", which was close to the previous guideline that "economic construction has to depend on science and technology and scientific and technological work must help economic construction, and new heights have to be scaled," but it put more stress on the industrialization of high technologies and made higher requirements on technological innovation. It was a transitional S&T plan that reflected the three S&T planning philosophy in the transition period, namely "technology comes first in S&T plan," "utilitarian tendency" and "reinforcing economic function", but it began to reflect on the excessive emphasis on pursuing economic benefits. In this plan, the idea of building a national innovation system had already been formed, but "people centrism" appeared for the first time, meaning the CPC had reflected on the purpose of economic development. Nevertheless, the utilitarian philosophy had a strong inertia and this plan didn't mention what to reflect on and how.

Despite the tendency of reflection, utilitarian ideas still had a strong presence in the 10th five-year S&T plan and utilitarian practices were obvious too. In basic research, for instance, the plan expected to make breakthroughs in a short period of time regardless of the laws governing scientific development, so it was only transitional.

As we came to the twenty-first century, the political, economic and social environment in China has seen tremendous changes. When the new administration came into office, it took the initiative to reflect on the old development mode of pursuing economic benefits and GDP only. "Three Represents" and the Scientific Outlook on Development were successively adopted as the guiding thought for the CPC, exerting far-reaching influence on building socialism with Chinese characteristics. In this book, when we begin to discuss each S&T plan, we first analyze the political and economic background at home and abroad against which our S&T plans came into being. It is a national action, so we have to study the corresponding political, economic and military environment, which gives rise to the underpinning philosophy of the plans. Compared with the situation in foreign countries, the relation between S&T and society is much more complicated in China due to its special national conditions and special circumstances in S&T development.

The changing political and economic environment in the twenty-first century has urged the CPC to reflect on our achievements and deficiencies during reform and opening-up in the past few decades. Great hopes are pinned on science and technology to achieve the third-step strategic goal of modernization. "We need science and technology more than ever before"—this implies the higher position of S&T in China. What should we do to put S&T into better use, so that it will not only promote economic development but also mitigate the negative impacts caused by fast development, such as environmental aggravation, widening poor-rich gap, and weak innovation capability? To answer that question, we have to reflect on and adjust the philosophy behind previous S&T programs.

The adjustment of philosophy will take a process and there is no government document to explicitly speak it out. It was the scholars that generalized the concepts, which corresponded to guidelines in S&T plans. The guideline of the 2006 S&T program—"indigenous innovation, breakthroughs in key areas, supporting the development and leading the future"—showcased the great changes in the philosophy of the first long-term S&T program of the new century.

When the 2006–2020 S&T program first began to be formulated, the S&T circle and economic circle argued about it. Some economists were inclined to develop S&T according to the "comparative advantage" theory, but the S&T circle insisted on the path of indigenous innovation, to which the economic circle had come to agree based on thorough discussions. The global financial crisis in 2008 showed that the "indigenous innovation" theory was more suitable for China than "comparative advantage" theory. There were arguments about S&T plan within the S&T circle too regarding whether the "big science" mode was suitable for China's S&T development and issues about research funding. The arguments illustrated the special nature of Chinese S&T plans. We shouldn't analyze them simply from the perspective of the scientific community, but should also from the political perspective.

To sum up, S&T programs in the twenty-first century have three philosophical concepts —"technology comes first in S&T plan," "people centrism" and "indigenous innovation". The 2006 S&T program, continuing the philosophy of "technology comes first", listed 16 major projects, 62 priority subjects and 8 cutting-edge technologies, almost all technology-oriented. They were able to directly enhance national security, promote economic transformation, reinforce national strength and improve people's lives. As a result, this S&T plan was mostly a technological plan.

"Technology comes first in S&T plan" is the only planning philosophy that has been implemented in all three historical periods. That has something to do with Chinese history and traditions. China has a time-honored history of civilization and it once left the whole world far behind in science and technology, but the academia generally agrees that our technological tradition outweighs our scientific tradition, as evidenced by the fact that all four ancient inventions by China fall under the technological category. The development of technology includes but isn't driven by that of science. It takes a long paper to prove this viewpoint, so we'll just do it the simple way here. The "Joseph Needham's Question" is an acknowledged subject in the academic circle, and discussions about it have lasted a long time and yielded bountiful results. It is an open question without a standard answer, which, as a piece of counter-evidence, proved that ancient China didn't have a scientific system similar to that in the west, but China was indeed an ancient civilization, which meant that between science and technology that constituted civilization, at least one of them played an immense role in ancient China. If science was negated, or at least wasn't affirmed, then it must be technology. Therefore, China has a profound and long technological history and tradition that has lasted till the present day.

This concept that prioritizes technology is naturally reflected in S&T plan. From a realistic point of view, the priority to technology is good for meeting the pre-set goals. For S&T plan, which is a government or national action, it is unimaginable and unacceptable if it doesn't yield any result in the end or the result is hard to predict. Therefore, the S&T plans in China have always accepted and implemented the philosophy of "technology comes first", which will continue as long as we continue to formulate and carry out S&T plans and as long as they remain a government and national action. It must be admitted that this philosophy has taken positive effects and boosted substantial progress in China's economic, military, national defense, social and other sectors in a short period of time. This "pragmatic" approach consists of the CPC's "realistic" spirit. As long as wealth and resources keep piling up, they, when reaching a certain stage, are bound to come back and promote the massive development of basic science—a truth that has been proved in other countries. China has largely increased the input in basic science in recent years, as illustrated by the multiplication of funding at NSFC. We should understand the concept of "technology comes first in S&T plan" as well as other concepts in a positive light.

After the "Three Represents" and Scientific Outlook on Development became the guiding thought of China, they affected the S&T planning philosophy. With the development of the world economy, individuals were given more importance, and their independent choice was considered equally or even more important than the choice imposed upon them by the collective. The sustained high-speed economic development in China has brought about a lot of problems, so it was right to implement the "people centrism" concept in S&T programs of the new century because that conformed to the historical trend. In other words, the importance of S&T development didn't just lie in boosting economic construction and national security on the macro level but also in promoting the independent development of individuals. That was why the 2006 S&T program and 11th five-year S&T plan added many scientific programs and projects that benefited the people and promoted individual development.

Having carried out reform and opening-up for so many years, especially after the "market for technology" strategy didn't lead to expected effects, the philosophy of "independent development based on domestic conditions" established in the 1963 S&T program came back to the picture. In the twenty-first century, innovation and national innovation system became the trend of the time, which, coupled with China's reality, gave rise to the S&T planning philosophy of "indigenous innovation". S&T programs in the new century put it in a prominent position and made deployments for building the national innovation system across the board. The 60 auxiliary measures rolled out later also revolved around those two centers. Both the 11th and 12th five-year S&T plans took "indigenous innovation" as one of the central concepts. As "people centrism" and "indigenous innovation" are still in process, let's wait and see their final effect.

This chapter expounds on the comments and suggestions on China's S&T development and plans made by three China-themed supplements to *Nature*, a top scientific journal in the world. The papers in those supplements are all written by scientists familiar with the western and Chinese culture. On the one hand, this means China's S&T development has reached a new stage and drawn close attention from the world scientific community; on the other hand, it reveals some problems in our S&T development and in the formulation and implementation of S&T plans. There are some papers that directly discuss the S&T plans. These discussions and reflections will help us understand more deeply the special nature of Chinese S&T plans and the historical evolution of planning philosophy.

References

2008 Statistical Bulletin of National Economic and Social Development of the PRC. 2009. *China Statistics* 3.

Campbell, Philip. 2003. *Supplement to Nature*, vol. 426, no. 6968, 18/25 December 2003, A3.

Chen, Zhu. 2004. *Supplement to Nature*, vol. 432, no. 7015, 18 November 2004, A24. CAS Academician.

Chien, Kenneth R., and Luther C. Chien. 2003. *Supplement to Nature*, vol. 426, no. 6968, 18/25 December 2003, A9.

Department of Development Planning of MOST. 2008. *Statistical Report on Science and Technology*, 22.

Feng, Zhenggang. 2004. Thoughts About the Argument Between Economic Circle and Science and Technology Circle. *Invention and Innovation* (9).

Guanhua, Xu. 2004. Comprehensively Implementing Scientific Outlook on Development, Creating New Situation in Science and Technology Work. *Hangzhou Keji* (2).

He, Shigang. 2004. *Supplement to Nature*, vol. 432, no. 7015, 18 November 2004, A30.

Ho, David D. 2004. *Supplement to Nature*, vol. 432, no. 7015, 18 November 2004, A6.

Hsiao, James C., and Kenneth Fong. 2003. *Supplement to Nature*, vol. 4: 26, no. 6968, 18/25 December 2003, A29.

Huang, Alice Shih-hou. 2003. *Supplement to Nature*, vol. 426, no. 6968, 18/25 December 2003, A39.

Ji, Weizhi, et al. 2004. *Supplement to Nature*, vol. 432, no. 7015, 18 November 2004, A46.

Jintao, Hu. 2006. *Adhering to the Path of Independent Innovation with Chinese Characteristics, Striving to Build an Innovative Country—Speech at the National Science and Technology Conference*. Beijing: People's Publishing House.

Jintao, Hu. 2007. Holding up High the Great Banner of Socialism with Chinese Characteristics; Fighting for New Victory in Building a Moderately Prosperous Society in All Respects—Report at the 17th CPC National Congress. *People's Daily*, October 25.

Lin, Yifu. 2005. Comparative Advantages and China's Economic Development. *Forward Position in Economics* (11).

Mingzheng, Teng. 2013. Study on the Collective Studies of the Political Bureau of CPC Central Committee since the 16th CPC National Congress. *Journal of Tianjin Administration Institute* 4.

Outline of National Medium and Long-term Science and Technology Development Program (2006–2020). 2006. Bulletin of the State Council of PRC (2).

Poo, Mu-ming. 2003. *Supplement to Nature*, vol. 426, no. 6968, 18/25 December 2003, A5.

Poo, Mu-ming. 2004. *Supplement to Nature*, vol. 432, no. 7015, 18 November 2004, A18.

Ren, Wenwei, and John Meligrana, et al. 2004. *Supplement to Nature*, vol. 432, no. 7015, 18 November 2004, A50.

Shen, Jianbei, and Long Yongtu. 2005. Chinese Auto Industry Should Develop Steadily and Buy Volkswagen in the Future. *CBN Daily*, September 8.

Tso, T.C. 2003. *Supplement to Nature*, vol. 426, no. 6968, 18/25 December 2003, A21.

Wu, Chung-I, Shi Suhua, and Zhang Yaping. 2003. *Supplement to Nature*, vol. 426, no. 6968, 18/25 December 2003, A25.

Wu, Ray. 2003. *Supplement to Nature*, vol. 426, no. 6968, 18/25 December 2003, A35.

Wu, Shanchao, Hui Meng, and Cunhao Zhang. 2014. *Supplement to Nature*, vol. 432, no. 7015, 18 November 2004, A36.

Xia, Wang. 2004. New Situations and Requirements for China's Technological Introduction after Joining the WTO. *China Science and Technology Achievements* (1).

Xiaojuan, Jiang. 2008. Review and Outlook of China's 30-year Reform and Opening-Up. *Social Sciences in China* (6).

Xiao, Ruiping, and Heping Cheng, et al. 2004. *Supplement to Nature*, vol. 432, no. 7015, 18 November 2004, A42.

Xin, Hao, and Gong Yidong. 2006. China Bets Big on Big Science. *Science*, 311 (March 17): 1548–1549.

Yan, Kunlun. 2007, May. *Study of Science and Technology Award System in China*. Doctoral thesis paper, University of Science and Technology of China.

Yang, Xiangzhong. 2003. *Supplement to Nature*, vol. 426, no. 6968, 18/25 December 2003.

Yi, Rao, Bai Lu, and Chen-Lu Tsou. 2004. *Supplement to Nature*, vol. 432, no. 7015, 18 November 2004, A12.

Yuan, Chunlin. 2004. 'Big Science' Mode Helps Accelerate Life Science Study. *China Youth Daily*, August 10.

Zemin, Jiang. 2006. Speech at the Conference Celebrating the 80th Anniversary of the Founding of the Communist Party of China. In *Selected Works of Jiang Zemin*, vol. 3. Beijing: People's Publishing House.

6 Analysis of Scientific and Technological Planning Philosophy and Case Studies

Abstract

This chapter expounds on the reasons for the historical evolution of the Medium and Long-Term Science and Technology Program and its philosophy, from politicization to utilitarianism, and then to humanization. Science and technology planning has a profound impact on contemporary Chinese academy. The Large Aircraft Project in Medium and Long-Term Science and Technology Program shows the complex factors and interest game in the process of formulating and implementing China's science and technology planning.

Keywords

Academia-officialdom affiliation · Large aircraft project · Technology trajectory · Philosophy historical change · Humanization

6.1 Adaption to Historical Dynamics

As a critical vying aspect, scientific and technological planning has become an endeavor to perpetuate advancement. This is particularly true for China under ideological guidance and Marxism. The endeavor has been a significant state affair since its founding. The preceding writing has analyzed the historical progress of China in the first fifty years. This section will reveal the historical trail of China's philosophy in planning and adaption to historical dynamics of science and technology.

I. China's Ideology in Scientific and Technological Planning

From China's founding and onward to 2020, China has made ten mid- and long-term scientific and technological programs in which the planning ideology has been the core and soul. As programs sprout from the core and soul, the philosophy has been the vital force that steers directly the orientation of China's scientific and technological plans, giving the process a historical trail that echoes in great depth the political, economic environment and China's realization of its own track in the advancement planning of science and technology. To study the conceptual changes to adapt to the dynamics will help further our understanding of the mid- and long-term programs and contemplate them philosophically.

The philosophy concept is a general principle, guiding ideology, fundamental idea, and prospective goal formulated and implemented for the development and planning of science and technology. A system can be divided into ideal and realistic planning concepts per external interference. The ideal planning concept allows unrestraint and self-guided development without considering social application and contamination

Z. Chen, *China's Medium and Long-Term Science and Technology Program*,
https://doi.org/10.1007/978-3-030-75146-3_6

of science and technology. This category includes study programs or protocols of the scientific community. Whereas, the realistic planning concept emphasizes the instrumental and social nature of science and technology in which the progress is administered to achieve the preset objective. The development is swayed more by human understanding rather than by the nature of the advancing science and technology underpinning the burgeoning economy, society and national defense. In general, the influence and interference are from the government. China's scientific and technological planning is definitely not protocols among scientific communities but a governmental act that must echo the will of state. Therefore, from the overall perspective, China's scientific and technological planning focuses on short-term accomplishments and the political and economic derivatives generated during different stages with distinctive targeted orientation. Basic sciences sometime are not under the spotlight amid the emphasized undertakings to boost economy and serve the society. China's scientific and technological planning belongs to the realistic planning concept.

II. Politically Inclined Planning Before Reform

To decipher China's scientific and technological planning must be from a historical perspective of different stages. As mentioned in the preceding text, there are three evolving stages. During the pre-reform planned economy, three core ideas of "for technology over science," "for ultimate national core interests" and "for political function emphasis" drove the planning. To recapitulate them in one core term is "politicized propensity". The academia has reached unanimity with China's politicized science. Ma Huidi believes China's first scientific and technological planning manifested the politicized propensity.[1] She pointed out two aspects—the great censure of science and the economic disarray in technology —in her paper *Theoretical Deliberation on China's First Science and Technology Development Plan*. Science and technology have been used as means to cement political interests and swirling centripetally on politics. "The consequence of politicized science is to equal science to technology. In an ultimate political society, the importance of science becomes applications of technology."[2] Gong Yuzhi wrote the political contention between Deng Xiaoping and the "Gang of Four" during the last stage of the Cultural Revolution, where the core deliberation was "Had Chairman Mao ever mentioned 'science and technology is productivity'." The "Gang of Four" tried to smash Deng with the bout. During the bash exertion, this report was wrongly accused as "one of the three poisonous weeds." The "Gang of Four" launched fiercely a great censure on Deng. Besides the "far-right cassation frenzy" accusation, they threw in two more major charges, rumor-mongering Chairman Mao's quotations and tampering the translation of the true meaning of Marxism, and claimed Marxism's productivity did not mention "first".[3] Revealed here is the fact that science and technology turned into means of political contention. During the pre-reform era, politicized propensity became a feature of scientific and technological planning or even a conception.

The first 12-Year Program of Science and Technology (1956–1967) of China since its founding made great achievement that laid the foundation for China's science and technology, "a formulation of a complete science and technology system that has a great deal of influence on China's scientific and technological advancement."[4] Through analysis on the context of the 12-Year Program of science and technology, many political expressions of seeking quick success and instant benefits can be found. For example, "The blank sectors of science must be quickly filled; the better-established sectors must be rapidly enhanced and strengthened; the backward situation of China's science and technology must be cast off immediately and improved to catch up or close in with the world

[1]Ma (1995, 13–15).

[2]ditto.

[3]Gong (2004: p. 7).

[4]Hu (2006: p. 209).

level within twelve years. This is a must-accomplish historical mission."[5] Despite the extreme enthusiasm and vigor of China's science and technology devotees, it was unrealistic to race for the advanced world level right after the founding of China. It was more a political gesture. During the preparation of the guiding principle of the 1956–1967s scientific and technological planning, deliberation between "to develop the focused, proceed to be the spearhead" and "to develop the focused, progress across the board, solidify the fundamental and proceed to be the spearhead" arose. The planning outline proposed the former as the principle that was challenged by a few. Some suggested changing it to the latter and some had opinions to make other changes. We are for the former."[6] The deliberation exposed the contradiction between a program-oriented and academic-oriented approach. The politicization of scientific and technological planning prevailed at the national founding stage. The marriage of the political passion and good will of scientists during this stage accomplished an extraordinary result five years earlier for the first five-year plan. However, the social effect and historical legacy of the 12-Year Program of Science and Technology remain pondering-worthy. "In overall, the system of scientific and technological planning before the 'Cultural Revolution' was beneficial to China at a cost of becoming state-controlled science."[7] Under the circumstances, the planning was not just to develop the science and technology but also to strengthen the regime and build the basic functionality of national economy. The politicized propensity of the planning was evident.

The Eight-Year Program of Technology spanning 1963–72 clearly stated "Scientific and technological work must better serve class conflict and production struggle." This indicates the objective of the plan was to secure the regime and serve politics not the scientific community. The propensity began during the Yan'an era. The then-founded Yan'an Institute of Natural Science discouraged unrealistic advanced researches. Li Fuchun, the first incumbent dean of the institute, criticized in 1943 some researchers' unrealistic work by saying "On the other hand, the Natural Science Society has never studied or discussed these problems, so has not the institute and the science journal, but the universe and earth, instead, at great length. Why?"[8] The natural science research mission assigned to us by Xu Teli was to "solve the natural resources problems for the production during the Anti-Japanese Invasion, regardless of profound or basic technology."[9] To use science and technology as means to strengthen the political regime and serve politics had its significance during the time. However, it had become a convention carried forth even after the founding of China. "The guiding principle for the development of science and technology in the early stage of the national founding was a continuation of the Yan'an legacy or the same emphasis on serving reality."[10] Before the national founding, the service was for the core political mission of winning the war, and after that, for securing the regime. Therefore, the philosophy of the scientific and technological planning before the reform was the three ideas mentioned in the preceding or the planning with politicized propensity.

III. Utilitarianized Science and Technology Planning During Transition Stage

The ideas of the planning philosophy during the transition stage changed to "for technology over science," "Utilitarianized propensity" and "bolstering economic function" or just Utilitarianized propensity to be inclusive by one core term. Becoming means to brace economic development, the scientific and technological advancement in China was interpreted as being utilitarian for a very long period. Under the context, the

[5] 1956–1967 *Scientific and Technological Development Planning*, Ministry of Science and Technology: http://www.most.gov.cn/kjgh/lskjgh.

[6] *Selected Works* of Nie (1999: 11).

[7] [US]Richard (1989: 41, p. 231).

[8] He et al. (1990: 407).

[9] Xu (1991: p. 108).

[10] Yang (2002: 1, p. 30, 31).

mid- and long-term planning was infiltrated by strong Utilitarianism. The concept was utilitarian. Richard P. Suttmeier, an expert of China's scientific and technological problems pointed out the goal of China's scientific and technological advancement was utilitarian in nature. "Since the founding of the People's Republic of China, we have seen perpetuated utilitarian interests and objectives in science and technology. To be short, these include: serving the economic construction, national defense construction, cultural reformation through broadened propaganda and promotion of science and technology."[11]He made his point on the planning during the transition stage very appropriate. Before the reform and opening-up, "the cohesion of science and politics" was apparent for political purposes in the planning; while afterward and onward to the early twenty-first century, the planning turned to serve economy. Despite of the scientific emphasis in the planning during the transition stage, in general, and especially in the implementation perspective, the scientific development had been ignored and overrun by Utilitarianized concept. Guangbi exposed the "the utilitarianized propensity in China's scientific goal" in his treatise, *Thesis of the History of Modern and Contemporary Science and Technology in China*," "coherence between utilitarian requirements and law of autonomous development of science cannot always be seamless."[12]

After the "Cultural Revolution", the toiling objectives of the 1978–85 scientific and technological planning were to "in the twentieth century, automate all possible departments and places, electrify the cities and rural areas, automate production processes in major industrial production departments to boost labor productivity, accelerate societal productivity, uplift China's societal economy, strengthen national defense and secure the regime under proletarian centralization. In science and technology, a loyal and dedicated team of proletariat must be built across China to include the first class scientists and technological experts and to have the most advanced scientific experimental equipment for reaching as close as possible the world-class science and technology, partially matching or even leading the contemporaneous world-class level or fields." The repression and anxiety inflicted by the ten-year Cultural Revolution burst into urgency and eagerness to catch up with and suppress the world advanced countries in science and technology, rushing the planning to become an ambitious and aggressive one." The Eight-Year Program was prepared right after the "Cultural Revolution". The zest to chase or exceed the word advanced level was completely exposed in this plan. The plan laid out a series ambitious objectives that later were found unrealistic. The projects included were changed to "National Scientific and Technological Breakthrough Plan." The Eight-Year Program was criticized as an 'aggressive plan'."[13] Many reasons were behind the aggressiveness and recklessness. The high expectation of science and technology was the most critical one that uncorked a very strong utilitarianism in the concept of scientific and technological planning.

After the reform and opening-up, we have bettered our understanding of the development law of science and technology and added weight to the law. However, the effort was overwhelmed by the much emphasis on the economic efficiency and short-term benefits, making the planning much more utilitarian during execution. This is attested by a nominal proportional change of R&D/GDP. In the "9th Five-Year Plan", the proportion was targeted at 1.5% by 2000. It turned out to be a failed projection. The scientific and technological planning does not have legal binding power, which gives governmental agencies of all levels to pay attention to the short-term benefits over the long-term ones, rendering much more utilitarianized propensity into the implementation of the plan.

Weighing on technological planning during the transition stage, the science and technology were instruments for economic development and revenue driving vehicles. The emphasis on the

[11][US]Richard (1989: 41, p. 231).

[12]Guangbi (1991: p. 112).

[13]Hu (2003: p. 66).

short-term results of science and technology characterized the planning with utilitarianized propensity.

Besides, the sloganeering language and text add impurity into this paper with obvious sigils of the time. Utilitarianism plus political motives of the times characterizes the scientific and technological planning, indicating too much expectation and meanings are attached to the planning. Science and technology have been more means and tools but a development objective. China's scientific and technological plans are attached with more marginal connotations than they should.

The planning philosophy of the new century is through introspection to bring forward the "people-oriented" and "independent innovation" ideas. To condense the ideas into one term, it may be the "scientific and technological planning with humanity-oriented propensity." Apparently, this remains in its formation, which is not ready to be categorized.

IV. **Factors of Philosophy Evolution Steering**

The formation and evolution of scientific and technological planning conception do not happen overnight. The conception evolution of the mid- and long-term planning since the founding of China has been cradled by swaying factors and a complicated background. The half-century long evolution and possible perpetuation into the future reveal a certain objective regularity.

(1) Correlation between science, technology, and society: After the 1930s and 1940s, science was transformed from a cliquey knowledge luxury to popular learning necessity, taken from a handful of the privileged to become a societal infrastructure, blurring the distinction between basic science and applied science. Fusion and division among sciences are refined and deepened. Donald Stokes introduced "Pasteur's Quadrant" and the concept of "applied basic research". He advocated the reestablishment of the contractual relationship between the government and the scientific community. "Breakthroughs acquired from basic research induced by applications can stimulate further pure research, and vice versa. This confirms Pasteur's belief, 'There is no pure science and applied science, only science and applied science'."[14] This suggests the scientific development must be associated with social interests, which make planned scientific programs a necessary undertaking. The same situation holds true for the technological development that has gone from an "insignificant technological focus" through "fairly significant technological development" to the "societal technological advancement". As the impact of science and technology increases, the need to steer the advancement of science and technology becomes more urgent. A variety of planning formats have emerged in many countries. The evolution of the scientific and technological planning philosophy is the result of the correlation between science, technology and society.

(2) Resources factor: Resources available for scientific and technological development are not in a forever supply. The case is truer for China, so is it a fundamental factor for the planning. "Where" and "how" to allocate the resources are matters of decision criteria and value evaluation. During the period of "fairly significant scientific development" and "fairly significant technological development," the owners of resources, the state, interest group and people all demanded returns from their sponsorship. Scientific communities and the purity of scientific development were challenged. The sponsors' demands have become one of the headsprings of the conception of scientific and technological planning. This was particularly true for underdeveloped countries like China. After the founding of the new China, elevating China's scientific and technological competence became a burning desire to scientists, technological workers and state leaders. There was an outburst of long-suppressed national emotion and outcry of

[14][US]Donald (1999: p. 87).

external pressure-imposed frustration of scientific incompetence. Vesting limited resources on technological plans that could produce short-term results was an undertaking with pertinence. Particularly, under the specific international context during the early stage of PR China's founding, cementing the regime and developing economy were the ultimate urgent missions of science and technology. Therefore, the planning of scientific and technological development with a politicized propensity from start and later utilitarianized propensity as a consequential result is apprehensible. When mulling over the strong utilitarianized propensity in scientific and technological planning in the new century, we can say the conception was correlated with limited resources.

(3) Science and technology with Chinese characteristics: It is obvious that the concept and guiding principle for China's mid- and long-term scientific and technological planning are bred with Chinese characteristics, so conspicuous are the will of government and national steering, a result of politicians' political insights and scientific opinions of scientists or the scientific communities. It is a model of "politicians + scientists". The outcome of the combination varies by the contemporaneous environment and the understanding of law of scientific development. The injection of politics has brought resource availability for the scientific and technological planning, which in turn introduced other political elements into the planning for the decision makers must consider political risks. Consequentially, in successive plans, the anticipated objectives have a large technological proportion. This also explains the philosophy, "for technology over science," that has become a common adherence to the planning.

(4) CCP's guiding principle and planning conception: Marx believed science and technology are the actual productivity. The "Chief Architect" Deng Xiaoping stated the same in the implementation of the reform and opening-up and emphasized science and technology are the primary productivity." Deng's view of science and technology as the primary productivity at the early stage of the reform and opening-up should not be construed as just a slogan for economic development but a fresh new ideology sprouting under the contemporaneous societal reality in China. It is another theoretical expression for China's modernization march."[15] This new attitude branded science and technology as an instrument to upgrade merchandise and meliorate the welfare of the masses in actual work; the undue pursuit of national economic development turned science and technology into appendage of the movement. The "action for some and non-action for others" exposed the actual interference from the will of leaders and the state, not the needs of science. Habermas also stated science and technology were the primary productivity in the late 1960s, and confirmed science and technology had created a new ideology. China's scientific and technological planning is the manifestation of an ideology of state. However, Marx considered science and technology being productivity not ideology. We might misinterpret Marx's "science and technology are productivity," resulting in excessive utilitarianized propensity in our concept of scientific and technological planning.

Although the conceptual utilitarian element was a realistic choice with rationality and ineluctability for China's scientific and technological planning, during the transition stage, this element was an unhealthy one or unsuitable as a guiding ideology for a long-term development, basic research, in particular. The long-neglected basic research and limited free scientific exploration have been two major factors keeping China short in Nobel prizes. For all the great achievements through scientific and technological planning, China's independent innovation capability remains weak, giving much less success in original innovation and basic research.

[15]Wu (2005: 3, p. 107).

"Quick production and fast returns have been emphasized by Chinese scientists. The mentality of racking one's wits in doing research has dragged down the forwarding speed of China's devotion in high-level, leading-edge science."[16] The recognition of China's accomplishment in science and technology is not in parallel with its economic power. This triggers the introspection on the past planning conception as we enter the new century, therefore, the emergence of "independent innovation" initiative.

As the formation and evolution are an objective process, the philosophy of scientific and technological planning requires examination from a philosophical perspective, rather than just analyzing the planning concept through planning. For a nation in whole lacking rational thinking, the study of the philosophy of scientific technological planning must be philosophized to avoid undermining significantly the innovation process. The most fundamental issue is to make progress through the development law of science and technology. During the fairly significant science era, to a country with rapid economic growth, a long-term emphasis on building basic science infrastructure to amplify the overall national prowess through technology under sagging per capita resources and less competent basic science is necessary.

The historical evolution process of the philosophy of scientific and technological planning from "politicized propensity → utilitarianized propensity → humanization propensity" will continue. The study on the planning philosophy will benefit the deep understanding of the planning and the development law of science and technology.

6.2 Planning Philosophy Influence on Contemporary Academics

The preceding context mentioned the planning of science and technology is the core steering mechanism commanding China's science and technology system. The planning has great significance to China's science and technology system and academic environment. To deepen and expand the study, this section tackles the importance of the planning and its philosophy to the scientific and technological advancement and academic studies from the perspective of the impact of the planning to contemporary academics. The economic circle considers the period from the reform and opening-up onward to the sprout of the socialist market system as the transition stage. Per this, the planning remains in the transition stage. The earlier definition of the new century as another stage is based on the periodization of the history of scientific and technological planning.

I. Utilitarianized Academics and Academia-Officialdom Affiliation

Current issues with academic studies in China remain a concern of the academia. One typical aspect exposed by the treatise "China's Academics and Education: Deliberations on Issues and Remedies" is published in *Journal of Dialectics of Nature*. From 2006 to 2007, as the leading publication of humanities, the Journal of Dialectics of Nature published a series of treatises that explore issues and remedies of China's academics and education, where Xu and Jianwei pointed out that the overemphasized quantity, formalism and uniformity under the empiricist management model for scientific studies are a hindrance for academic innovation.[17] Li Hongwei's discourse on Western postmodern theory also said the destruction of traditional intellectuals is to be blamed mostly on Chinese themselves, and "the current urgent mission is to rectify academic morality and modulate academics."[18] Wang Bolu was more explicit. He indicated China's academic and education issues have been the malpractice in the system and intensifying competition.[19] These discourses have disclosed the issues. The major cause has been the academia's inclination to satisfy

[16]Cong and Suttmeier (2006).

[17]Xu and Jianwei (2006: 1, pp. 1–2).

[18]Li (2006: 2).

[19]Wang (2006: 1, pp. 1–2).

political needs by currying academic resources with political circles, or the "academic utilitarianism", to be short.

The "utilitarianized conception" is the core philosophy during the transition stage of scientific and technological planning, according to the preceding analysis. Onward with the implementation of the planning and execution of scientific projects, the concept spreads across the entire development system of the science and technology or in a general sense, the entire academic environment, making utilitarianism one of the characteristics of China's academic studies.

The academic utilitarianism is induced by that of the planning of science and technology. The manifestation of the academic utilitarianism is the "academic-officialdom affiliation". The affiliation or even alliance of academia and political circles facilitates the injection of technology funds and acquisition of academic resources, giving the planning a "utilitarian sigil". This has been conspicuous and different from other countries' academic environment and that of ancient China. It is anticipated to be different from China's future academic environment centuries later after crossing the underdeveloped stage. The utilitarianized planning of science and technology at the state level avails the accumulation of wealth and boost of national strength. However, to academic studies, the utilitarianized academics do not deliver benefits.

The affiliation of academia and political circles exists in developed countries of the West, only the one in China has its own contemporary characteristics. The circumstances for coalition are complicated in China. The social phenomenon commonly known is the common occurrence of dual identities of "an officer and a scholar". Scholars actively seek the status of "scholar + leader". The public in society prefer the status of "professor + doctoral advisor + director" or similar title, despite scholars are the majority and only a handful of leader seats available. Therefore, the arrangement of "scholar + sinecure" is acceptable by most scholars. A typical example is the ranking-standardization. In an academic organization, a scholar is given a sinecure title like director, consultant, or member to indicate its academic position and significance in a specific academic field. An academic organization is for scholars to exchange and encourage each other for academic purposes, one of the formats in a scientific community. It is neutral from the competition of academic resources. However, under China's contemporary research environment of "utilitarianized academics", the scholar's position becomes a convenience to convince the public of its academic achievements and extend influence to acquire academic resources and portions of the projects of the scientific and technological planning. It is also an indirect path to seize power and hold rights. Therefore, scholars incline to hold positions in academic organizations besides the duty of academic faculty as one of the means to obtain academic resources. The social phenomenon is an expert's business card populated with multiple titles of academic organizations in addition to "Professor". Scholars and experts hold many positions in society, as consultants of many governmental agencies, as independent directors of enterprises or member of think tanks, etc. This background of the phenomenon comes from the philosophy of utilitarianized academics induced by the philosophy of utilitarianized planning of science and technology. Similarly, politicians can easily hold positions in academia as a guest professor or leaders of various academic groups.

Robert K. Merton believed that science, as a societal system, while playing a critical role in societal development and life, is constantly being restrained and influenced by political, economic, societal and cultural factors. From Merton's sociological perspective, China's science has many complicated correlations with political and economic factors. The interactions among the factors have many characteristics. In the contemporary academic environment, the swaying power of politics on science is stronger than the opposite. This may be deemed as sociology of science with Chinese characteristics. Harriet Zuckerman, in his famous book *Scientific Elite: Nobel Laureates in the United States*, pondered on the relationship between scientists and interests. His Nobel Laureate's authoritative voice in many public realms has changed scientists' societal roles. For example, Watson and et al. switched themselves as technocrats and scientific administrators. Some turned their authority in science into influential power of politics and society. For instance, the "pressure group" pokes into political affairs.[20] China's indigenous scientists, lots of them seems like acquire more resources, including political interests no less than their foreign counterparts, with their academic achievements, despite few of them has been awarded a Nobel.

A status of scholar-leader has its advantages in acquiring and accumulating academic resources. It adds convenience to applying for a funded project, publishing a treatise or a book, participating contest of achievements, establishing a curriculum, etc. Under China's contemporary academic context, the circumscribed academic resources add impetus to competitions inside and outside the academia. Regardless of the case, the identity of "scholar + leader" has its advantages. On the other aspect, an "officer-scholar" status can benefit greatly the career path in the officialdom. This makes the academia and officialdom affiliation an inevitable phenomenon in China, confirming the breed of utilitarianized scholarship from a utilitarianized environment.

China's academia and officialdom affiliation is a phenomenon. It is beyond judgment. This is a bridge between science and power. From China's contemporary reality, using science to access power and resources is more common. For example, scientists after making achievements are easier to redirect their focus to politics and become high-rank officials. This is not necessarily a misdoing. It helps to gain more resources and qualify officials for the state. This, however, gives uncertainty if the interference of politics is removed from the equation for the imperfect structure of China's scientific community. Many places require political coordination, involvement and balance.

China's affiliation of academia and officialdom is an unavoidable outcome in contemporary academic research. The affiliation is necessary and rational under a specific circumstance (regardless of legitimacy and rationale). The necessity for the affiliation is a better way for the academia to acquire scientific and technological resources. To political circles, the sponsorship to the academia involves many aspects. The affiliation becomes a proper and acceptable way for the political circles to support the scientific and technological planning and programs. The rationality is the lowered difficulty and cost for scholars to accumulate knowledge needed and obtain necessary academic resources for the scientific research to accomplish more. "The bridge between academic and political circles" increases in number and widens in size.

The result of utilitarianized academic environment is an irrational evaluation system of colleges and universities and the nonstandard evaluation system of academics. The phenomenon of the entire evaluation system of society is to convince the investor or sponsor—the government, a rewarding upfront investment in the initial stage that returns the academia with more resources. This encourages much adoption of SCI and SSCI that are known to be deficient by many university presidents in China. The most important achievement of scholars is knowledge production. Indexing of knowledge production weighs more on quantity and ignores quality, resulting in splitting one treatise into three. Of increasing number of published papers, the number of heavyweight treatises is only a few. Instructors are busy in filling out annual evaluation forms, coping with evaluation pressure, applying for projects of the scientific and technological programs, even with deception. In many cases, scholars do not delve into knowledge but brew knowledge, or the so-called "making up stories or vying for projects."

To an individual scholar, the utilitarian bout is unwillingly engaged. To change an academic environment, adapting to it becomes the primary. Presidents of universities are often involuntarily flexed by pressure to maintain the vicious practice and only be able to make some improvement within their power in some areas. As presidents of universities, they are required to amass academic resources for not only themselves but also the schools. The most feasible solution for them

[20][US]Harriet (1979).

is to enhance the affiliation of academia and officialdom to meet the needs of political circles, and at the same time to step up the collaboration with the education and industry to cater the demands of industries.

II. External and Historical Factors for Utilitarianized Academics

1. Puzzledom of Academic Amassment

Chinese scholars shoulder more social responsibility than their foreign counterparts. As a common Chinese citizen, he must strive to make his ends meet. Scholars use their knowledge to serve the society to get remuneration. According to the wage definition of Western economics, "Wages are the monetary form of labor productivity or the price of labor." In other words, scholars sell their labor, primarily headwork capacity to acquire physical goods and metal remuneration. However, from economic practice, scholars at all times and all places are at the secondary distribution tier in the greater society distribution for their indirect merchandise production. Therefore, scholars in most cases get non-equivalent returns for their toil with the inclination of more giving.

This is true for contemporary Chinese intellectuals. Scholars as citizens must work to survive. The academic studies by Chinese scholars are driven not only by interest but also or more importantly by the survival means to make a living mostly. The utilitarianized propensity in academia is understandable. The population of contemporary Chinese scholars is large. In late 2014, state-owned enterprises had 55.5 million specialists and technicians in total. The number did not include other scholars. A general scholar, given no resources, is hard to make any achievement. Chinese scholars must find a way to amass resources from amid limited total academic ones in a societal context with averagely less resources available, including economic, social, scientific and technological ones like projects, programs, equipment, payroll, publication, title, authoritative voice, etc.

"Karoshi", frequently seen in Japanese society, is emerging in China in another form, "Die Young". Young professors in their 30 s of some prestigious universities passed away because of severe stress. "In the second half of January in 2005, in a short period of four days, two young instructors of Tsinghua University died successively. On October 20, 2004, Zhang Yingqing, a Nobel hopeful, dean of Institute of Holographic Biology of Shandong University, died young. Xiao Liang, 32, an academician of the Center for Studies of Border Area History and Geography of Chinese Academy of Social Sciences died in his sleep at 4 AM on January 5, 2005."[21] The nowadays general public laments the fretting surroundings. Scholars are similar, resulting in fewer masters. The "forced-fidget" is not from scholars but the adversity and stress from the hard-to-get resources for their scientific research. "The current evaluation system overemphasizes quick success and instant benefits. Regardless of private or peer reviews, the weight is on quantity not quality. To become a prestigious figure in the academia, quality in achievement is necessary. Innovation is a criterion. However, high quality research results are not short-term achievable."[22] The philosophy of utilitarianized planning of science and technology is the instigator of this mischievous academic tout. The overemphasized utilitarian goal of pursuing economic efficiency misguides academicians to hound excessively academic resources. Maybe we could make a bold assumption that China's scholars mired in asperity of survival and development decide to undertake the utilitarianized objective for them to overcome the plight. Therefore, the transition stage of scientific and technological planning is utilitarianized. The propensity spreading to the academia has brewed the "utilitarianized academics" in the academic environment.

2. Reality and History

The affiliation of academics and politics or "politicized academics and academicized

[21] Xiao (2005).

[22] Wang (2006: 4, 1–2).

politics" has become a global fad. The state government has evolved from empirical governance to technocracy. From the supersedure of national leaders post-World War II, leaders with military background have been gradually replaced by highly-educated politicians with an impressive diploma. Many countries have been governed by experts (including natural scientists, sociologists, artists, etc.). A certain percentage of state leaders in the world possess a doctorate. The proportion of China's new round of governmental officials (2008) with higher education background has climbed up. "In 270 governors (mayors, chairmen of autonomous regions) and lieutenant governors (vice mayor, vice chairmen of autonomous regions), the number of officer with graduate school background is 210 or 77.8% of the total, of which 58 have a doctorate, or an impressive 21.5%."[23] Their background facilitates the much easier infusion of academia and political circles, so does the influence of utilitarianized concept on the academia.

The affiliation of academics and politics has been a historical phenomenon since the ancient days. Scholars of the history of science claim: China's ancient science was categorized in utility technology, which was used to secure the imperial supremacy that was included in the ruling feudalism. Many known ancient scientists were officials in the imperial court. The official academics was much influential than that of private schools. For example, Zhang Heng had served as the imperial astronomer for 14 years. Zu Chongzhi served in many official positions. Shen Kuo was appointed the head of the Three Bureaus (or the Ministry of Treasury). Guo Shoujing was the head of astrological commission for life. Xu Guangqi was the Chief of the Ministry of Rites, just to name a few.

China's traditional affiliation of academics and politics led to the effort on solving realistic problems rather than establishing an axiomatic system of China's ancient science. Searching for answers to natural origins remains a conflict between scholars' curiosity and the investors' or ruling class' interests. This tradition has become an inheritance as an inescapable yet understandable historical reason, to some extent, responsible for today's utilitarianized academics. Through deepened analysis, in today's scientific research environment, the emphasis on solving short-term and realistic problems (including the survival of intellectuals and scholars) continues, ignoring researchers' curiosity toward nature and motivation of pursuing scientific principles. In the overall context, the mentality of quick success preferred by the investor or the government and academicians is a rational outcome. Ancient tradition is imperceptibly found in the preparation and implementation of scientific and technological planning of China. The utilitarianized philosophy in the planning is not just a practice in modern times but also an inheritance from ancient utilitarianism of technology over two millennia.

6.3 Case Study: Large Aircraft Project of Scientific and Technological Program

The second scientific and technological planning of the twenty-first century, *2006–2020 National Mid- and Long-term Scientific and Technological Development Program*, listed 16 key projects of which one was the research and development of large aircraft. In the past successive plans, elaboration and deployment of aircraft R&D were given. For example, the *National Medium- and Long-Term Science and Technology Development Program (1986–2000)* stated to "develop technologies for the design, manufacture and mass production of domestic trunk route aircraft, and strengthen research into flight safety technologies." In the *10-Year Program for Development of Science and Technology of the People's Republic of China* and *Eighth Five-Year Plan Outline (1991–1995–2000),* "In civil aviation, focus on breakthroughs in the design and manufacturing of trunk route aircraft and helicopters. By the end of the century, the ability to design and manufacture 150–200 trunk route aircraft and light helicopters shall be

[23]High Academic Diploma of 80% Provincial Leaders, Beijing News, Feb. 1, 2008.

substantiated" was spelled out. The *Outline of National Mid- and Long-term Scientific and Technological Program (1990–2000–2020)* put forward, "We will develop design, manufacture and mass production technologies for domestic trunk route aircraft." In the *10th Five-Year Plan of National Economic and Social Development—Special Development Plan for Science and Technology Education Science and Technology Development Plan (2001–2005)*, the goal to "focus on research and development of the world's round-trip transportation system technology, integrated air-ground data network technology and medium and large transport aircraft system technology" was clearly given.

The National "11th Five-Year" Plan for Scientific and Technological Development elaborated on the large aircraft projects, "Large aircraft projects, driven by the key technical needs of contemporary large aircraft, shall carry out key technology pre-research and examination. Focusing on the design, development and manufacturing of system integration, power system and test system for large-scale domestic aircraft, we will break through the core key technologies and make technical reserves for the development of large passenger aircraft." From these planning, China's attention on R&D of large aircraft has been clear in the scientific and technological plans of the past 20 years from 1986 and onward.

During the study of scientific and technological planning documents, we were enlightened by this case. It can serve as an exemplar to elaborate the complexity of the preparation and implementation of projects in scientific and technological planning process and deepen the exploration of China's scientific and technological planning. As the result of interactions of interest groups and various relevant factors, the scientific and technological planning, especially, during the transition stage, this case delivers a complicated and dynamic landscape.

Large aircraft, as an excellent practical area for independent innovation, can propel advanced leap of technologies in many related fields besides boosting national economy and defense strength. This section will give a brief overview of the development history of large aircraft.

China's endeavor in the development of large aircraft or the large passenger aircraft began in the 70s of the twentieth century. ("Large aircraft" means jumbo jets weighing over 100 tons with a capacity over 100 passengers and a range over 5,000 km.) This kind of aircraft started serving national chairman (Chairman Mao) and premier (Zhou Enlai). The project, named the "708" Project according to the contemporaneous naming convention of China for large-scale projects, was kicked off in August 1970. Phonetically, "708" is close to the "707" of Boeing, which many mistook it a mimic but on the contrary, a result of independent innovation. The project focused on the R&D work of the "Y-10" (or Yun 10/Y-10), which was presided over by Shanghai City. Its operations were administrated by the Third Ministry of Machine-Building of the PRC under the direct command of the central government and collaboration of central ministries, military, 21 provinces and cities and autonomous regions of each jurisdiction with a total of 262 units. From its commencement, debate on the undertaking of domestic or foreign advanced technology had clouded the research and design. This case demonstrates the collective decision-making process under pressure from all aspects, including politics, science and technology and economy, for a scientific and technological planning.

I. "R&D Success and Technology Debate" of Y-10[24]

At the began for design of the Y-10, disputes arose within the professional field between supporters of imitationism and those who supported innovationism,[25]※ centered largely in the Department of Industry on Aviation and Spaceflight. At that time, some Chinese authorities lacked confidence about the ability of Chinese

[24]Chen (2009: 31(4), pp. 414–418).

[25]In this book, *imitationism* and *innovationism* are new words created by the author. *Imitationism* refers to copying foreign airplane technologies and includes little innovation from Chinese professionals; *innovationism* means that innovation is the core of development and specific use is made of indigenous Chinese innovation(s).

manufacturers on design a large airplane. They consequently adhered to the imitationism, which they defined as copying technologies and techniques from existing planes produced by developed countries. However, others scholars argued that China should promote domestic indigenous innovations through the adoption of relevant technologies. That debate was never stopped throughout the process of development of China's large airplanes.

While in the early period, there were three differing examples on how to install an airplane engine: the Soviet aerofoil model, the European aerofoil model, and the U.S. aerofoil model. The people who were promoters of innovation eventually chose the U.S. model as most optimal for the Y-10 based on wind-tunnel experiments with all three models. However, this decision was severely against by imitationists because the new development did not follow the previous model, and ultimately the Civil Aviation Administration of China (CAAC) refused to purchase it. Despite the controversy, it was subsequently proved that the U.S. aerofoil model was most suitable finally.

Most of the adherents of imitationism worked in the government, while promoters of innovationism explore more practical positions in airplane design and were supported by state leaders. For this reason, innovationists can had the means to reject interference from imitationists and innovationism dominated, especially when the Y-10 project began and indigenous innovations were successfully applied to the development of the Y-10.

Fortunately, the general airplane design group decided to adopt the suggestions of vice-chief designer Cheng Bushi (Fig. 6.1), who aimed for a breakthrough from 10 to 100 tons in the airplane's design that would directly advance the Chinese design. After a decade of hard effort and with 166 large-scale ground experiments, the main body of the Y-10 was produced at the comparatively low cost of RMB 350 million. Only two examples were built. The first prototype (01) was used for static testing and while the other (02) was used for flight testing. The plane had its first flight on September 26, 1980.

Before 1984, Y-10 made 130 flights requiring 170 h of flying time to locations such as Beijing, Harbin, Urumqi, Zhengzhou, Hefei, Guangzhou, Kunming, and Chengdu (Figs. 6.2 and 6.3). It also was even used to fly to Lhasa in Tibet seven times for specific missions during snow hazards.

From Table 6.1 we can see the comparative specifications between the Chinese Y-10 and the U.S. Boeing 707–020. The table shows that most of the specifications of the Y-10 were better than the Boeing 707–020 at that time. However, the controversy between imitation versus innovation never ceased, and the forces of imitation had long prevailed in the field of Chinese aviation.

From the tabulate comparative data, except speed, the "Y-10" shows better data than that of Boeing. The "Y-10" had made an initial success and a good start.

II. Contention of Innovation and Imitation After Reform and Opening-Up

In the 80s of twentieth century, China was in a perspiring reform. Many circumstances had changed dramatically and different from the time of designing the "Y-10" that ended prematurely after its creation. In December 1980, the 65% complete of "Y-10" 03 required 30 M RMB to finish up its remaining 35% construction. The death of Chairman Mao and Zhou Enlai dimmed the fate of the "Y-10" as the leader's plane.

As the only client, the Civil Aviation Administration of China (CAAC) slashed the possibility of procurement of the "Y-10". On January 8, 1981, the CAAC submitted its report (81) MINHANGGONGZI No. 2, *Views on Domestic "Y-10" Aircraft*, to the Central

Fig. 6.1 Vice-chief of Y-10 designer Cheng Bushi

Fig. 6.2 "Y-10" Test Flight Routes[26]

Fig. 6.3 "Y-10" Maiden Flight[27]

Financial and Economic Affairs Commission. In the report, it stated, "...The CAAC now has ten Boeing 707 s that are not frequently used. They can serve for another two decades...The CAAC in last year's Ten-Year Planning (1980) did not include any acquisition for the Y-10..." In fact, a few months later, China's official media disclosed the news on the procurement of foreign large aircraft. Since 1980, China had spent over 15 billion US dollars purchasing 500 large aircraft. In October 1983, a meeting called by the State Development Planning Commission (SDPC), was attended by Shanghai, representatives of the Third Ministry of Machine-Building of the PRC, where Shanghai City argued to finish up the Y-10 03 would require over 30 M RMB, and CAAC clearly stated the rejection of the idea. After the meeting, the state did not invest more in the "Y-10" R&D. The meeting did not have resolution minutes. However, because of lacking funding, the "Y-10" program had to stop.

All those ruses of discrepant technologies brought up by the pro-imitation clique could have been refuted today by sufficient facts, for example, the untenable reasons, such as the weight efficiency, outdated technologies, far-stretched national capacity, not to finish an unfinished program, etc. In the early 80 s of the twentieth century, the pro-imitation clique's failure to convince the innovation advocates but winning through funding has exposed the power wrestling behind doors and bouts of other unspoken factors.

The debate and contention were continued. On July 30, 1986, Hu Xitao, the former director of Aircraft Bureau of the Ministry of Aerospace Industry of the PRC, along with Shen Yuan, Zhang Azhou and Ji Wenmei, three presidents of Beihang University, Nanjing University of Aeronautics and Astronautics and Northwestern University of Technology submitted a petition to Comrade Deng Xiaoping, *Provide and Use Domestic Trunk Route Aircraft at All Costs*. This was the famous "Petition of Four Gentlemen". In the petition, the four top-notch aerospace experts elaborated the urgency of the development of trunk route aircraft from the needs of

[26]*"Y-10" Successful Test Flight*, Cheng Bushi Blog, http://chengbushi.blog.china.com/photo.

[27]ditto.

transportation development and the aviation industry. This was the background for the inclusion of the R&D program of large aircraft in the 1986 Scientific and Technological Planning, to "develop technologies for the design, manufacture and mass production of domestic truck route aircraft, and strengthen research for flight safety technologies"[28] However, the planning was not legislation. The program was not implemented.

China's aviation industry decided on a three-stage development strategy, Imitation-Collaboration-R&D, amid the restless debate. According to the strategy, the first stage was to assemble and partially manufacture trunk route aircraft. McDonnell-Douglas Corporation's 80/90 (MD-80/MD-90) series aircraft were the assembly targets, while technology was offered by the corporation; followed by a collaboration with foreign countries to design and develop 100-passenger seat aircraft; before building self-reliantly 180-passenger seat aircraft by 2010. The path undertaken was referred to as the "Santana" development path of aircraft, which was to make collaborated production with foreign corporations before shifting to domestication.

Adhering to the three-stage strategy, China had assembled 35 MD-82s from March 1985 to October 1994. In 1996, Airbus SAS signed a collaboration agreement with the HQs of China Aviation Industry Corporation (AVIC) to jointly design and manufacture 100-passenger seat AE-100. In August 1997, after acquiring McDonnell-Douglas Corporation, Boeing announced the termination of making MD-90 aircraft. Raw materials of 20 MD-90s being assembled in China were already procured and warehoused. By the then price over 30 M US dollars per aircraft the material cost 500 M USD. Only two MD-90s were sold domestically by AVIC. In 1996, Airbus Asia, Singapore Technologies and AVIC came to an agreement to jointly design and develop 100-passenger seat AE-100. In 1998s Farnborough International Air Show, Airbus released the end of the AE-100 contract and disclosed its plan to invest 500 M to R&D its own 107-passenger seat trunk route aircraft A-318. The "Three-Stage Strategy" concluded with a complete defeat. China failed to acquire any critical technology and opened large share of domestic market at a cost of a large sum of technology transfer fee and damages (breach of contract) caused by Airbus.

During the deliberation of imitation and innovation, China's top leader once announced to support the innovation idea. On October 13, 1981, Comrade Deng Xiaoping set the tone to "consider manufacturing domestic aircraft by ourselves" in his keynote speech delivered in an Enlarged Meeting of the Political Bureau of the CPC Central Committee. On December 30 of the same year, he added, "From now on, all domestic aircraft must be made in China." However, the development path of China's large aircraft did not follow Deng Xiaoping's instruction after the successful test flight of the "Y-10" in 1980. Digging into the reasons, the intertwined and complicated politics, technologies, economy, public relations, and power influence were unveiled. The study of this specific case might deserve another book. This book is set to use this case to ponder on the complexity of large-scale scientific and technological programs. Items, included in the planning of science and technology but not implemented by the execution of the planning, are variables. In 2006s scientific and technological planning of the new century, large aircraft program was listed as a critical program that was prepared specifically according to the previous planning experience.

III. Critical Program Planning in New Century

Riding into the twenty-first century, the Chinese government realized again the importance of large aircraft to the politics, economy, and technologies of a nation and its people. The innovation advocates on the same topic were once again winning the bid. In March 2003, Wang Daheng, an academician of CAS (Chinese Academy of Sciences), allied over 20 other

[28]*Outlines of National Mid- and Long-term Development Plan of Science Technology*, Ministry of Science and Technology, http://www.most.gov.cn/kjgh/lskjgh.

Academicians of the academy to submit a petition to Premier Wen Jiabao to give their suggestions on the jumbo jet transport aircraft. In May 2003, Premier of the State Council Wen Jiabao visited Wang Daheng to ask for his suggestions on the mid- and long-term scientific and technological planning of the new century. The 88-year old "Pioneer of China's Two-Bomb and One-Satellite" Wang Dayan said, "China must R&D its own large aircraft." In November 2003, the Ministry of Science and Technology was appointed by the State Council to organize the "Large Aircraft Program Team". Guan Qiao and Liu Daxiang, two Academicians of the China Academy of Engineering, earnestly submitted Suggestions on Immediate Research and Development of Large Aircraft on 2004s Two Sessions of the National People's Congress and the Chinese People's Political Consultative Conference (NPC & CPPCC), which triggered a social frenzy of self-reliant R&D of large aircraft. Professor Lu Feng of Peking University wrote and published his *Strategic Study on China's Large Aircraft Development* in the early 2005. Business Watch Magazine spent large space and a cover story to open discuss the large aircraft program on the magazine's March 2004, March 2005, June 2006 and January 2007 issues.

On February 9, 2006, the State Council issued the *Outline of National Mid- and Long-term Scientific and Technological Development Program*, which set a deterministic tone on the research and development of large aircraft as one of the 16 critical programs in the following 15 years. On March 14 of the same year, The Fourth Session of the Tenth National People's Congress unanimously passed the 11th Five-Year Plan of national economy and societal development in which the large aircraft emphasis, especially, in critical programs of high-tech industrial projects, was focused on the trunk route aircraft. The state investment to the entire large aircraft program was 50–60 billion RMB to cover large transport aircraft, large passenger aircraft and engines. After the inauguration, the program was divided into the large transport aircraft and the passenger aircraft projects and carried out simultaneously.

"This is the first time the R&D and manufacture of trunk route aircraft approved by the National People's Congress," said Hu Xitao.[29] In the early twenty-first century, supported by the national leader, the innovation advocates claimed victory of the bout. On February 26, 2007, the State Council during an execution meeting approved the critical program of the R&D of large aircraft in principle and agreed to establish a corporation for the large aircraft program. On May 11, 2008, China Commercial Aircraft Company Limited was founded in Shanghai, marking a new chapter of China's self-reliant R&D of large aircraft. However, the specific design and development of large aircraft were handled by related departments to curb the rooted problems in China's scientific and technological system and the inveteracy of pro-imitation mentality in the aviation industry, avoiding being trapped in the same rut like the previous "Y-10". The successively proposed R&D and manufacture of large aircraft in the past did not become a reality until the twenty-first century. The "three on, three off and reactivation" of large aircraft program in the history of scientific and technological planning has been a rare case. Other planning did not encounter such setback. It is the unusual characteristic of this case that has enlightened our understanding of the intertwined complexity between science, technology, politics and society of the planning of science and technology. The planning of science and technology has been the result of the complicated influential factors. Technology is not the only deterministic factor for the success and failure of a critical scientific and technological program. The system also plays a steering role. Although the significance of the successful test flight of the "Y-10" twenty years ago was not given correct attention, it was a valuable set of materials for people who study China's contemporary history of science in the future. The fault was the price that "our country paid in the forwarding process." To lower the possibility of making a similar mistake and the price paid to correct it and to sublime the scientific conception and encourage independent

[29]Cao (2006).

innovation deserve further contemplation of scientists, technological workers, and policy makers of science and technology. This book attempts to borrow the specific case to deepen the understanding the complexity of the planning of science and technology and its implementation, as well as the philosophy of planning.

IV. **Conclusion and Thought**

Analyzing the debate and contention between imitation and innovation that have been continuing for 30 years, we are able to make a preliminary conclusion.

First, ideas of imitation and innovation have been co-existing in aircraft design and manufacturing since the founding of China and highly possible sustaining. In the early stage of China's founding, the pro-imitation clique had a large share of market and many successful cases. The idea, indeed, shortened the technological gap between China and foreign powerful countries, primarily the USSR, in a short period. However, with the increase of China's self-reliant R&D competence, switching from imitation to innovation should have taken place timely. The inertia of the deep-rooted imitation thought dragged on to engage in fierce on–off debates with the *Innovationism*. During the transition stage after the reform and opening-up, the interest groups and interest channels had changed. The debate between the pro-imitation clique and innovation advocates transformed into interest contention, turning the planning of science and technology, the typical and special case, into a contention platform of interest row.

Second, ideas of imitation and innovation have been common in other technological areas of China, for instance, automobile. A judgment is difficult to make if possible. The two thoughts served different functions at different historical stages. Generally, in the early stage of the founding of the Republic, the technology and science infrastructure was weak. The imitation approach was a feasible choice. However, after a period of amassment of technologies, the independent innovation became more appealing, making it a favorable choice. A determination to break the propensity of imitation should be made to turn merits of others into our own dexterity for independent innovation movement. General technologies available through acquisition can be imitated to reduce R&D cost. The idea of independent innovation should be substantiated for high-tech areas like large aircraft with the capacity to drive national economy.

Third, the special case unraveled the complexity of planning preparation and implementation of large scientific and technological programs. As a state behavior, the long-term scientific and technological development should be planned at the top state level to outline a clear blueprint. The change of time requires adaptivity of scientific and technological planning, which has become the core of China's science and technology after the reform and opening-up. To make this core change to function, the planning must be substantive in implementation. The "three on, three off and reactivation" of the large aircraft case suggests the importance of the top level's will in decision-making. The persistence and determination of the execution level is more critical. The coherence of the decision-making level and the execution level, particularly, under the context of market economy with socialistic characteristics, has been an issue that the 2006s mid- and long-term program and science and technology and must resolve in the future.

When discussing the more than 30-year history of Chinese large airplane development, several questions must be considered: Will the debates between *imitationism* and *innovationism* continue for a long period in China? Are there similar debates in other engineering and technology areas? In the "Big Science" stage, the relationship between science, technology, and society is becoming much closer than before, so how should China deal with the complex relations among them? Why can't the efforts of Chinese scientists, engineers, and experts have practical effects on the process of developing the Y-10? How does the relationship between Chinese scientists and politicians differ from that of other countries? And to what degree has this difference influenced Chinese governmental policy making? How will the conflicts between decision makers and implementers affect the development of Chinese science and technology?

I believe that an in-depth discussion of these and other questions will promote greater understanding of the history of China's science and technology.

6.4 Recap and Inspiration

The planning of science and technology has been through a half-century after the founding of China. The planning has evolved with time, so has the planning philosophy changed with history.

The planning conception is a complicated and dynamic process. No document specifically addresses the topic. It lurks in the background hardly detected yet affects and restrains the preparation and implementation of the scientific and technological planning. During the planned economic period before the reform, China's political activities were bustling. Politics drove almost societal behavior. The planning of science and technology, as an extension of governmental acts, echoed the will of the state. Therefore, it served politics. Ostensible politicized propensity defined the attitude of the planning of science and technology of this period. The success of the first planning of science and technology after the founding the R.P.C. set the base tone for China's science and technology and the system of fairly significant science characteristics of China. It became a successful political movement. The accomplishment of the "Two-Bomb and One Satellite" secured the new regime and respect of the world. Another politicized propensity was the science and technology frequently drawn into class conflict actively or passively, turning scientists who possessed the science and technology into the source and target of political movements.

After the reform and opening-up, politics, economy and international environment had changed a great deal. The practical mentality of the Chinese Communist Party reaffirmed itself. This needed to be echoed by the science and technology and the planning of them to return to an according "practical" state. Everything must be centered on the economic construction, swaying the planning from obeying the secured regime to serving the economic construction. Before the Republic's founding, the scientific and technological practice of the CCP weighed on management experience with practical efficiency. Onward, it learned the opposite from experience. This made the economic function among many societal functions of science and technology stood out, almost replacing entirely other societal functions. Chasing short-term interests, relying on only GDP to gauge the economy, ignoring intentionally or unintentionally basic sciences, the utilitarianized propensity inundated the planning of science and technology of the transition stage. It can be called as utilitarianized planning of science and technology. The utilitarianism must be perceived from the positive side. It can help gain impressive achievement and bring many negative issues into the process, especially, the sluggishness of the overall development level of basic science.

Moving into the twenty-first century, China began its introspection of the overwhelming utilitarian concept. The planning philosophy of science and technology started showing some humanistic features, emphasizing more on the development of individual entity and social justice and harmony. It can be anticipated that this trend will be sustained.

The planning of science and technology is the core of China's science and technology system, while the planning philosophy is the core of the planning. Therefore, the planning philosophy as the core of the core has a great impact to China's overall science and technology system. This chapter uses the characteristic of the transition stage—utilitarianized philosophy—to prove the impact on scientific and technological environment and academic studies.

The utilitarianized philosophy brewed the utilitarianized academic environment and academic studies. A typical issue is the "affiliation of academic and officialdom." The two sides can gain quick short-term interests, and the long-term benefits are not concerned by individual scholars and officers. Under the utilitarianized context, to maximize the short-term interests and achieve

Table 6.1 Comparison of the Chinese Y-10 and the U.S. Boeing 707–020

	Y-10	Boeing707-020
No. of passengers	178	140
Takeoff weight	110,227 kg	100,800 kg
Empty weight	58,120 kg	46,785 kg
Operating range	8,300 km	6,800 km
Cruising speed	639 km/h	999 km/h
Length	42.93 m	41.25 m
Wingspan	42.24 m	39.90 m
Height	13.42 m	12.65 m
Powerplants (4x)	Pratt & Whitney JT3D	Pratt & Whitney JT3C

raw amassment are the primary goals. The cause of the phenomenon is rather complicated. China's unique national situations toughened the survivability of scholars. The tradition of the historical "scholar-officer" subservient to the ruling class and the external diving force of utilitarianism have made scholars to incline toward the affiliation with officialdom a legitimate choice.

The utilitarianized concept is advantageous to bolster the economic development at the state level, but it becomes a demerit at the microscopic tier. This proves the formation and impact of the planning of science and technology and their historical evolution are an extremely complicated process. Academic autonomy and independent innovation are the same idea with different expressions. The scientific and technological planning steered by the state may propel the formation of a self-reliant research environment, such as the private academic foundation encouraged by the state.

The planning of science and technology is a product created by the interactions of politics, economy and international environment. What to develop, how to develop, how to configure it involve bouts of all aspects. Particularly, after moving into the transition stage, the diversification of interest groups had inconsistent voices from different classes and channels, forcing the planning to be substantially compromised. How to trace out the landscape is a challenge. The large aircraft program is a special and persuasive case. The "three on, three off and reactivation" experience of the "Y-10" appeared in many planning works in decades. An objective description of this period of history will benefit the ones who want to understand and discern the complex and dynamic landscape amid rows of various factors.

In the twenty-first century scientific and technological planning, the large aircraft is included as one of the sixteen key special projects. Once again, a departmental endeavor ascends to a national undertaking, another exemplification of scientific and technological planning driven by the will of state and governmental steering. The study to maximize the benefits for the state, scientists, government and scientific community and to explore the philosophy of scientific and technological planning to China's academia remains ahead a challenge to scholars.

References

Cao, Haidong. 2006. Decision on Made-in-China Large Aircraft. *Southern Weekly*, Apr 13.

Chen, Z.H. 2009. A brief history of China's Y-10: Imitation versus innovation. *Technology in Society* 31 (4): 414–418.

Cong, Cao, Richard P. Suttmeier, and Denis Fred Simon. 2006. China's 15-year science and technology plan. *Physics Today*.

Donald, E. 1999. *Stokes, Pasteur's Quadrant: basic science and technological innovation*. Beijing: Science Press.

Guangbi, Dong. 1991. *Thesis of the history of modern and contemporary science and technology in China.* Changsha: Hunan Education Press.

Gong, Yuzhi. 2004. Historical case and a fresh outset—preface: mao zedong thought on science and technology. *Chinese Cadres Tribune*. Issue 1.

Hu, Weijia. 2006. Preparation effect and inspiration of the twelve-year plan for science and technology. *Policy and Management Research*. Issue 3.

Zhiping, He, Yin Gongcheng, and Zhang Xiaomeo. 1990. *China Science and Technology Group*. Shanghai: Shanghai Scientific and Technical Publishers.

Hu, Weijia. 2003. Overview of studies on china's successive science and technology plans. *Studies in the History of Natural Sciences.*

Harriet, Zuckerman. 1979. Scientific elite: nobel laureates in the United States. Beijing.

Li, Hongwei. 2006. China's academics and education: issues and remedies (II) introspection on the death of intellectuals. *Journal of Dialectics of Nature*. Issue 2.

Ma, Huidi. 1995. Theoretical deliberation on China's first science and technology development plan. *Studies in Dialectics of Nature*. Issues 6.

Richard, P. Suttmeier. 1989. Research and revolution—science policy and societal change in China, translated by Yuan Nansheng et al. Beijing: University of National Defense Science and Technology Press.

Nie, Rongzhen. 1999. *Selected Works of Nie Rongzhen on Science and Technology*. Beijing: National Defense Industry Press.

Wu, Weiyi. 2005. The role of science and technology in human society and its limitations—thought on Jürgen Habermas' technology and science as ideology. *Forum on Science and Technology in China*. Issue 3.

Wang, Bolu. 2006. China's academics and education: issues and remedies (IV) ,On Source of Issues. *Journal of Dialectics of Nature*. Issue 4.

Xiao, Fang. 2005. *Law and Life*. Issue 9, 26;27.

Xu, Teli. 1940. On natural science study, *Chinese Culture*, Issue 4, cited from Dong Guangbi. 1991. Thesis of the history of modern and contemporary science and technology in China. Changsha: Hunan Education Press.

Xu, Hongzhuan, and Meng Jianwei. 2006. Advantages and disadvantages under the empiricist management model for scientific studies. *Journal of Dialectics of Nature*. Issue 1.

Yang, Lifang. 2002. Guiding ideology for scientific and technological development: from Yan'an era to new China. *Studies In The History Of Natural Sciences.* Issue 1.

7 Conclusions

Abstract

This chapter expounds on the basic point of the whole book: China's science and technology planning has achieved great success, and at the same time, it needs philosophical reflection, so as to better promote the development of science and technology in China in the future. It also expounds on seven specific arguments, including the priority of technology, the implementation of national will, the deepening of understanding, the balance of game, science, and political consultation, the role of historical accumulation, and the paradox of planning and freedom.

Keywords

Rejuvenating the country through science and education · National will · China's national conditions · Interest game · Science and politics consultation · Collective cooperation

So far in this book, we have given a panoramic view of science and technology (S&T) planning in the more than 60-odd years since 1949 when the People's Republic of China was founded, as well as an analysis of the concepts behind those S&T programs in different periods. On this basis we can put forth some basic viewpoints: The historical evolution of and changes in the philosophy of S&T planning in China are the result of the combined action of a variety of factors—political, economic, and social—in particular settings; S&T planning as a national action incorporates numerous factors that are beyond the scope of S&T; S&T planning is not only a macro-vision and layout for the development of science and technology but also a political action taken to fulfill the will of the state; while the previous S&T plans were quite fruitful, we should ponder retrospectively over the deficiencies and weaknesses of those plans, with a view to promoting S&T development in China more effectively.

This book is roughly divided into two parts, the first being primarily a historical account and the second an analytical review. In dealing with historical facts, the writing is guided by a definite historical view: the basic concept underlying the S&T plan for each period is used to form the framework within which the plan and the changes in the planning concept were analyzed. Research on these plans is of great significance for the study of contemporary S&T history and policies. From 1949 to 2010, China made ten medium- and long-term S&T programs in total. This book divides the history of China's S&T planning into three periods, that is, the planned-economy period, the economic transition period, and the new century. As each of these periods is characterized by distinct S&T planning concepts, the book tries to explore the changes in the

Z. Chen, *China's Medium and Long-Term Science and Technology Program*,
https://doi.org/10.1007/978-3-030-75146-3_7

concepts while providing an overview of historical development. To sum up, we may draw some specific conclusions as follows.

Conclusion 1: The appearance of S&T programs after the People's Republic of China was founded in 1949 was both a historical continuation and a realistic choice. Technological traditions were inherited and strengthened, and S&T programs were naturally inclined to prioritize technology.

It is undeniable that China made resplendent scientific achievements in ancient times, especially technological achievements. Studies on the science history of ancient China show that the country has a long and strong tradition of putting more emphasis on practical technologies than on scientific research. In a sense, the history of Chinese civilization that has lasted thousands of years is also a history of technological civilization, which explains the country's profound technological foundation as well as technological awareness among the general public. Subconsciously, people have an inherent capability of understanding and creating technologies, and they are well accustomed to and would readily accept the organization of technical engineering projects. Therefore, the appearance of S&T plans after 1949 was a natural continuation of history. In the meantime, the understanding of technology was elevated to an unprecedented height, and the management and organization of applied science and technical engineering reached a higher level.

The Communist Party of China (CPC) began S&T practices as early as when it was at the revolutionary bases. The long and hard military struggles gave the CPC a profound understanding and grasp of applied sciences represented by military technology, and S&T practices before the People's Republic of China was founded in 1949 were characterized by an emphasis on timeliness, technical practicality, S&T personnel, and S&T management in a military manner. Due to limited resources, the achievements of natural science studies, especially of basic research, were much inferior to those of technological research in terms of scale and quality. No wonder the Chinese Workers' and Peasants' Red Army, most of whose members were workers and farmers, and leaders from this group, did not fully understand the importance of the "useless" basic sciences. This inclination to stress technology over science was naturally reflected in the S&T planning after the People's Republic of China was founded, and gave birth to a potential atmosphere and presumption that technology was more effective than science for China's revolution and state establishment. As a result, the S&T plans of China had a strong technological inclination from the very beginning, which affected its subsequent S&T plans. Technology often replaced science or science was mixed with technology. Most senior leaders of the CPC derived from revolutionary struggles and only a few had studied abroad or received higher education, so their recognition of the importance of science and basic sciences was inadequate or seriously inadequate. Zhou Enlai and Deng Xiaoping, thanks to their experiences of studying in France, had a deeper understanding of the importance of basic sciences than other leaders, and arrangements for the research on basic sciences were made in the first S&T program exactly because of Zhou Enlai's support.

The nascent people's government of New China had to build the country basically from scratch, so how to consolidate the state and quickly lay a foundation for S&T and education was the top consideration in S&T planning. Therefore, no argument like the one between Polanyi and Bernal appeared in China. To advance S&T in a short period, it was obviously necessary to make plans for S&T development, and because of the general understanding of socialism at that time, it was believed that science and technology, including basic sciences, should develop in a well-planned way. Therefore, the appearance of S&T programs after the People's Republic of China was founded in 1949 was both a historical continuation and a realistic choice.

The large number of senior S&T cadres that CPC fostered in war times, such as Wang Zheng and Wu Heng, played an important leading role in formulating and implementing S&T plans after 1949. China's scientific foundation was not only laid by CPC's S&T cadres but also by scientists in Kuomintang-controlled areas, who, along with the scientific community in those areas, contributed a lot to basic sciences in particular. They were a critical supportive force for the implementation of S&T plans after 1949. Due to the change of regime, most scientists that represented the world advanced level then did not become leaders in the S&T field, so their right opinions could not be heard, which undermined the development of basic sciences to some extent.

Conclusion 2: China's S&T programs are neither research agreements between scientific communities nor research outlines for scientists. They are a government action and a national decision to fulfill the nation's will and to secure and increase national interests.

China's S&T programs are created in specific settings with Chinese characteristics. They are special in many ways, of which the most important one is that S&T programs are issued in the form of government documents and carried out as national actions, so they have to consider things from a national perspective. At the International Congress of Mathematicians (ICM) held in Paris in 1900, Hilbert delivered a famous speech in which he put forth 23 mathematical problems based on the achievements and trend of mathematical research in the nineteenth century. Covering most areas of modern mathematics and being applauded by many scientists, those problems could be regarded as the research outline for scientific communities, leaving a profound influence on the 20th-century mathematics. The research agreements between scientific communities or the research visions of individual scientists can be completely based on scientists' research interest without considering national interests, and they do not necessarily need to use government resources.

But S&T planning in China is basically a government action. Although scientists participate in formulating and implementing the plans and the contents are about science and technology, they play their role within a national framework and in accordance with government documents. The system of "national economic plan—S&T plan—S&T program" analyzed earlier in the book is an example in point. Guaranteeing and increasing national interests and improving the efficiency of national survival are the major premise and starting point of S&T planning in China. S&T programs are endowed with many extensions and connotations far beyond the research outline of scientific communities, and they can be regarded as political plans and actions about science and technology from this point of view. That is why China's S&T plans can use a lot of government resources in the name of the nation and some projects that twist the original intention of science and technology become reasonable. From consolidating the state to promoting economy and to pushing national rejuvenation, S&T plans have carried and will continue to carry many historical missions. These plans and the historical changes in the concepts behind them reflect the changes of national will with the passage of time.

As the first S&T program in China, the one for the period of 1956–1967 created many traditions for S&T planning, laid the foundation for China's S&T development and resulted in the big scientific development system with Chinese characteristics. One of the core concepts of this plan was "national core interests are paramount." The S&T plan is a national action rather than a voluntary action of the scientific community, so it represents and centers on national core interests, which, when the People's Republic of China was just founded, meant consolidating the new state and proving its reasonableness and legitimacy. Another core concept of the plan was "intensifying political function", which meant S&T's political function outweighed its S&T function. This concept was reflected by the fact that as the Chinese government was newly established, everything including science and technology must serve political purposes as the central task, and that the S&T plan must endorse the reasonableness and legitimacy of the nascent government.

S&T plans and planning concepts in the economic transition period were also characterized by the fulfillment of national will and the efforts to gain more space for national survival. In that period, the sole reliance on S&T plans to push China's S&T development was replaced by the reliance on the "S&T plan—S&T program" system. If we take into account China's S&T development before 1949, it will show a clear historical trajectory "from diverse and multi-regional development to development promoted by S&T plans and to multi-layered development through S&T programs." Such a historical trajectory denoted the evolution of state power and the mainstream ideology, and also the change in national will and interests in step with the times.

The core of national will in the economic transition period was to boost economic development and enhance the overall national strength centered on economic strength. Everything centered on economic development then and almost all social sectors served that purpose. At that time, the philosophy of S&T planning was influenced by the national will of promoting economic development, so the S&T program in 1986 appeared mature and well-proven and reflected the utilitarian planning concept. The outline of S&T development (1990–2020) continued such a concept although it wasn't put into practice in the end. The 8th and 9th five-year S&T development plans were matched with the corresponding five-year economic development plans, reflecting the idea of better serving the central task of economic development.

The economic centrism in the economic transition period changed the role of science and technology from supporting state consolidation to supporting economic development, and a guideline was gradually formed that believed "economic development must rely on S&T, S&T work must serve economic development and efforts must be made to scale S&T heights." Such a guideline implied a utilitarian attitude of pursuing short-term economic benefits. The positive effects of the "utilitarian tendency" must be affirmed as it quickly expanded the S&T scale in China, strongly boosted economic development, and considerably enhanced the overall national strength. Otherwise, the country couldn't have developed so rapidly and many serious problems would have occurred. In particular, after the People's Republic of China was founded, the nascent state didn't pay due attention to population control, and resources were in extremely short supply, so the utilitarian attitude was a reasonable choice. S&T planning in the economic transition period stressed its economic function, which outweighed its other functions. S&T was deemed as a productive force and later the No. 1 productive force. Science and technology didn't just develop for their own sake, but, more importantly, for economic development. Its economic function was highlighted as paramount, which was a reflection of the national will that made economic development the center of all work.

Conclusion 3: Changes in the philosophy of S&T planning over time indicate the Chinese government's ever deeper understanding of S&T development. The reflection on S&T planning philosophy and the proposition and development of a national innovation system in the new century shows that the CPC's understanding of S&T development has reached a new height and S&T planning has entered a new historical stage.

From establishing the Party to military struggles and then to founding and developing the country, the CPC has continuously deepened its understanding of the development of things and extracted the laws thereof. Just as we have to combine the rationale of Marxism with the actual situations of Chinese revolution and development, China's S&T programs are the results of combining the rationale of S&T development with China's development and its S&T reality. Whether the result is good depends on whether the components are good. Obviously, the CPC has a better understanding of China's actual conditions and its S&T reality, but its understanding of the rationale of S&T development has undergone a process.

For instance, the extensions and connotations of science and technology are confused most of the time, which gives rise to *keji* (a combination of science and technology), a word with typical Chinese characteristics. In the S&T plan formulated in 2006, namely the *Outline of the National Program for the Medium and Long-term Development of Science and Technology (2006–2020)*, science was differentiated from technology for the first time. This indicated the CPC's deeper recognition of the difference between science and technology, and that opinions from the scientific circle were more readily accepted by the political circle. Another example is that since the People's Republic of China was founded, the country has fully realized S&T's positive effect and how it benefits the society, but is unclear about or even blind to its negative effect and how it harms the society. In a way, the pure pursuit for GDP during the transition period was a reflection of the belief that "there is only good science, no bad science." That's why the country has begun to reflect in the new century on the pure pursuit for GDP, including the utilitarian attitude in the transition period.

Several S&T programs have been implemented or are being implemented in the twenty-first century, including the 10th five-year S&T plan, the 2006–2020 S&T plan, and the 11th five-year S&T plan, and so on. The first one is a transitional S&T plan that fully reflects the utilitarian attitude in spite of the tendency of reflection. There have been tremendous political, economic, and social changes since the beginning of the new century. After the new administration came into power, it reflected on the old development model that only pursued economic benefits and GDP. The "Three Representatives" and the "Scientific Outlook on Development" became the CPC's guiding thought in succession, and the old S&T planning concepts had to be adjusted. As the former Soviet Union cut off its assistance to China, the 1963–1972 S&T program added an extra philosophy—"develop independently based on national conditions," which can be viewed as the origin of the planning philosophy of "independent innovation" in the twenty-first century.

S&T programs in the twenty-first century also implement the philosophy of "people centrism", which conforms to the trend of historical development. The 2006 S&T plan and the 11th five-year S&T plan include a number of science programs and projects for public welfare. The "independent innovation" philosophy is put in a prominent position in S&T plans of the new century, and deployments are made for comprehensively building a national innovation system. The world's top science magazine *Nature* issued three China specials, which commented on China's S&T development and plans and gave suggestions. On the one hand, this means China's S&T development has come to a new stage and captured close attention from the global scientific community. On the other hand, it also points out some problems in our S&T development and the formulation and implementation of S&T plans, which is where S&T planners need to reflect on and improve. In sum, the CPC's understanding of S&T development has reached a new height and its S&T planning has entered a new historical stage.

The utilitarian attitude toward S&T has led to tremendous benefits. The Chinese economy and society progressed substantially, its international standing was elevated to a height never reached before, and science and technology themselves developed by leaps and bounds, especially in quantity and scale. The planning concepts in the past 50-plus years have a strong realistic and utilitarian inclination. While they proved extremely fruitful and made great contributions to economic development, we must ponder over the deficiencies in China's science development and in the scientific community. The long-term neglect of basic sciences and the restriction on free explorations are an important reason why few Chinese scientists had received the Nobel Prize so far.

Conclusion 4: The formation and implementation of S&T programs and the changes in the philosophy of S&T planning are the result of the combined action of a variety of factors—political, economic, military, and others—in particular settings. The finalization of the S&T programs is the result of various gaming and balancing among various relations and interests.

When analyzing every S&T program, this book would study its background, including political, economic, military, diplomatic, and other factors. Such a research approach aims to clarify that every S&T program is the result of the combined effect of a string of factors. But exactly because of this, the synergy of multiple parallelograms sometimes results in the local distortion of the S&T plan and some phenomena that deviate from the original intention and objective laws of S&T development.

For example, early S&T plans were required to serve class struggle and eliminate illiteracy, which was the task of schools; projects were often included in S&T plans whose objectives were impossible to achieve. This was attributed to the interference from non-science factors. Scientists, especially famous Chinese scientists of foreign nationality, often criticize China's S&T plans. Their criticism is reasonable and does point to the deficiencies of our S&T plans, but they don't consider the special national conditions of China and the various factors underlying the formation of S&T plans. This is not to find excuses for S&T planning but is a realistic statement. Only by finding the true reasons for the problem can we truly solve it.

Our study on the big plane case shows that the S&T program is the result of balancing the interests of different parties and the historical evolution of planning concepts is a very complicated dynamic process. Especially in the economic transition period when reform and opening-up are deepened and stakeholders are diversified, all social classes are looking for their own spokespeople, as evidenced by the variety of spokespersons and agents popular in the society today. Spokespersons of different classes have widely different or even opposite interest demands, and the S&T plan is a national plan that can adjust and balance these demands. As stated before, S&T plans in China are not research agreements among scientific communities that have no major conflicts of interests but are a government action that can freely or easily mobilize and use public resources. Therefore, S&T planning in a sense is the gaming process of various interests.

As these interests involve national interests, class interests, group interests, and even personal interests, the finalization of the S&T program is the result of various gaming and balancing relations and interests. For instance, to prove whether a discipline or project is important and how to prove it and make arrangements for it in future S&T plans involves not only national interests but also the immediate interests of different parties. To put it bluntly, a research project is a matter of huge amounts of research funds ranging from dozens to hundreds of millions of RMB. In the widely recognized system featuring "no evaluation for massive fund, easy evaluation for medium fund, and tedious evaluation for minor fund," stakeholders can secure more resources through the national S&T plan, which makes the interesting game even more complicated. When making S&T plans, the standard observed by the scientific community is not the only one. There are also political, economic, and cultural standards or even hidden rules that must be taken into account.

The big plane project is perhaps the most special case in all S&T programs. The author discovered this problem during research and singled it out as a case study to prove the conclusion. Studies show that due to the combined effect of multiple factors, S&T plans in China carry many connotations beyond their original boundary, and the interest game is not good for China's healthy S&T development in the long term, particularly the development of basic sciences. Since the S&T plan is in the central position of China's overall S&T system, the concepts behind it are naturally brought to the system and the general environment of academic research. If the S&T plan is the result of interest game, the overall S&T system and the general environment of academic research will be subject to negative impacts too, which we should be alert to. The "utilitarian research" analyzed earlier is one of the negative impacts.

Conclusion 5: An S&T plan in China is the outcome of "science + politics" and the result of the consultations between "scientists and politicians". The individual role of political leaders and the collective opinions of scientists command different positions and play different roles in S&T planning, and the two sides strive for the common goal of maximizing both scientific and national interests.

In modern times, it seems to be a historical trend for science and politics to get closer and even merged. The presidents of some western countries attach great importance to S&T development, but as their scientific communities are well developed, the political circle seldom intervenes in scientists' work directly or designates specific projects for scientists to accomplish, and the sponsor barely interferes in the specific scientific research work after providing the fund. However, S&T plans in China are the outcome of "science + politics", a combination that is both common and special in China.

It is common because the combination of science and politics is almost a common sense in China. The author analyzed earlier that the class of "scholarly officials", who were the intellectuals in ancient times, had the inclination for and tradition of being attached to the ruling class. It is special because in modern and contemporary China, the scientific community isn't well developed and doesn't have the spirit or habit of independence. The unstable policies toward intellectuals after 1949 make it difficult for the Chinese intellectual circle, including the scientific circle, to become a highly independent class. Externally, it cannot resist the interferences from the political circle; internally, it tends to accept and seek political protection and endorsement, especially given the limited resources for academic research.

Both the scientific circle and political circle are layered, and high-level figures in each circle have their way of communication, so correct opinions and suggestions from the scientific circle can be passed on by top scientists to high-ranking government officials to secure their support. There are many examples in point. In May 1983, Qu Bochuan, in conjunction with the honorary president of Nanjing University Kuang Yaming, honorary president of Tianjin University Li Shusen and honorary president of Zhejiang University Liu Dan, submitted a proposal to the Secretariat of the CPC Central Committee to "list 50 or so institutions of higher learning as major national construction projects." The proposal suggested the central government allocate RMB 5 billion to 50 universities to help them reach the world advanced level regarding instruments and equipment in the 1990s. This proposal received close attention from Deng Xiaoping and other leaders of the central government and was generally accepted later, leaving positive impacts on the "211 Program" and "985 Program" afterward. The proposal submitted by Qu Bochuan and others was called the "835 proposal" in the circle of higher education. There are other successful cases of Chinese intellectuals making suggestions directly to top leaders, such as the "863 Program" put forth by four old-generation scientists including Wang Daheng and supported by Deng Xiaoping, the suggestion on developing China's own big plane made in 1986 by "four scholars" (referring to Hu Xitao, head of the plane bureau of the former Ministry of Aviation Industry, president of Beihang University Shen Yuan, president of Nanjing University of Aeronautics and Astronautics Zhang Azhou, and president of Northwestern Polytechnical University Ji Wenmei), and the "knowledge innovation program" proposed by the Chinese Academy of Sciences in the 1990s. There are also negative cases alongside these positive cases of science-politics combination.

China's S&T planning is a typical field where politics and science are well combined, and every S&T plan receives the attention and direct support from top leaders, although to varying degrees. For instance, the first S&T program wouldn't have been so successful without the direct leadership of and support from top leaders including Mao Zedong and Zhou Enlai. Hu Jintao and Wen Jiabao attached great importance to the medium and long-term S&T program formulated in 2006.

Owing to their different understanding of science and technology, scientists and politicians play different roles and occupy different positions in S&T planning. Generally speaking, politicians come up with the direction and goal from a national political perspective, while the scientific circle works out specific S&T projects according to the goal. The two circles sometimes have disagreements, so consultation is important. When formulating the S&T development program for 2006–2020, the scientific circle had arguments both internally and externally, which indicated the special nature of S&T plans in China. It shouldn't be analyzed from the perspective of the scientific community alone, instead by combining political and scientific views. The two sides strive for the common goal of maximizing both national and scientific interests.

Conclusion 6: China's S&T planning has been quite successful in general, which has not only enhanced China's overall national strength and international standing but also promoted the development of S&T. Such success results from the combined action of the communist production mode of collective cooperation and the historical legacy of feudal centralization over long years.

China's S&T planning has been quite successful in general. In particular, the planning for technological categories has played a critical and irreplaceable part in fortifying national strength, modernizing the country, and rejuvenating the nation. Although other countries have similar science development outlines or technology progress plans, they don't seem able to achieve such tremendous success within such a short period as in China. The great success in science and technology, especially in the high-tech field and some basic sciences, reflects the special historical position of S&T planning in China. The reason for such success, according to the author, is the combined action of the communist production mode of collective cooperation and the historical legacy of China's feudal centralization over long years.

According to Marx's original definition of communism and socialism, the production mode of collective coordination is an ideal for communists. The CPC was created in the 1920s, when communism was on the rise in the global stage. After long years of revolutionary struggle, the CPC established the national government, and the production mode of collective coordination was implemented across the country. Ownership by the whole people and collective ownership are the forms of collective cooperation, but how to put it into practice in the field of science and technology? The answer is through centralized use of resources, coordinated efforts to tackle S&T difficulties and completion of large-scale scientific projects.

In the initial period after the CPC took over the political power, collective cooperation was considered the best way to liberate and develop productive forces and the integral characteristic of communism or socialism. The former Soviet Union also had successful and fruitful central S&T plans, which were highly advocated by Bernal, founder of the sociology of science. Economically speaking, collective cooperation can generate scale merit, raise the efficiency of resource utilization and is, therefore, one of the indispensable modes of large-scale social production. The human society has evolved from agricultural production featured by individual workshops to collective production of the industrial society. As a result, the series of social movements launched by the CPC after the People's Republic of China was founded were partially aimed to explore the best mode of collective cooperation in China. Collective cooperation requires members of a group to give up some of their personal thoughts and the free and undisciplined research approach and adopt the unified research model. Obviously the denial of collective cooperation would be associated with the denial of the newborn state, which would easily escalate to the "contradiction between the people and the enemy." As science and technology are two different fields, which one of them is more suitable for the unified production mode of collective cooperation? The answer is obvious—technology. This explains why technology takes a large proportion in China's S&T plans. S&T plan, as a national action, costs a great deal of capital, and failure to achieve the expected results would be unimaginable and not allowed, but goals set in technological plans are easy to achieve.

A deeper reason is that S&T plans are more suitable for the production mode of collective cooperation, but this mode cannot be taken as the only reason why China's S&T plans have been so fruitful. There are of course other reasons; otherwise, why cannot the communist parties in other countries consistently formulate and implement S&T plans? Another reason is more profound and more important, which is China's long history of feudal centralization.

China is a time-honored and civilized oriental power where the idea of centralization is deeply rooted. This idea is quite pronounced even today and reflected in academic research from time to time. For instance, there are always scholars who hope to establish a "dominant" theory or occupy an academic high ground or "rule the academic world." But in fact, academic research should be something resembling a sphere, featuring equality and free development. The idea of centralization stems from the long years of feudal centralization in ancient Chinese society. The ancient S&T history of China shows that the ruling class always attached great importance to large-scale social and technological planning, such as the massive construction of the Great Wall, digging of the Grand Canal and measurement of sky and earth. The ruling class was usually capable of mobilizing the people and controlling the social situation and development. This capability formed under centralization was accumulated and reinforced over time and passed on from one dynasty to another, and was naturally inherited by the People's Republic of China, the newly born state of the people. The easy launch of many social movements in China after 1949 and the amazing power of collective social action, which was rarely seen in the world, was the result of a long historical legacy. During the opening ceremony of the Beijing 2008 Olympic Games, the unified movements of thousands of performers astonished the western world. The collective action or unconscious collective action in current China can be viewed as deriving from the long historical legacy.

Therefore, the great success of China's S&T plans should be attributed to the combined effect of the communist production mode of collective cooperation and the historical legacy of feudal centralization over long years. Other countries and other historical periods of China don't have such special historical conditions, so China's S&T programs are very special, because of which they are underpinned by different philosophy in different historical periods. The evolution of China's medium and long-term S&T programs and the changes in the concepts behind them are a representation of this beautiful combination. In view of this, we may infer that the future evolution of Chinese S&T plans will be related with collective cooperation and historical legacy. In other words, if our understanding of the production mode of collective cooperation changes or its specific model changes, or the historical legacy of centralization gradually dissolves or a new form of the rule appears, then our S&T plans will change substantially too.

Conclusion 7: The paradox of planned S&T and free research exists in China's S&T plans.

Throughout the world, completely independent and free research seems to have become more difficult. The S&T planning model in China provides a reference model of S&T development and scientific research for the world scientific community. Going forward, the big scientific development model may be a combination of planned S&T development and market-based free research both in China and the world at large.

Ultimately the purpose of S&T development is to serve human development, but what's the essence of human development? It's to seek freedom for human beings, including external and physical freedom and internal and mental freedom. S&T plans are made to pursue such freedom, but the freedom of human beings cannot be planned; otherwise, how can it be free? Likewise, if human beings are allowed to stop the planned S&T development freely, how can we arrive at the freedom that can only be achieved through collective efforts?

This is obviously a paradox between planning and freedom, as evidenced by the ceaseless arguments about S&T plans in the past few decades and the flourishing technologies and the relatively weak basic sciences. How should we look at this kind of paradox? What should we do to make our S&T plans strike a reasonable balance as well as a necessary tension between planned development and free research?

Although China's S&T planning is a collective action, the intervention from politicians usually gives it a political touch. How does a politician's personal viewpoint, especially personal psychology, affect S&T plans? A psychological analysis of the game between decision makers and executors will steer the study of S&T planning into the social-psychological paradigm of quantitative analysis.

No similar S&T plans in foreign countries are analyzed for comparison in this book because western countries, especially developed ones, represented by the United States, have neither the communist collective cooperation nor the long-term historical legacy of centralization, so conclusions drawn from a comparative analysis are not necessarily of reference value. All in all, S&T plans and the history of them, as a field of academic study, can deepen people's understanding of China's science history and S&T policies in contemporary times. The "research on S&T planning history" may be a grand goal to be considered in the future.

Appendix
A Brief Introduction to the Guidelines of the 10 Science and Technology Planning[1]

Since the founding of the People's Republic of China, the central government has formulated ten programs for the development of science and technology, which determined the strategies, guidelines, and key tasks for the development of science and technology in different periods, and put forward corresponding policies and measures. Every science and technology planning has a clear guideline, which is a reflection of the philosophy on science and technology planning in the text.

1. Science and Technology Development Program (1956–1967)

   This was the first science and technology plan since the founding of the People's Republic of China. Hundreds of scientists from various fields and disciplines had participated in the preparation of this plan. The guiding principle was "focus on development and catch up the front." 57 major scientific and technological tasks and 616 central issues were proposed from 13 aspects, and 12 key tasks were further proposed from them. It was implemented in 1956 and completed five years ahead of schedule in 1962.

2. Science and Technology Development Program (1963–1972)

   Around 1962, the CPC Central Committee decided to draw up a ten-year plan for the development of the national economy, with which it decided to work out a ten-year scientific and technological plan at the same time, as the support of the national economic plan. It was finalized in June 1963. At the end of that year, it was officially implemented with the approval of the CPC Central Committee and the State Council. The ten-year plan has established the policy of "self-reliance, catching up the peak" with the development of science and technology. Due to the interference of the Cultural Revolution, it basically stopped in 1966. However, some individual actions of plans were continued.

[1]This part is based on the medium- and long-term science and technology planning texts and relevant documents published by the China government. The 12th and 13th Five-Year S&T Plans would be brief introduced in the book.

Z. Chen, *China's Medium and Long-Term Science and Technology Program*,
https://doi.org/10.1007/978-3-030-75146-3

3. Science and Technology Development Program (1978–1985)

   It was formulated in June 1977 and adopted at the National Science Conference from March 18 to 31, 1978. In October 1978, the CPC Central Committee formally forwarded the outline of the national plan for the development of science and technology from 1978 to 1985 (referred to as the Outline of the Eight Year Plan for short) and began to implement it, which put forward the guideline of "overall arrangement, highlighting key points."

4. National Medium- and Long-Term Science and Technology Development Program (1986–2000)

   At the end of 1982, the China State Council approved the report of the State Planning Commission and the State Science and Technology Commission on the preparation of the fifteen year (1986–2000) science and technology development plan, and established the science and technology leading group of the State Council to lead the formulation of the long-term plan for science and technology. The planning was completed from March to October 1984 and implemented in 1986. The guideline of this plan is "science and technology must face economic construction, and economic construction must rely on science and technology" (for short, "Facing and Relying").

5. National Medium- and Long-Term Science and Technology Development Program (1990–2000–2020)

   It was formulated around 1990. This plan was basically the same as the 1986–2000 medium- and long-term plan for science and technology and the later ten-year plan for the development of science and technology of the People's Republic of China (1991–1995–2000), so it had not been implemented in practice. This plan had some guidelines for different fields, but it continues to adhere to the basic policy of "economic construction must rely on science and technology, and scientific and technological work must face economic construction."

6. Ten-Year Program & the Eighth Five-Year Plan for the Development of Science and Technology in the People's Republic of China (1991–1995–2000)

   It was formulated in March 1991. On March 8, 1992, the State Council officially approved the outline of the plan and issued it to the whole country for implementation. This science and technology plan continued to implement the guideline that "economic construction must rely on science and technology, and scientific and technological work must face economic construction."

7. Ninth Five-Year Plan for National Science and Technology Development (until 2010)

   It was formulated in 1994 and implemented in 1995. The guiding principle of this scientific and technological plan is to continue to implement "economic construction must rely on science and technology, scientific and technological work must face economic construction and strive to climb the peak of science and technology" (hereinafter referred to as "Facing, Relying on and Climbing the peak")

8. Tenth Five-Year Plan for National Economic and Social Development—Science, Technology and Education

   On October 28, 1998, the second meeting of the National Science and Education Leading Group decided to prepare the Tenth Five Year Plan for science and technology and completed the drafting of the plan in June 2000. In May 2001, the State Planning Commission and the Ministry of Science and Technology officially issued the special plan for the development of science and technology education in the Tenth Five Year Plan for national economic and social development (Science and technology development plan). Based on "facing, relying on and climbing the peak," the Tenth Five Year Plan for science and technology put forward the guidelines of "doing some important thing and following up the whole picture, making breakthroughs in key points and developing high technology and realizing industrialization, improving the ability of continuous innovation of science and technology, and realizing leapfrog development of Technology (hereinafter referred to as" innovation, industrialization").

9. National Medium- and Long-Term Program for Science and Technology Development (2006–2020)

On March 22, 2003, at the first plenary meeting of the new State Council, it was decided to start to study and formulate the national medium—and long-term scientific and technological development plan. The formulation of this science and technology plan was divided into three stages: the first stage was strategic research (from June to December 2003); the second stage was the drafting of the outline of national medium and long term science and technology development plan (2006–2020) (from January to June 2004); the third stage was the examination and approval of national science and technology development plan for 2006–2010 (hereinafter referred to as the 11th Five Year Plan for science and Technology) (July September 2004). On December 30, 2005, the State Council issued the outline of the national medium and long term plan for science and technology development (2006–2020). The guidelines of it were: indigenous innovation, key leapfrogging, supporting development, and leading the future.

10. Eleventh Plan for National Science and Technology Development

The 11th Five Year Plan for science and technology was formulated in accordance with the 11th Five Year Plan for national economic and social development and the 2006 medium and long term plan for science and technology. It was implemented in 2005. Adhere to the guidelines of "indigenous innovation, key leapfrogging, supporting development and leading the future."